SOFTWARE DEFINED MOBILE NETWORKS (SDMN)

Wiley Series in Communications Networking & Distributed Systems

Series Editors: David Hutchison, *Lancaster University, Lancaster, UK*
Serge Fdida, *Université Pierre et Marie Curie, Paris, France*
Joe Sventek, *University of Oregon, Eugene, OR, USA*

The "Wiley Series in Communications Networking & Distributed Systems" is a series of expert-level, technically detailed books covering cutting-edge research, and brand new developments as well as tutorial-style treatments in networking, middleware and software technologies for communications and distributed systems. The books will provide timely and reliable information about the state-of-the-art to researchers, advanced students and development engineers in the Telecommunications and the Computing sectors.

Other titles in the series:

Wright: *Voice over Packet Networks*, 0-471-49516-6 (February 2001)
Jepsen: *Java for Telecommunications*, 0-471-49826-2 (July 2001)
Sutton: *Secure Communications*, 0-471-49904-8 (December 2001)
Stajano: *Security for Ubiquitous Computing*, 0-470-84493-0 (February 2002)
Martin-Flatin: *Web-Based Management of IP Networks and Systems*, 0-471-48702-3 (September 2002)
Berman: *Grid Computing. Making the Global Infrastructure a Reality*, 0-470-85319-0 (March 2003)
Turner: *Service Provision. Technologies for Next Generation Communications*, 0-470-85066-3 (April 2004)
Welzl: *Network Congestion Control: Managing Internet Traffic*, 0-470-02528-X (July 2005)
Raz: *Fast and Efficient Context-Aware Services*, 0-470-01668-X (April 2006)
Heckmann: *The Competitive Internet Service Provider*, 0-470-01293-5 (April 2006)
Dressler: *Self-Organization in Sensor and Actor Networks*, 0-470-02820-3 (November 2007)
Berndt: *Towards 4G Technologies: Services with Initiative*, 0-470-01031-2 (March 2008)
Jacquenet: *Service Automation and Dynamic Provisioning Techniques in IP/MPLS Environments*, 0-470-01829-1 (March 2008)
Gurtov: *Host Identity Protocol (HIP): Towards the Secure Mobile Internet*, 0-470-99790-7 (June 2008)
Boucadair: *Inter-Asterisk Exchange (IAX): Deployment Scenarios in SIP-enabled Networks*, 0-470-77072-4 (January 2009)
Fitzek: *Mobile Peer to Peer (P2P): A Tutorial Guide*, 0-470-69992-2 (June 2009)
Shelby: *6LoWPAN: The Wireless Embedded Internet*, 0-470-74799-4 (November 2009)
Stavdas: *Core and Metro Networks*, 0-470-51274-1 (February 2010)
Gómez Herrero: *Network Mergers and Migrations: Junos® Design and Implementation*, 0-470-74237-2 (March 2010)
Jacobsson: *Personal Networks: Wireless Networking for Personal Devices*, 0-470-68173-X (June 2010)
Minei: *MPLS-Enabled Applications: Emerging Developments and New Technologies, Third Edition*, 0-470-66545-9 (December 2011)
Barreiros: *QOS-Enabled Networks*, 0-470-68697-9 (December 2011)
Santi: *Mobility Models for Next Generation Wireless Networks: Ad Hoc, Vehicular and Mesh Networks*, 978-1-119-99201-1 (July 2012)
Tarkoma: *Publish/Subscribe Systems: Design Principles*, 978-1-119-95154-2 (July 2012)

SOFTWARE DEFINED MOBILE NETWORKS (SDMN)

BEYOND LTE NETWORK ARCHITECTURE

Edited by

Madhusanka Liyanage
Centre for Wireless Communication, University of Oulu, Oulu, Finland

Andrei Gurtov
Helsinki Institute for Information Technology HIIT, Aalto University, Espoo, Finland

Mika Ylianttila
Centre for Internet Excellence, University of Oulu, Oulu, Finland

WILEY

This edition first published 2015
© 2015 John Wiley & Sons, Ltd

Registered Office
John Wiley & Sons Ltd, The Atrium, Southern Gate, Chichester, West Sussex, PO19 8SQ, United Kingdom

For details of our global editorial offices, for customer services and for information about how to apply for permission to reuse the copyright material in this book please see our website at www.wiley.com.

Library of Congress Cataloging-in-Publication Data

Liyanage, Madhusanka.
  Software defined mobile networks (SDMN) : Beyond LTE network architecture / Madhusanka Liyanage, Wireless Communication, University of Oulu, Finland, Andrei Gurtov, for Information Technology HIT, Aalto University, Finland, Mika Ylianttila, Centre for Internet Excelience, University.
    pages cm.
  Includes bibliographical references and index.
  ISBN 978-1-118-90028-4 (hardback)
1. Software-defined networking (Computer network technology)   I. Gurtov, Andrei.   II. Ylianttila, Mika.   III. Title.
  TK5105.5833.L59 2015
  004.6′5–dc30

                                   2015004425

A catalogue record for this book is available from the British Library.

Set in 10/12pt Times by SPi Publisher Services, Pondicherry, India

1   2015

Contents

Editors   xv

Contributors   xvii

Foreword   xxvii
Ulf Ewaldsson

Foreword   xxix
Lauri Oksanen

Preface   xxxi

Acknowledgments   xxxvii

Abbreviations   xxxix

PART I   INTRODUCTION
1   Overview   3
Madhusanka Liyanage, Mika Ylianttila, and Andrei Gurtov
    1.1   Present Mobile Networks and Their Limitations   4
    1.2   Software Defined Mobile Network   5
    1.3   Key Benefits of SDMN   7
    1.4   Conclusion   9
    References   9

2   Mobile Network History   **11**
Brian Brown, Rob Gonzalez, and Brian Stanford
   2.1   Overview   11
   2.2   The Evolution of the Mobile Network   12
         2.2.1   Sharing Resources   13
         2.2.2   Orchestration   14
         2.2.3   Scalability   15
   2.3   Limitations and Challenges in Current Mobile Networks   15
   2.4   Requirement in Future Mobile Networks   18
   Reference   19

3   Software Defined Networking Concepts   **21**
Xenofon Foukas, Mahesh K. Marina, and Kimon Kontovasilis
   3.1   Introduction   21
   3.2   SDN History and Evolution   23
         3.2.1   Early History of Programmable Networks   23
         3.2.2   Evolution of Programmable Networks to SDN   25
   3.3   SDN Paradigm and Applications   28
         3.3.1   Overview of SDN Building Blocks   28
         3.3.2   SDN Switches   30
         3.3.3   SDN Controllers   31
         3.3.4   SDN Programming Interfaces   34
         3.3.5   SDN Application Domains   37
         3.3.6   Relation of SDN to Network Virtualization and Network
                 Function Virtualization   38
   3.4   Impact of SDN to Research and Industry   39
         3.4.1   Overview of Standardization Activities and SDN Summits   40
         3.4.2   SDN in the Industry   41
         3.4.3   Future of SDN   41
   References   42

4   Wireless Software Defined Networking   **45**
Claude Chaudet and Yoram Haddad
   4.1   Introduction   45
   4.2   SDN for Wireless   47
         4.2.1   Implementations: OpenRoads and OpenRadio   49
         4.2.2   SDR versus SDN   50
   4.3   Related Works   50
   4.4   Wireless SDN Opportunities   51
         4.4.1   Multinetwork Planning   51
         4.4.2   Handovers and Off-Loading   53
         4.4.3   Dead Zone Coverage   55
         4.4.4   Security   55
         4.4.5   CDN and Caching   56
   4.5   Wireless SDN Challenges   56
         4.5.1   Slice Isolation   56
         4.5.2   Topology Discovery and Topology-Related Problems   56
         4.5.3   Resource Evaluation and Reporting   57

     4.5.4   *User and Operator Preferences*   57
     4.5.5   *Nontechnical Aspects (Governance, Regulation, Etc.)*   58
  4.6   Conclusion   59
  References   59

5 Leveraging SDN for the 5G Networks: Trends, Prospects, and Challenges   61
  Akram Hakiri and Pascal Berthou
  5.1   Introduction   61
  5.2   Evolution of the Wireless Communication toward the 5G   62
     5.2.1   *Evolution of the Wireless World*   62
  5.3   Software Defined Networks   64
  5.4   NFV   65
  5.5   Information-Centric Networking   67
  5.6   Mobile and Wireless Networks   68
     5.6.1   *Mobility Management*   68
     5.6.2   *Ubiquitous Connectivity*   69
     5.6.3   *Mobile Clouds*   70
  5.7   Cooperative Cellular Networks   71
  5.8   Unification of the Control Plane   73
     5.8.1   *Bringing Fixed–Mobile Networking Together*   73
     5.8.2   *Creating a Concerted Convergence of Packet–Optical Networks*   74
  5.9   Supporting Automatic QoS Provisioning   75
  5.10   Cognitive Network Management and Operation   76
  5.11   Role of Satellites in the 5G Networks   77
  5.12   Conclusion   79
  References   79

PART II   SDMN ARCHITECTURES AND NETWORK IMPLEMENTATION

6 LTE Architecture Integration with SDN   83
  Jose Costa-Requena, Raimo Kantola, Jesús Llorente Santos,
  Vicent Ferrer Guasch, Maël Kimmerlin, Antti Mikola and Jukka Manner
  6.1   Overview   83
  6.2   Restructuring Mobile Networks to SDN   84
     6.2.1   *LTE Network: A Starting Point*   84
     6.2.2   *Options for Location of the SDMN Controller*   86
     6.2.3   *Vision of SDN in LTE Networks*   88
  6.3   Mobile Backhaul Scaling   91
  6.4   Security and Distributed FW   95
     6.4.1   *Customer Edge Switching*   97
     6.4.2   *RG*   97
  6.5   SDN and LTE Integration Benefits   98
  6.6   SDN and LTE Integration Benefits for End Users   100
  6.7   Related Work and Research Questions   103
     6.7.1   *Research Problems*   104
     6.7.2   *Impact*   104
  6.8   Conclusions   104
  References   105

7   EPC in the Cloud   107
James Kempf and Kumar Balachandran
7.1   Introduction   107
   7.1.1   Origins and Evolution of SDN   108
   7.1.2   NFV and Its Application   109
   7.1.3   SDN and Cross-Domain Service Development   112
7.2   EPC in the Cloud Version 1.0   115
7.3   EPC in the Cloud Version 2.0?   117
   7.3.1   UE Multihoming   117
   7.3.2   The EPC on SDN: OpenFlow Example   119
7.4   Incorporating Mobile Services into Cross-Domain
    Orchestration with SP-SDN   123
7.5   Summary and Conclusions   125
References   126

**8   The Controller Placement Problem in Software
Defined Mobile Networks (SDMN)   129**
Hakan Selvi, Selcan Güner, Gürkan Gür, and Fatih Alagöz
8.1   Introduction   129
8.2   SDN and Mobile Networks   130
8.3   Performance Objectives for SDMN Controller Placement   132
   8.3.1   Scalability   133
   8.3.2   Reliability   133
   8.3.3   Latency   134
   8.3.4   Resilience   135
8.4   CPP   136
   8.4.1   Placement of Controllers   137
   8.4.2   Number of Required Controllers   143
   8.4.3   CPP and Mobile Networks   145
8.5   Conclusion   146
References   147

9   Technology Evolution in Mobile Networks: Case of Open IaaS Cloud Platforms   149
Antti Tolonen and Sakari Luukkainen
9.1   Introduction   149
9.2   Generic Technology Evolution   150
9.3   Study Framework   152
9.4   Overview on Cloud Computing   153
9.5   Example Platform: OpenStack   154
   9.5.1   OpenStack Design and Architecture   155
   9.5.2   OpenStack Community   156
9.6   Case Analysis   156
   9.6.1   Openness   157
   9.6.2   Added Value   157
   9.6.3   Experimentation   158
   9.6.4   Complementary Technologies   158

	9.6.5	Incumbent Role	159
	9.6.6	Existing Market Leverage	160
	9.6.7	Competence Change	160
	9.6.8	Competing Technologies	160
	9.6.9	System Architecture Evolution	161
	9.6.10	Regulation	161
9.7	Discussion		162
9.8	Summary		164
Acknowledgments			165
References			165

PART III   TRAFFIC TRANSPORT AND NETWORK MANAGEMENT

10   Mobile Network Function and Service Delivery Virtualization and Orchestration   **169**

Peter Bosch, Alessandro Duminuco, Jeff Napper, Louis (Sam) Samuel, and Paul Polakos

10.1	Introduction		169
10.2	NFV		170
	10.2.1	The Functionality of the Architecture	170
	10.2.2	Operation of the ETSI NFV System	174
	10.2.3	Potential Migration and Deployment Paths	177
	10.2.4	NFV Summary	182
10.3	SDN		182
10.4	The Mobility Use Case		183
10.5	Virtual Networking in Data Centers		185
10.6	Summary		186
References			186

11   Survey of Traffic Management in Software Defined Mobile Networks   **189**

Zoltán Faigl and László Bokor

11.1	Overview		189
11.2	Traffic Management in Mobile Networks		190
11.3	QoS Enforcement and Policy Control in 3G/4G Networks		191
	11.3.1	QoS for EPS Bearers	193
	11.3.2	QoS for Non-3GPP Access	195
	11.3.3	QoS Enforcement in EPS	195
	11.3.4	Policy and Charging Control in 3GPP	195
	11.3.5	Policy Control Architecture	196
11.4	Traffic Management in SDMNs		198
	11.4.1	Open Networking Foundation	198
	11.4.2	The OF Protocol	199
	11.4.3	Traffic Management and Offloading in Mobile Networks	200
11.5	ALTO in SDMNs		201
	11.5.1	The ALTO Protocol	202
	11.5.2	ALTO–SDN Use Case	202

11.5.3   The ALTO–SDN Architecture   204
11.5.4   Dynamic Network Information Provision   205
11.6   Conclusions   206
References   206

12   Software Defined Networks for Mobile Application Services   209
Ram Gopal Lakshmi Narayanan
12.1   Overview   209
12.2   Overview of 3GPP Network Architecture   210
12.3   Wireless Network Architecture Evolution toward NFV and SDN   212
12.3.1   NFV in Packet Core   212
12.3.2   SDN in Packet Core   213
12.4   NFV/SDN Service Chaining   215
12.4.1   Service Chaining at Packet Core   215
12.4.2   Traffic Optimization inside Mobile Networks   217
12.4.3   Metadata Export from RAN to Packet CN   221
12.5   Open Research and Further Study   222
Acknowledgments   223
References   223

13   Load Balancing in Software Defined Mobile Networks   225
Ijaz Ahmad, Suneth Namal Karunarathna, Mika Ylianttila, and Andrei Gurtov
13.1   Introduction   225
13.1.1   Load Balancing in Wireless Networks   226
13.1.2   Mobility Load Balancing   227
13.1.3   Traffic Steering   227
13.1.4   Load Balancing in Heterogeneous Networks   227
13.1.5   Shortcomings in Current Load Balancing Technologies   227
13.2   Load Balancing in SDMN   229
13.2.1   The Need of Load Balancing in SDMN   230
13.2.2   SDN-Enabled Load Balancing   233
13.3   Future Directions and Challenges for Load Balancing Technologies 244
References   244

PART IV   RESOURCE AND MOBILITY MANAGEMENT
**14   QoE Management Framework for Internet Services in
SDN-Enabled Mobile Networks   249**
Marcus Eckert and Thomas Martin Knoll
14.1   Overview   249
14.2   Introduction   250
14.3   State of the Art   251
14.4   QoE Framework Architecture   252
14.5   Quality Monitoring   254
14.5.1   Flow Detection and Classification   254
14.5.2   Video Quality Measurement   255

	14.5.3	Video Quality Rating	255
	14.5.4	Method of Validation	257
	14.5.5	Location-Aware Monitoring	259
14.6	Quality Rules		259
14.7	QoE Enforcement (QEN)		260
14.8	Demonstrator		261
14.9	Summary		263
References			264

15 Software Defined Mobility Management for Mobile Internet **265**
Jun Bi and You Wang
15.1	Overview		265
	15.1.1	Mobility Management in the Internet	265
	15.1.2	Integrating Internet Mobility Management and SDN	267
	15.1.3	Chapter Organization	267
15.2	Internet Mobility and Problem Statement		268
	15.2.1	Internet Mobility Overview	268
	15.2.2	Problem Statement	271
	15.2.3	Mobility Management Based on SDN	273
15.3	Software Defined Internet Mobility Management		274
	15.3.1	Architecture Overview	274
	15.3.2	An OpenFlow-Based Instantiation	275
	15.3.3	Binding Cache Placement Algorithm	277
	15.3.4	System Design	281
15.4	Conclusion		285
References			285

16 Mobile Virtual Network Operators: A Software Defined Mobile Network Perspective **289**
M. Bala Krishna
16.1	Introduction		289
	16.1.1	Features of MVNO	291
	16.1.2	Functional Aspects of MVNO	292
	16.1.3	Challenges of MVNO	293
16.2	Architecture of MVNO: An SDMN Perspective		294
	16.2.1	Types of MVNOs	294
	16.2.2	Hierarchical MVNOs	294
16.3	MNO, MVNE, and MVNA Interactions with MVNO		296
	16.3.1	Potential Business Strategies between MNOs, MVNEs, and MVNOs	299
	16.3.2	Performance Gain with SDN Approach	300
	16.3.3	Cooperation between MNOs and MVNOs	300
	16.3.4	Flexible Business Models for Heterogeneous Environments	301
16.4	MVNO Developments in 3G, 4G, and LTE		303
	16.4.1	MVNO User-Centric Strategies for Mobility Support	303
	16.4.2	Management Schemes for Multiple Interfaces	304
	16.4.3	Enhancing Business Strategies Using SDN Approach	304

16.5   Cognitive MVNO   305
        16.5.1   *Cognitive Radio Management in MVNOs*   305
        16.5.2   *Cognitive and SDN-Based Spectral Allocation*
                 Strategies in MVNO   306
16.6   MVNO Business Strategies   307
        16.6.1   *Services and Pricing of MVNO*   308
        16.6.2   *Resource Negotiation and Pricing*   309
        16.6.3   *Pushover Cellular and Service Adoption Strategy*   309
        16.6.4   *Business Relations between the MNO and MVNO*   310
16.7   Conclusions   310
16.8   Future Directions   311
References   311

PART V   SECURITY AND ECONOMIC ASPECTS
17   Software Defined Mobile Network Security   **317**
        Ahmed Bux Abro
17.1   Introduction   317
17.2   Evolving Threat Landscape for Mobile Networks   318
17.3   Traditional Ways to Cope with Security Threats
        in Mobile Networks   318
        17.3.1   *Introducing New Controls*   318
        17.3.2   *Securing Perimeter*   319
        17.3.3   *Building Complex Security Systems*   320
        17.3.4   *Throwing More Bandwidth*   320
17.4   Principles of Adequate Security for Mobile Network   320
        17.4.1   *Confidentiality*   321
        17.4.2   *Integrity*   321
        17.4.3   *Availability*   321
        17.4.4   *Centralized Policy*   321
        17.4.5   *Visibility*   322
17.5   Typical Security Architecture for Mobile Networks   322
        17.5.1   *Pros*   323
        17.5.2   *Cons*   325
17.6   Enhanced Security for SDMN   325
        17.6.1   *Securing SDN Controller*   325
        17.6.2   *Securing Infrastructure/Data Center*   325
        17.6.3   *Application Security*   326
        17.6.4   *Securing Management and Orchestration*   326
        17.6.5   *Securing API and Communication*   326
        17.6.6   *Security Technologies*   326
17.7   SDMN Security Applications   327
        17.7.1   *Encryption: eNB to Network*   327
        17.7.2   *Segmentation*   327
        17.7.3   *Network Telemetry*   329
        References   329

18 Security Aspects of SDMN   **331**
Edgardo Montes de Oca and Wissam Mallouli
18.1 Overview   331
18.2 State of the Art and Security Challenges in SDMN Architectures   331
       18.2.1 Basics   332
       18.2.2 LTE-EPC Security State of the Art   332
       18.2.3 SDN Security in LTE-EPC State of the Art   334
       18.2.4 Related Work   339
18.3 Monitoring Techniques   344
       18.3.1 DPI   347
       18.3.2 NIDS   348
       18.3.3 Software Defined Monitoring   349
18.4 Other Important Aspects   351
       18.4.1 Reaction and Mitigation Techniques   351
       18.4.2 Economically Viable Security Techniques for Mobile Networks   352
       18.4.3 Secure Mobile Network Services and Security Management   353
18.5 Conclusion   354
References   355

19 SDMN: Industry Architecture Evolution Paths   **359**
Nan Zhang, Tapio Levä, and Heikki Hämmäinen
19.1 Introduction   359
19.2 From Current Mobile Networks to SDMN   360
       19.2.1 Current Mobile Network Architecture   360
       19.2.2 Evolutionary SDMN Architecture   361
       19.2.3 Revolutionary SDMN Architecture   363
19.3 Business Roles of SDMN   364
19.4 Industry Architectures of Evolutionary SDMN   366
       19.4.1 Monolithic MNO   366
       19.4.2 Outsourced Subscriber Management   368
       19.4.3 Outsourced Connectivity   370
19.5 Industry Architectures of Revolutionary SDMN   371
       19.5.1 MVNO   371
       19.5.2 Outsourced Interconnection   372
       19.5.3 Outsourced Mobility Management   374
19.6 Discussion   374
References   376

Index   **379**

Editors

Madhusanka Liyanage

Centre for Wireless Communication, University of Oulu, Oulu, Finland

Madhusanka Liyanage received the B.Sc. degree in electronics and telecommunication engineering from the University of Moratuwa, Moratuwa, Sri Lanka, in 2009; the M.Eng. degree from the Asian Institute of Technology, Bangkok, Thailand, in 2011; and the M.Sc. degree from University of Nice Sophia Antipolis, Nice, France, in 2011.

He is currently a research scientist with the Centre for Wireless Communication, Department of Communications Engineering, University of Oulu, Oulu, Finland. In 2011–2012, he was a research scientist at I3S Laboratory and Inria, Shopia Antipolis, France. His research interests are SDN, mobile and virtual network security. He is a student Member of IEEE and ICT.

Madhusanka is a coauthor of various publications including books chapters, journals, and research papers. He is also one of the work package leaders of Celtic call 2012 SIGMONA (SDN Concept in Generalized Mobile Network Architectures) project.

Andrei Gurtov

Helsinki Institute for Information Technology HIIT, Aalto University, Espoo, Finland

Andrei Gurtov received his M.Sc. (2000) and Ph.D. (2004) degrees in computer Science from the University of Helsinki, Finland. He is a principal scientist at the Helsinki Institute for Information Technology HIIT, Aalto University.

He is also adjunct professor at Aalto University, University of Helsinki and University of Oulu. He was a Professor at University of Oulu in the area of Wireless Internet in 2010–2012. Previously, he worked at TeliaSonera, Ericsson NomadicLab, and University of Helsinki. He was a visiting scholar at the International Computer Science Institute (ICSI), Berkeley in 2003, 2005, and 2013.

Dr. Gurtov is a coauthor of over 130 publications including two books, research papers, patents, and five IETF RFCs. He is a senior member of IEEE.

Mika Ylianttila

Centre for Internet Excellence, University of Oulu, Oulu, Finland

Mika Ylianttila received his doctoral degree in communications engineering at the University of Oulu, in 2005. He has worked as a researcher and professor at the Department of Electrical and Information Engineering.

He is the director of the Centre for Internet Excellence (CIE) research and innovation unit. He is also docent at the Department of Computer Science and Engineering. He was appointed as a part-time professor at the Department of Communications Engineering for a 3-year term from January 1, 2013. The field of the professorship is broadband communications networks and systems, especially wireless Internet technologies.

He has published more than 80 peer-reviewed articles on networking, decentralized (peer-to-peer) systems, mobility management, and content distribution. Based on Google Scholar, his research has impacted more than 1500 citations, and his h-index is 19. He was a visiting researcher at Center for Wireless Information Network Studies (CWINS), Worcester Polytechnic Institute, Massachusetts, and Internet Real Time Lab (IRT), Columbia University, New York, USA. He is a senior member of IEEE, and editor in *Wireless Networks Journal*.

Contributors

Ahmed Bux Abro (CCDE, CCIE Security, CISSP) is a technologist, strategist, and contributor for multiple technology fronts such as Digital Transformation, IoE, ITaaS, Cloud, SDDC, and SDN. Bringing change through the technology innovation and being distinguished in developing new technologies, architectures, frameworks, solutions, and services to solve complex and challenging business problems have remained his area of focus. He has introduced various new frameworks, architectures and standards around Cloud, Network Function Virtualization, and SDN. He is author of a book, multiple drafts, and papers with some patents on the waiting list. Playing an influential and mentor role for the professional community by leading a program that enables senior architects and experts, and help them in attaining necessary professional skillset and certifications. He has played technology leader role in Fortune 50 in diverse markets (North America, EMEA, Asia) and industry sectors. He is also a frequent public speaker in multiple industry events.

Ijaz Ahmad received his B.Sc. degree in computer systems engineering from University of Engineering and Technology (UET), Peshawar, Pakistan, in 2008. He completed his M.Sc. (Technology) degree of wireless communications engineering with major in telecommunications engineering from University of Oulu, Oulu, Finland, in 2012. Currently, he is a doctoral student with the Department of Communications Engineering, University of Oulu, Oulu, Finland, and a research scientist in Centre for Wireless Communications (CWC), Oulu, Finland. His research interest includes software defined mobile networks, network security, and network load balancing.

Fatih Alagöz is a professor in the Department of Computer Engineering, Bogazici University. He received the B.Sc. degree in electrical engineering from Middle East Technical University, Turkey, in 1992, and the M.Sc. and D.Sc. degrees in electrical engineering from the George Washington University, USA, in 1995 and 2000, respectively. His current research interests are in the areas of wireless/mobile/satellite networks. He has contributed/managed to ten research projects for the US Army of Intelligence Center, Naval Research Laboratory, UAE

Research Fund, Turkish Scientific Research Council, State Planning Organization of Turkey, BAP, etc. He has published more than 150 scholarly papers in selected journals and conferences.

M. Bala Krishna received the B.E. degree in computer engineering from Delhi Institute of Technology (presently Netaji Subhash Institute of Technology), University of Delhi, Delhi, India, and the M.Tech. degree in information technology from University School of Information Technology (presently University School of Information and Communication Technology), GGS Indraprastha University, Delhi, India. He had received his Ph.D. in computer engineering from JMI Central University, New Delhi, India. He had earlier worked as Senior Research Associate and Project Associate at Indian Institute Technology, Delhi, India, in the areas of Digital Systems and Embedded Systems. He has worked as Faculty Member and handled projects related to Networking and Communication. He is presently working as Assistant Professor in University School of Information and Communications Technology (formerly University School of Information Technology), GGS Indraprastha University, New Delhi, India. His areas of interest include computer networks, wireless networking and communications, mobile and ubiquitous computing, and embedded system design. He has publications in International Journals, Conferences and book chapters. His teaching areas include wireless networks, mobile computing, data and computer communications, embedded systems, programming languages, etc. His current research work includes wireless ad hoc and sensor networks, green networking and communications, cognitive networks and smart grid communications. He is member of IEEE and ACM Technical Societies.

Kumar Balachandran received his B.E. in electronics and communications engineering with honors in 1986 from the Regional Engineering College, Tiruchirapalli, India, and his masters and doctorate in computer and systems engineering from Rensselaer Polytechnic Institute, Troy, NY, in 1988 and 1992, respectively. His postgraduate work was on convolutional codes and trellis-coded modulation. He started his career at PCSI where he helped specify and build Cellular Digital Packet Data (CDPD). Since 1995, he has worked at Ericsson Research on a broad range of topics in wireless communications. He is well published and frequently invited as speaker and panelist. He is named on 54 US patents. He is currently Principal Research Engineer at Ericsson Research at San Jose. His recent work is on spectrum issues and 5G Systems.

Pascal Berthou is an associate professor of computer science in the University of Toulouse, France, and Senior Research Scientist at LAAS-CNRS French research Labs, Toulouse, France. His current research focuses on network QoS-support for autonomic applications, wireless sensor networks, multinetwork communication architecture, software defined networking, network virtualization, multimedia applications over QoS-based broadband satellite systems, 5G broadband wireless networks, middleware-based communication, and network emulation.

Jun Bi received the B.S., M.S., and Ph.D. degrees in computer science from Tsinghua University, Beijing, China. He was a postdoctoral scholar at Bell Laboratories Research, and a research scientist at Bell Labs Research and Bell Labs Advanced Communication Technologies Center. Currently, he is a full professor and director of Network Architecture & IPv6 Research

Division at Institute for Network Sciences and Cyberspace, Tsinghua University. His research interests include Internet architecture and protocols, future Internet (SDN and NDN), Internet routing, and source address validation and traceback. Please visit http://netarchlab.tsinghua.edu.cn/~junbi for more information. Jun Bi is the corresponding author of this chapter.

László Bokor graduated in 2004 with M.Sc. degree in computer engineering from the Budapest University of Technology and Economics (BME) at the Department of Telecommunications. In 2006, he received an M.Sc.+ degree in bank informatics from the same university's Faculty of Economic and Social Sciences. He is a Ph.D. candidate at BME, member of the IEEE, Multimedia Networks Laboratory, and Mobile Innovation Centre of BME where he works on advanced mobility management related projects (i.e., FP6-IST PHOENIX and ANEMONE, EUREKA-Celtic BOSS, FP7-ICT OPTIMIX, EURESCOM P1857, EUREKA-Celtic MEVICO, EUREKA-Celtic SIGMONA, and FP7-ICT CONCERTO).

Peter Bosch currently addresses cellular and noncellular packet routing combined with in-line service routing for 5G mobile systems. Before Peter designed and implemented the MobileVPN routing system used to manage mobility between cellular and non-cellular access systems by way of IP VPN/MP-BGP, codesigned and implemented the world's first UMTS HSDPA systems with MIMO technology, codesigned and implemented the UMTS base station router (BSR), designed and implemented LTE mobility solutions before these were standardized and worked on cloud operating systems and virtual radio access networks. Peter worked at Bell Labs, Alcatel-Lucent and Juniper before joining Cisco.

Brian Brown, CCIE #4805, is a solutions architect in the service provider group within the Cisco Advanced Services organization. In his 14 years at Cisco, he has supported large mobile and wireline networks around design and implementation of new technology. Brian earned his B.S. degree in computer engineering from Virginia Polytechnic and State University, and has worked in the networking industry for 20 years.

Claude Chaudet received his Ph.D. in 2004 from Insa de Lyon, France. He is now Associate Professor at Telecom ParisTech, where he conducts his research in the domains of wireless infrastructureless networks and embedded communicating devices. He published several articles on ad hoc, sensors and body area networks. Besides, he is also interested in large-scale networks where global properties can emerge from local behaviors and published papers related to intelligent transportation systems, smart grids and critical infrastructures protection.

Jose Costa-Requena is a research manager in the COMNET department. He has participated in several EU projects since 2002. Dr. Costa-Requena has been working both in AALTO and Nokia Ltd. for more than 10 years. He has been working as System Program Manager at Nokia with the development of mobile terminals, network infrastructures, and home appliances. He has managed large multisite projects with 10+ subcontractors and over 100 developers. He was in Nokia patenting board for 7+ years and holds 15+ granted patents and 40+ filed applications. He has several journal and conference publications. He is also a coauthor of IETF RFCs, and he has been delegate in 3GPP standardization forum working in architecture CN1, Security SA3 and Services SA2 working groups. Besides 10+ of experience in industry, he has large experience in managing EU projects such as MobileMan, DC2F, SCAMPI, MEVICO, SIGMONA, and PRECIOUS.

Alessandro Duminuco received a bachelor degree in 2003 and a master degree in 2006 in computer science engineering from Politecnico di Torino (Turin, Italy). In 2009, he received a Ph.D. in computer science from Telecom ParisTech at Eurecom (Sophia Antipolis, France), where he worked on distributed systems and defended a thesis on peer-to-peer backup technologies. From 2009 to 2011, he worked for Alcatel-Lucent as researcher in the system infrastructure group, where he worked on cloud technologies applied to telcos focusing on distributed elastic MME for LTE and cloud operating systems. From 2011 to 2012, he worked for Accenture as senior consultant in the IGEM Cloud offering team. Since 2012, Alessandro is part of the mobility CTO team in Cisco, where he works on software defined networks and network function virtualization focusing in particular on in-line services for 4G and 5G mobile systems.

Marcus Eckert studied Information and Communications Systems at TU-Chemnitz, Germany. During the studies, he specialized in communication networks as well as high frequency technics. In April 2011, he received the engineering diploma (Dipl.-Ing.) in information and communications science. In May 2011, he started his work as a research assistant, and since then he have been a member of the research staff at the chair of communication networks within Chemnitz University of Technology. The chair is led by Professor Bauschert. There he is working on my Ph.D. in the field of QoE estimation and management for different Internet services (like Video Services).

Zoltán Faigl received his M.Sc. degree in telecommunications from Budapest University of Technology and Economics (BUTE), Hungary, in 2003. Currently, he works on his Ph.D. at Mobile Innovation Centre, BUTE. His fields of interests are communication protocols, network architectures, information security, mobile and wireless networks, network dimensioning, and decision making. He has participated in several international projects such as FP6-IST PHOENIX and ANEMONE, EURESCOM P1857, EUREKA-Celtic MEVICO, and EUREKA-Celtic SIGMONA.

Xenofon Foukas holds a diploma in computer science from the Department of Informatics in the Athens University of Economics and Business (AUEB) in 2012, and an M.Sc. in advanced computing from the Department of Computing at Imperial College London in 2013, where he was awarded a departmental scholarship on academic distinction. He has collaborated with various institutions including the R&D department of Intracom Telecom S.A. and the Informatics Laboratory (LIX) of École Polytechnique in France. Currently, he has a joint appointment, with NCSR "Demokritos," as a Marie-Curie research fellow for the GREENET project, and with the University of Edinburgh where he is pursuing a Ph.D. degree in wireless networks. His research interests include software defined mobile networks, energy-efficient wireless communication networks, wireless network management, and distributed algorithms.

Rob Gonzalez (CCDE 20130059, CCIE 20547) is a solutions architect at Cisco Systems, working in advanced services with service provider customers. He has worked on large-scale projects from physical layer optical networks to the latest IP NGN and Mobile backhaul networks for many of the largest ISP's in the United States. His influence ranges from the planning and design phase of a project through implementation and support. He has earned two CCIE certifications and the CCDE certification. Rob received his B.S. degree in electrical engineering from Georgia Tech in Atlanta, GA, in 1993 and has worked in the telecommunications and networking industry for 20 years.

Vicent Ferrer Guasch received his M.Sc. in communications engineering from Aalto University, Finland, in 2014, with major in networking technology. He has worked as IT support for HP in Spain. In Finland, he has worked as a research assistant and Ph.D. student at the Department of Communications and Networking in Aalto University since September 2014. His research interests include Future Internet architectures, SDN, and LTE networks.

Selcan Güner is a master degree student in Bogazici University in Computer Engineering department. Her research activities for master thesis has a focus on software defined networks. She received her B.Sc. degree in computer engineering from Bogazici University in 2011. She received her BA degree in management from Anadolu University in 2009. Outside the academia, she has been working in industry as a full-time software developer. She has been specialized in the mobile operating systems as a senior developer.

Gürkan Gür received his B.S. degree in electrical engineering in 2001, and Ph.D. degree in computer engineering in 2013 from Bogazici University. Currently, he is a senior researcher at Provus—A Mastercard Company. His research interests include cognitive radios, green wireless communications, small cell networks, network security, and information-centric networking.

Yoram Haddad received his B.Sc., Engineer diploma and M.Sc. (Radiocommunications) from SUPELEC (leading engineering school in Paris, France) in 2004 and 2005, and his Ph.D. in computer science and networks from Telecom ParisTech in 2010 all awarded with honors. He was a Kreitman postdoctoral fellow at Ben-Gurion University, Israel, between 2011 and 2012. He is actually a tenured senior lecturer at the Jerusalem College of Technology (JCT)-Lev Academic Center in Jerusalem, Israel. Yoram's published dozens of papers in international conferences (e.g., SODA) and journals. He served on the Technical Program Committee of major IEEE conferences and is the chair of the working group on methods and tools for monitoring and service prediction of the European COST action on Autonomous Control for a Reliable Internet of Services (ACROSS). He is the recipient of the Henry and Betty Rosenfelder outstanding researcher award for year 2013. Yoram's main research interests are in the area of wireless networks and algorithms for networks. He is especially interested in energy-efficient wireless deployment, Femtocell, modeling of wireless networks, device-to-device communication, wireless software defined networks (SDNs) and technologies toward 5G cellular networks. He has been recently awarded a grant as a PI on SDN "programmable network" within the framework of the MAGNET NEPTUNE consortium from the Office of Chief Scientist of Israel.

Akram Hakiri, Ph.D., is an associate professor of computer science at the University of Haute-Alsace, and Research Scientist at LAAS-CNRS. His current research focuses on developing novel solutions to emerging challenges in wireless networks, QoS-based broadband satellite systems, network virtualization, middleware-based communication, Cloud networking and software defined networking for heterogeneous networks, 5G broadband communication, and multi-technologies communication systems.

Heikki Hämmäinen is a professor of network economics at Department of Communications and Networking, Aalto University, Finland. His main research interests are in technoeconomics and regulation of mobile services and networks. Special topics recently include measurement and analysis of mobile usage, value networks of cognitive radio, and diffusion of Internet protocols in mobile. He is also active in International Telecommunications Society, national research foundations in Finland, in addition to several journal and conferences duties.

Raimo Kantola holds an M.Sc. in computer science from St. Petersburg Electrotechnical University (1981), and Dr. of science degree from Helsinki University of Technology (1995). He has 15 years of industry experience in digital switching system product development and architecture, product management, marketing, and research. Since 2005, he is a full professor of networking technology at Aalto University, Finland. He has held many positions of trust at the Helsinki University of Technology, now Aalto University, among them being the first Chairman of the Department of Communications and Networking. He has done research in routing, peer-to-peer, MANET protocols, traffic classification, etc. His current research interests are new networking paradigms, network security, privacy, and trust.

James Kempf graduated from University of Arizona with a Ph.D. in systems engineering in 1984, and immediately went to work in Silicon Valley. Prior to his current position, Dr. Kempf spent 3 years at HP, 13 years at Sun Microsystems, primarily in research, and 8 years at Docomo Labs USA as a research fellow. Dr. Kempf worked for 10 years in IETF, was chair of three working groups involved in developing standards for the mobile and wireless Internet, and was a member of the Internet Architecture Board for 2 years. He is the author of many technical papers and three books, the latest of which, *Wireless Internet Security: Architecture and Protocols* was published by Cambridge University Press in 2008. Since 2008, Dr. Kempf has worked at Ericsson Research in Silicon Valley on Software Defined Networking (SDN)/ OpenFlow and cloud computing.

Maël Kimmerlin received his B.Sc. in information technology from Télécom SudParis in 2012. He is working in the Department of Communications and Networking at Aalto University as a research assistant.

Thomas Martin Knoll studied electrical engineering at the Chemnitz University of Technology, Germany, and at the University of South Australia, Adelaide. He specialized in information technology and earned the Dipl.-Ing. degree in 1999. From then on, he has been a member of the research staff at the chair of communication networks within Chemnitz University of Technology. Besides ongoing teaching tasks in data communications, he is focusing on carrier network technologies for QoS support within and between AS, as well as the resulting QoE of Internet services. In 2009, he received a Ph.D. from Chemnitz University for his thesis about a coarse grained Class of Service based interconnection concept. Thomas Martin Knoll is member IEEE, ACM, ITU, and VDE.

Kimon Kontovasilis holds a 5-year diploma and a Ph.D. in electrical engineering from the National Technical University of Athens, and an M.Sc. in computer science from North Carolina State University. He is with the Institute for Informatics & Telecommunications of the National Center for Scientific Research "Demokritos" since 1996, currently ranking as a research director and serving as the Head of the Telecommunication Networks Laboratory. His research interests span several areas of networking, including modeling, performance evaluation and resource management of networks, management and optimization of heterogeneous wireless networks, energy-efficient wireless networks, and mobile ad hoc and delay-tolerant networks. He has participated in a considerable number of research projects at a national and international scale and has been involved in the organizing or technical program committees of numerous international conferences in the field of networking. He is a member of IFIP WG6.3.

Ram Gopal Lakshmi Narayanan is a Principle architect, Verizon, USA. For the past 20 years, he has been working on various research projects in the area of wireless networking, Internet security and privacy, network analytics and video optimization, and machine learning. He is one of the authors for ForCES requirement and protocol specification in IETF standards. He has contributed to several standards working groups including Internet Engineering Task Force, Network Processing Forum, Service Availability Forum, trusted computing group and held NPF high availability task group chair position. He holds more than 25 patents and published several papers. He received the B.S. degree in information systems from Birla Institute of Technology and Science, Pilani, India, M.S. degree in computer science from Boston University, and Ph.D. from the University of Massachusetts.

Tapio Levä received his M.Sc. in communications engineering from Helsinki University of Technology (TKK), Finland, in 2009, with major in networking technology and minors in Telecommunications Management and Interactive Digital Media. He is currently finalizing his dissertation in the Department of Communications and Networking at Aalto University concerning the feasibility analysis of new Internet protocols. His research interests include techno-economics of Internet architecture evolution, Internet standards adoption, and Internet content delivery.

Sakari Luukkainen took his D.Sc. at Helsinki University of Technology. In the 1990s, he worked in the Technical Research Centre of Finland, where he directed the Multimedia Communications research group. He also has practical managerial experience in technology companies of the telecommunications industry. Today, he works as a senior research scientist in the Department of Computer Science and Engineering at Aalto University, and is responsible for the Networking Business education program, which combines business and technology studies in the telecommunications field. Mr. Luukkainen's research interests include technology innovation management and commercialization of new network services. Currently, a part of his research effort is directed to studying the effects of virtualization and cloud computing in mobile networking business.

Wissam Mallouli, senior R&D engineer at Montimage, has graduated from the National Institute of Telecommunication (INT) engineering school in 2005. He received his masters degree from Evry Val d'Essonne University also in 2005, and his Ph.D. in computer science from Telecom and Management SudParis, France, in 2008. His topics of interest cover formal security testing of embedded systems and distributed networks. He is expert is testing methodologies. He has a strong background in monitoring and testing network security (protocols and equipment). He also has a solid experience in project and R&D management. He is involved in several projects such as the FP6/FP7 IST calls, CELTICPlus, ITEA projects, and national ones. He also participates in the program committees of numerous national and international conferences. He published more than 20 papers in conference proceedings, books, and journals.

Jukka Manner received his M.Sc. (1999) and Ph.D. (2004) degrees in computer science from the University of Helsinki. He is a full professor (tenured) of networking technology at Aalto University, Department of Communications and Networking (Comnet). His research and teaching focuses on networking, software, and distributed systems, with a strong focus on wireless and mobile networks, transport protocols, energy-efficient ICT, and cyber security. He was the academic coordinator for the Finnish Future Internet research programme 2008–2012. He is an active peer reviewer and member of various TPCs. He was the local co-chair of Sigcomm 2012 in Helsinki. He has contributed to standardization of Internet technologies in the IETF since 1999, and was the co-chair of the NSIS working group. He has been principal investigator and project manager for over 15 national and international research projects. He has authored over 90 publications, including 10 IETF RFCs. He is a member of the ACM and the IEEE.

Mahesh K. Marina is a reader in the School of Informatics at the University of Edinburgh, UK. Prior to joining Edinburgh, he had a 2-year postdoctoral stint at UCLA. During 2013, he was a visiting researcher at ETH Zurich and Ofcom (the UK telecommunications regulator) Head Office in London. He received his Ph.D. from the State University of New York at Stony Brook. His current research interests include wireless network management (including network monitoring and dynamic spectrum access), mobile phone sensing, next-generation mobile cellular networks, and software defined networking. His research has received recognition in the form of awards/nominations for best papers/demos at IEEE SECON 2013, IEEE IPIN 2013, IEEE WiNMee 2012, and ACM WiNTECH 2010. His work on the Tegola rural wireless broadband project won the NextGen Challenge 2011 Prize for its positive community/societal impact. A co-founder of the ACM MobiCom WiNTECH workshop focusing on experimental wireless networking, he has chaired two ACM/IEEE international workshops in wireless networking and mobile computing areas, and also served on 30+ technical program committees of international conferences/workshops in those areas. He is an IEEE senior member.

Antti Mikola received his B.Sc. in communications engineering from Aalto University in 2013 with a major in networking technology. He has worked as a research assistant at the Department of Communications and Networking in Aalto University since April 2014.

Edgardo Montes de Oca graduated as engineer in 1985 from Paris XI University, Orsay. He has worked as research engineer in the Alcatel Corporate Research Centre in Marcoussis, France, and in Ericsson's Research centre in Massy, France. In 2004, he founded Montimage, and is currently its CEO. He is the originator and main architect of Montimage Monitoring Tool (MMT). His main interests are network and application monitoring and security;

detection and mitigation of cyberattacks; and, building critical systems that require the use of state-of-the-art fault-tolerance, testing, and security techniques. He has participated and participates as company and WP leader of several EU and French national projects (e.g., CelticPlus projects MEVICO and SIGMONA, CIP-PSP projects ACDC and SWEPT). He is also member of the APWG (Anti-Phising Working Group).

Suneth Namal Karunarathna received the B.Sc. degree in computer engineering from the University of Peradeniya, Peradeniya, Sri Lanka, in 2007. He completed M.Eng. degree in information and communications technologies from the Asian Institute of Technology, Bangkok, Thailand, in 2010, and M.Sc. degree in communication network and services from Telecom SudParis, Paris, France, in 2010. Currently, he is a doctoral student with the Department of Communication Engineering, University of Oulu, Oulu, Finland, and a research scientist of Centre for Wireless Communications, Oulu, Finland. His research interest includes software defined wireless networks, communication security, mobile femtocells, fast initial authentication, and load balancing.

Jeff Napper currently works for Cisco developing proof-of-concept systems for mobile service providers. After graduating from the University of Texas at Austin, he worked on Grid Computing systems for VU University Amsterdam and on developing new operating systems for Bell Labs, Alcatel-Lucent before joining Cisco.

Paul Polakos is currently a Cisco fellow in the Mobility CTO Group at Cisco Systems where he is focusing on emerging technologies for future mobile networks. Prior to joining Cisco, he was a senior director of wireless networking research and Bell Labs Fellow at Bell Labs, Alcatel-Lucent in Murray Hill, NJ and Paris, France. He holds B.S., M.S., and Ph.D. degrees in physics from Rensselaer Polytechnic Institute and the University of Arizona.

Louis (Sam) Samuel recently joined Cisco, and is currently a director of engineering in the Chief Technical Office of Cisco's Mobility Business Group. Sam covers such topics as virtualization and orchestration of mobility products and small cell technologies. Prior to joining Cisco, Sam held several posts in Alcatel-Lucent. Among these were the posts of Chief Architect for Alcatel-Lucent's Software, Solutions and Services Business Group (S3G). In this role, Sam was responsible for the strategic technical leadership of S3G covering a wide remit of areas from software architecture innovation to innovation of Services.

Jesus Llorente Santos received his M.Sc. in communications engineering with honors from Aalto University, Finland, in 2012, with major in networking technology. He has worked as IT support for Telefonica in Spain. In Finland, he has worked as a Research Assistant and PhD student at the Department of Communications and Networking in Aalto University since April 2011. His research interests include Future Internet architectures, Network Address Translator, Realm Gateway, and Customer Edge Switching.

Hakan Selvi is a master degree student in the Department of Computer Engineering, Bogazici University. His current researches and thesis are related with software defined networks. He received his bachelor of science degree in computer engineering in 2010 from Bogazici University.

Brian Stanford is a solutions architect in the service provider group within the Cisco Advanced Services organization. His breadth of experience encompasses a wide variety of areas including IP NGN, IP RAN, Data Center, and Campus technologies. He joined the Wan Switching group in 1999 where he focused on Global Frame Relay & ATM networks. In his 15 years at Cisco, he has supported Global Enterprise customers as well as Global Service Providers. In his most recent role, his focus has been on promoting the adoption of a Network Architecture approach in US mobile carrier networks. As a solutions architect, He has acquired a thorough understanding of technology and how to apply those technologies to solve complex business problems. He has been CCIE certified since 1999 and currently holds four CCIE certifications (Routing & Switching, Wan Switching, SP, Security). He is also a project management professional (PMP) # 292000.

Antti Tolonen is a master's student at Aalto University School of Science studying in the master's programme in mobile computing—services and security. He has been working in both teaching and research assistant roles at Aalto University's Department of Computer Science and Engineering for the past several years during his studies. The research topics he has taken part in have consisted of peer-to-peer networking, cloud computing, and modern Web-based data transfer technologies. Currently, he is finishing his master's thesis on building and validating a testbed that allows studying the utilization of cloud computing in the LTE core network. Mr. Tolonen's future interests include software development in the fields of virtualization and networking.

You Wang received the B.S. degree in computer science from Tsinghua University, Beijing, China. He is now a Ph.D. candidate at Department of Computer Science, Tsinghua University. His research interests include Internet mobility management, identifier/locator split and software defined networking (SDN).

Nan Zhang received her M.Sc. (Tech.) in communications engineering from Aalto University School of Science and Technology, Finland, in 2010, with major in Data Networks and minor in strategy and international business. She is currently doing her postgraduate studies at the Department of Communications and Networking at Aalto University, Finland. Her research interests include techno-economics of Internet content delivery solutions, such as content delivery networks, information-centric based solutions, and software defined networks.

Foreword

Originally, the Internet was designed as a decentralized packet switching network that is multiply redundant, fault-tolerant and with mostly peripheral computational components. Today, the Internet has usurped the majority of all information and communications technology (ICT) functions in society, spanning the range of media access services such as television, music and video streaming, Web access, and interactive telecommunications (voice and video telephony), as well as supporting a diverse range of applications that connect machine devices in various environments to each other and to the network. In parallel, the relentless reduction in price, power consumption, and device size driven by Moore's law has led to computation, storage, and networking becoming so inexpensive that intelligence can be incorporated into almost all manufactured goods. Enterprise IT systems are also rapidly being centralized with cloud technologies, creating huge efficiencies in the way computation, networking, and storage are provisioned and deployed. The broad impact that the Internet and inexpensive computation are having on life is leading to a networked society, in which anything that would benefit from being connected, will be connected.

Central among the supporting technologies for a networked society are mobile networks which provide connectivity both for the devices that constitute the Internet of Things, as well as for devices used for communication by people. Mobile networks face increasing challenges going forward as the characteristics of the data transported over them will vary widely, from large volumes of small and periodic sensor readings to large high-definition video streams. New services such as vehicular communications and critical infrastructure for industrial applications of the Internet pose exacting requirements on security, latency, and reliability. Mobile operators must build and provision their networks to accommodate this flood of data, and simplify the way new services are defined and provisioned in accordance with their customers' needs. Vendors must in turn provide operators with systems that solve the posed challenges in a cost-effective manner.

Fortunately, new technologies in the pipeline will help the ICT industry rise to the challenge. Cloud computing and software defined networking (SDN), originally developed in enterprise networks, are now moving into operator networks, including mobile networks, through the European Telecommunication Standards Institute (ETSI) Network Function Virtualization

(NFV) effort. The articles in this book address important research questions in software defined mobile networking (SDMN), and represent the fruits of academic and industrial research over the recent years. Research in SDMN, such as that described in this volume, will play a critical role in defining NFV and in the 5G mobile network.

Ulf Ewaldsson
Senior Vice President, Chief Technology Officer
Ericsson

Foreword

Mobile networking is entering an exciting era of development, driven both by user needs and new technologies. Key new technologies are network function virtualization (NFV) and software defined networking (SDN). They are often considered together or even confused with each other. Simplified, NFV separates the network functions from the underlying hardware and software platforms, whereas SDN separates network control from the user data routing. They bring benefits in capacity scaling, cost reduction, and flexibility and speed in the introduction of new services. NFV concepts are already rather broadly agreed, for example, via the ETSI specification work and NFV-based products are entering into commercial service. Similar broad agreement on how to utilize SDN in mobile networks is missing, so this book is a timely contribution to that discussion. To get the full benefit of NFV and SDN in mobile networks, we need to take a new look at the network architecture as it now stands.

This book takes that look, both broad and deep. It introduces current state of the art and probes potential ways to evolve mobile networks to take better advantage of SDN and NFV. It discusses system wide and product architectures and key issues like network management, quality of service, and security. The technologies exposed in this book are central to the evolution of mobile networking and future definition of the 5G architecture. While there are no final answers, yet the book provides a good update on the leading edge of research in SDN and NFV for mobile networks.

Lauri Oksanen
VP, Research and Technology
Nokia Networks

Preface

The main objective of this book is to provide the cutting-edge knowledge about software defined mobile network (SDMN) architecture. SDMN is one of the promising technologies that are expected to solve the existing limitations in current mobile networks. SDMN architecture provides the required improvements in flexibility, scalability, and performance to adapt the mobile network to keep up with the expected future mobile traffic growth.

This book gives an insight into the feasibility of SDMN concept and its opportunities. It also evaluates the limits of performance and scalability of the SDMN architecture. The book discusses theoretical principles of beyond long-term evolution (LTE) mobile network architectures and their implementation aspects.

The SDMN architecture is based on software defined networking (SDN) and network virtualization principles. The book aims at evaluation, specification, and validation of SDN and network function virtualization (NFV) relevant to future SDMNs. The SDMN concept will change the network architecture of the current LTE mobile networks. It is foreseen that SDMN architecture will offer new opportunities for traffic, resource, and mobility management. Moreover, it will introduce new challenges on network security and impact on the cost of the network, value chain, business models, and the investments on mobile networks. This book presents a well-structured, readable, and complete reference of all these aspects of SDMN architecture. It contains both introductory level text as well as more advanced reference to meet the expectation of readers from various backgrounds and levels.

The Need for SDMN

The first mobile telecommunication network was introduced in the 1980s. During past four decades, the mobile communication technologies have achieved a significant development. The evolving mobile services, rapidly increasing broadband speed and inherent mobility support attract many subscribers. Thus, mobile communication is becoming the primary or even sole access method for more and more people. The present mobile networks support sophisticated network services such as Voice over IP (VoIP), high-density video streaming, high-speed broadband connectivity, and mobile cloud services. As a result, the mobile traffic volume is drastically increasing in each year. It is foreseen that the mobile data traffic usage is growing faster than the fixed Internet for the coming years. Thus, mobile networks must upgrade to keep up with the traffic growth and support rapidly evolving mobile services market. However, it is always challenging to increase the mobile network bandwidth due to the limited radio bandwidth resources and remarkably complex and inflexible backhaul devices.

On the other hand, the present business environment of telecommunication is changing rapidly. Usually, the telecommunication market is highly competitive. However, today's mobile operators must compete with a new class of competitors such as over-the-top (OTT) players, cloud operators, and established Internet service provider (ISP) giants. Thus, it is required to minimize CapEx of the network by reducing the cost of hardware and minimize OpEx by maximizing the utilization from hardware assets. In order to overcome these challenges, mobile networks have not only to go through architecture processes to optimize the current resources but also to add new components/technologies which increase the capacity.

On these grounds, SDN and NFV are promising technologies which are expected to solve these limitations in current mobile networks. SDN provides the required improvements in flexibility, scalability, and performance to adapt the mobile network to keep up with the expected growth. NFV offers a new way to design, deploy, and manage networking services. NFC allows decoupling the network functions from proprietary hardware appliances, so they can run in software. SDMN architecture is based on both SDN and NFV principles.

The adaptation of SDN concepts is expected to solve many limitations in current mobile networks. In SDN enables telecom networks, each operator has the flexibility to develop new networking concepts, optimize their network, and address specific needs of subscribers. Furthermore, software-programmable network switches in SDMN use modern agile programming methodologies. These software methodologies can be developed, enhanced, and upgraded at much shorter cycles than the development of today's state-of-the-art mobile backhaul network devices.

SDN and NFV Development

During the past few decades, many similar initiatives such as Telecommunications Information Networking Architecture (TINA), intelligent networks, and active networks which were proposed to solve the flexibility, scalability, and performance issues in telecommunication networks. These architectures were not finally realized due to failures at the implementation stage and the lack of support from industrial giants. However, the adaptation of SDN and NFV concept has extended up to commercial production level already. Almost all the device manufactures are working on the designing of devices to support SDMNs. Some telecommunication device manufactures already started to ship SDMN products for mobile network operators.[1]

---

[1] Josh Taylor, Telstra taps Ericsson for software defined networking, Technical report, 2014, URL http://www.zdnet.com/au/telstra-taps-ericsson-for-software defined-networking-7000032689/.

On the one hand, we have seen plenty of innovation in devices we use to access the network, the applications, and services. However, the network infrastructure has always consisted of remarkably complex and inflexible devices. The rapid traffic growth has demanded a significant change in these devices. The SDMN architecture changes the underlying infrastructure of telecommunication networks. Both SDN and NFV concepts are offering new ways to design, build, and operate the telecommunication networks.

The adaptation of SDN concepts propose to decouple the control plane from the data plane and make the control plane remotely accessible and remotely modifiable via third-party software clients. Thus, it is directing the current mobile network toward a flow-centric model that employs inexpensive hardware and a logically centralized controller. NFV concepts enable ubiquitous, convenient, on-demand network access to a shared pool of configurable network resources.

Thus, SDN and NFV concepts create a new telecommunication network environment which can support rapid service innovation and expansion. It also scales and optimizes the utilization of network resources more efficiently and cost effectively without compromising customer's experience.

SDMN Standardization

Many standardization bodies already started standardization efforts on SDMN concepts. Standardization is the first step toward the wide adoption of a new technology. Thus, the benefits of such standardization efforts are very significant for all stakeholders of the telecommunication field.

European Telecommunications Standards Institute (ETSI) is an industry-led standards development organization. It is the biggest telecommunication community which is working on the adaptation of NFV concepts for future mobile networks since 2012. Initially, seven of the world's leading telecoms network operators started Industry Specification Group for NFV (ISG NFV) at ETSI. During the past 2 years, this group has grown significantly. ISG NFV is now consisting of over 220 individual companies including 37 of the world's major service providers. This large community of experts are working intensely to develop the required standards for NFV as well as sharing their experiences of NFV development and early implementation.

On the other hand, Open Networking Foundation (ONF) is a non-profit organization dedicated to accelerating the adoption of open SDN, and it is the leading standardization organization in SDN domain. ONF was founded in 2011, and the membership of ONF as grown to over 100 company-members including telecom operators, network service providers, equipment vendors, and virtualization software suppliers. In 2014, ONF has formally launched one working group called Wireless & Mobile Working Group (WMWG). WMWG analyzes architectural and protocol requirements for extending SDN technologies to wireless and mobile domains include wireless backhaul networks and cellular Evolved Packet Core (EPC).

These two communities have started to share their standardization works since 2014. ETSI signed a strategic partnership agreement with ONF to further the development of NFV specifications by utilizing SDN concepts.

Furthermore, several Study Groups (SGs) of ITU (International Telecommunication Union)'s Telecommunication Standardization Sector (ITU-T) are already working on the adaptation of SDN concepts for public telecommunication networks. For instance, SG13 (Future networks) is focusing on functional requirements and architecture development for SDN enabled mobile networks. SG11 (signaling) works aligns with SG13 to develop signaling requirements and protocols for their SDN architectures.

Moreover, various other research communities such as Software Defined Networking Research Group (SDNRG) in Internet Engineering Task Force (IETF), Optical Internet Forum (OIF), Broadband Forum (BBF), and the Metro Ethernet Forum (MEF) are also working of the SDN and NFV deployment in various network scenarios.

It is always tricky to predict what will be the future of telecommunication networks beyond the LTE architecture. However, the benefits of applying the SDMN principles and the tremendous support by both the research community and the industry giants make SDMN a very promising candidate for future mobile networks.

Intended Audience

The book will be interesting for Industry/SMEs to design new telecommunication devices, network engineers implementing new technologies in the operator network, researchers working in the area of next-generation mobile network, and students working on master or doctoral theses in the areas of network security, mobility management, and Techno-economics.

This book provides a cutting-edge knowledge in areas such as network virtualization and SDN concepts relevant to next-generation mobile networks. This helps Industry/SMEs to create innovative solutions based on the research results in this book. It can be transferred to their products enabling the company to maintain its competitiveness in the challenging telecommunication market. Network operators can gain knowledge on latest network concepts such as SDN and virtualization. Also they can foresee the advantages of SDMN in terms of cost savings, scalability, flexibility, and security. The new possibilities of SDMN concept can help operators to cope with the increased cost pressure and competition.

Finally, this book facilitates the transfer of up-to-date telecommunication knowledge to university students and young scientists in research institutes and universities. Furthermore, the book will provide a complete introduction of SDMN concept for fresh researchers in the field. On the other hand, the book includes an overview of several related research areas such as virtualized transport/network management, traffic, resource and mobility management, mobile network security, and techno-economic modeling.

Organization of the Book

The book is organized into five parts: (I) Introduction, (II) SDMN architectures and network implementation, (III) Traffic transport and network management, (IV) Resource and mobility management, and (V) Security and techno-economic aspects.

The first part includes a general overview and background on SDN and present mobile network architectures. It presents the brief history and evolution of both SDN concepts and mobile network architectures. In the overview (Chapter 1), we present the SDMN architecture. Initially, we discuss the limitations of present-day mobile architectures. Then, we explain the advance features and key benefits of SDMN architecture. Chapter 2 presents the evolution of mobile networks. It analyzes the market trends and traffic projections of mobile communication.

Based on that, this chapter also explains requirements in future mobile networks. Chapter 3 explains the general SDN concepts. The chapter presents the history, the evolution and various application domains of SDN. It also provides a short summary of some of the prominent SDN related activities in industry and standardization bodies. Chapter 4 examines how the SDN concept and technologies could be applied in the wireless world. This chapter reviews opportunities and challenges which will happen during the adaptation of SDN concepts to wireless networks. Chapter 5 presents the role of SDN in the design of future 5G wireless networks. It also contains a survey on the emerging trends, prospects and challenges of SDN based 5G mobile systems.

The second part covers the basic of SDMN architectures and various implementation scenarios. It provides the state of the art in SDMN and describes changes to the current LTE architecture which is useful for wider SDMN deployment. Part II starts with Chapter 6 which provides an overview of LTE network architecture and its migration towards SDN based mobile network. This chapter describes how SDN based mobile networks benefit both mobile operators and end users. Chapter 7 presents a deployment model of the evolved packet core (EPC) where the EPC functions are deployed as services on a virtualized platform in a cloud computing infrastructure. It also explains the advantages of SDN which offers the potential to integrate the EPC services into the broader operator network using cross-domain orchestration. Chapter 8 discusses important aspects of the controller placement problem (CPP) in software defined mobile networks (SDMN). It presents the available solution methodologies and analyzes relevant algorithms in terms of performance metrics. Chapter 9 analyzes the factors influencing the future evolution of telecommunication clouds controlled by open-source platform software.

The third part discusses the impact of SDN concepts on traffic transport and network management functions of future mobile networks. It also explains the scalability of SDMNs and optimization of traffic transportation in SDMNs. Chapter 10 describes the European Telecommunications Standards Institute (ETSI) Network Function Virtualization (NFV) architecture and underlying support provided by SDN in the mobile network context. Chapter 11 introduces the main building blocks of traffic management in mobile networks, and provides an overview of quality of Servic (QoS) provisioning and dynamic policy control in 3G/4G networks. Then, it discusses the QoS enforcement features of OpenFlow switches and presents a future technology for improved resource selection using the application-layer traffic optimization protocol integrated into software defined networks. Chapter 12 describes the applicability of SDN to mobile applications by allowing dynamic service chaining inside the packet core, radio access networks and between user equipment (UE) for direct device-to-device communication. Chapter 13 provides platforms and technologies used for load balancing in software defined mobile networks. It also discusses the main challenges existing in current load balancing technologies and explains the requirements of novel load balancing technologies for software defined mobile networks.

The fourth part explains the various challenges on resource and mobility management of future mobile networks while adapting the SDN concepts. Chapter 14 presents the SDN enabled QoE monitoring and enforcement framework for Internet services. This framework augments existing quality-of-service functions in mobile as well as software defined networks by flow based network centric quality of experience monitoring and enforcement functions. Chapter 15 discusses SDN-based mobility management in the Internet. It reviews existing Internet mobility protocols and explains why SDN is beneficial to solve current Internet mobility problems. Chapter 16 reviews the existing MVNO architectures and explains the limitation in current architectures. Moreover, it explains SDN perspective of MVNOs which adds the reconfigurable mobile network parameters to the existing MVNO and enhances the features of the mobile network.

The fifth part covers the security and techno-economic aspects. It includes comprehensive literature review in security challenges in future mobile architectures and security management aspects in SDMNs. Moreover, it discusses the business cases in virtualized mobile network environments and presents both evolutionary and revolutionary industry architectures for SDMNs. Chapter 17 explains the limitation on traditional security models and presents the requirement to develop an inclusive and intrinsic security model across the SDMN. Chapter 18 presents the security issues introduced by SDN, NFV and future mobile networks that integrate these technologies to become software defined mobile networks (SDMNs). Finally, Chapter 19 defines the key business roles and presents both evolutionary and revolutionary industry architectures for SDMN, as well as discusses the perspectives of different stakeholders in the mobile network industry.

Acknowledgments

This book focuses on software defined mobile networks, which has been created by the joint effort of many people. First of all, we would like to thank all of the chapter authors for doing a great job!

This book would not have been possible without the help of many people. The initial idea for this book originated during our work in SIGMONA (SDN Concept in Generalized Mobile Network Architectures) project. Many of the partners not only motivated us but also contributed with various chapters. We would like to acknowledge the contributions of all the partners in CELTIC SIGMONA project.

Also we thank all the reviewers for helping us to select suitable chapters for the book. Moreover, we thank anonymous reviewers who have evaluated the proposal and gave plenty of useful suggestions for improving it. Ulf Ewaldsson (Ericsson) and Lauri Oksanen (Nokia) wrote nice forewords for the book and we really appreciate their effort. We thank Clarissa Lim, Sandra Grayson, Liz Wingett, and Anna Smart from John Wiley and Sons for help and support in getting the book published. Also, we would like to thank Brian Mullan (Cisco), Jari Lehmusvuori (Nokia), and James Kempf (Ericsson) for their continuous support on various occasions to complete this book.

Also, Madhusanka is grateful to the Centre for Wireless Communication (CWC) and University of Oulu for hosting the SDN mobile related research projects which helped him to gain the fundamental knowledge for this book. We also thank the Finnish Funding Agency for Technology and Innovation (Tekes) and Academy of Finland that funded research work at CWC and HIIT. Madhusanka thanks his wife Ruwanthi Tissera not only for her encouragement but also for help on proof reading and indexing tasks.

Last but not least, we would like to thank our core and extended families and our friends for their love and support in getting the book completed.

Madhusanka Liyanage, Andrei Gurtov, and Mika Ylianttila

Abbreviations

1G	First generation
2G	Second generation
3D	Three-dimensional
3G	Third generation
3GPP	Third-Generation Partnership Project
4G	Fourth generation
5G	Fifth generation
AAA	Authentication, authorization, and accounting
ACID	Atomicity, consistency, isolation, and durability
ACK	Acknowledgment
ADC	Application Detection and Control
ADC	Application delivery controllers
ADSL	Asymmetric digital subscriber line
AF	Application function
AKA	Authentication and key agreement
ALTO	Application-layer traffic optimization
AMPS	Advanced mobile phone system
AN	Access network
API	Application programming interface
APLS	Application label switching
APN	Access Point Name
APN-AMBR	Per APN aggregate maximum bit rate
APS	Access point services
APT	Advance persistent threat
AQM	Active queue management
ARP	Address Resolution Protocol
ARP	Allocation and retention priority
AS	Access stratum
AS	Autonomous system

ASS   Application Service Subsystem
ATM   Asynchronous Transfer Mode
BBERF   Bearer Binding and Event Reporting Function
BBF   Broadband Forum
BGP   Border Gateway Protocol
BR   Business revenue
BS   Base station
BSC   Base station controller
BSS   Base station subsystem
BSS   Business service subsystem
BT   Business tariff
BTR   Bit transfer rate
BTS   Base transceiver station
BYOD   Bring your own device
CAGR   Compound annual growth rate
CAM   Content-addressable memory
CAP   Cognitive access point
CapEx   Capital expenditure
CAPWAP   Control and Provisioning of Wireless Access Points
CBTC   Communication-Based Train Control
CCNs   Content-centric networks
CDF   Cumulative distribution function
CDMA   Code division multiple access
CDN   Content delivery network
CDNI   Content Delivery Networks Interconnection
CE   Control element
CES   Customer edge switching
CGE   Carrier-grade Ethernet
CG-NAT   Carrier-grade network address translation
CGW   Charging gateway
CIA   Confidentiality, integrity, and availability
C-MVNO   Cognitive mobile virtual network operator
CN   Core network
CoA   Care-of address
COTS   Commercial off the shelf
CP   Control plane
CPP   Controller placement problem
CPU   Central processing unit
C-RAN   Cloud—radio access network
CRM   Cognitive radio management
D2D   Device to device
DDoS   Distributed denial of service
DFI   Deep flow inspection
DHCP   Dynamic Host Configuration Protocol
DiffServ   Differentiated services
DL   Downlink

DMM	Distributed mobility management
DNS	Domain name server
DoS	Denial of service
DP	Data plane
DPDK	Data Plane Development Kit
DPI	Deep packet inspection
DSCP	Differentiated services code point
DSL	Digital subscriber line
DTLS	Datagram Transport Layer Security
EAP	Extensible Authentication Protocol
EAP-AKA	Extensible Authentication Protocol—Authentication and Key Agreement
EAP-SIM	Extensible Authentication Protocol—Subscriber Identity Module
EC2	Elastic Compute Cloud
ECMP	Equal-cost multipath
EDGE	Enhanced data rates for GSM evolution
EID	Endpoint identifiers
EM	Element manager
Email	Electronic mail
eNodeB	Enhanced NodeB
EPC	Evolved packet core
ePCRF	Enhanced Policy and Charging Rules Function
EPS	Evolved packet services
E-RAB	Evolved radio access bearer
ETSI	European Telecommunications Standards Institute
E-UTRAN	Evolved UMTS terrestrial radio access network
EVDO	Evolution-Data Optimized
FE	Forwarding element
FLV	Flash video
FM	Frequency modulation
FMC	Fixed–mobile convergence
ForCES	Forwarding and Control Element Separation
FQDN	Fully qualified domain name
FRA	Future radio access
FTP	File Transfer Protocol
FW	Firewalls
GBR	Guaranteed bit rate
GENI	Global Environment for Networking Innovations
GGSN	Gateway GPRS Support Node
GPL	General public license
GPRS	General Packet Radio Service
GRX	GPRS roaming exchange
GSM	Global System for Mobile Communications
GTP	GPRS Tunneling Protocol
GUTI	Globally Unique Temporary Identity
H2020	Horizon 2020
HA	Home agent

HA   High availability
HAS   HTTP Adaptive Streaming Services
HD   High definition
HeNB   Home eNodeB
HetNet   Heterogeneous network
HFSC   Hierarchical fair-service queue
HIP   Host Identity Protocol
HLR   Home location register
HLS   HTTP live streaming
HMAC   Hash message authentication code
HMIPv6   Hierarchical mobile IPv6
HoA   Home address
HRPD   High-rate packet data services
HSPA   High-speed packet access
HSS   Home Subscriber Server
HTB   Hierarchical token bucket
HTTP   Hypertext Transfer Protocol
I2RS   Interface to the Routing Systems
IaaS   Infrastructure as a service
IA-MVNO   Intramodular mobile virtual network operator
ICN   Information-centric networks
ID   Identifier
IDPS   Intrusion Detection and Prevention Systems
IDS   Intrusion Detection System
IEEE   Institute of Electrical and Electronics Engineers
IE-MVNO   Intermodular mobile virtual network operator
IETF   Internet Engineering Task Force
ILNP   Identifier/Locator Network Protocol
ILS   Identifier/Locator Split
IMS   IP Multimedia Subsystem
IMSI   International Mobile Subscriber Identity
IoT   Internet of Things
IP   Internet Protocol
IPsec   Internet Protocol security
IPTV   Internet Protocol television
IPX   Internetwork Packet Exchange
ISAAR   Internet Service quality Assessment and Automatic Reaction
ISP   Internet service providers
ISV   Independent software vendor
IT   Information technology
ITU   International Telecommunication Union
KASME   Key Access Security Management Entity
KPI   Key Performance Indicator
L1   Layer 1
L2   Layer 2
L3   Layer 3

L4	Layer 4
L7	Layer 7
LAN	Local area network
LISP	Locator Identifier Separation Protocol
LMA	Local mobility anchor
LR-WPAN	Low-rate wireless personal area networks
LSP	Label-switched path
LTE	Long-Term Evolution
LTE-A	Long-Term Evolution-Advanced
M2M	Machine to machine
MaaS	Mobility as a service
MAC	Media access control
MAC	Message authentication code
MAG	Mobile Access Gateway
MANO	Management and orchestration
MAP	Mobility anchor points
MBR	Maximum bit rate
MCP	Multiple controller placement
ME	Mobile equipment
MEF	Metro Ethernet Forum
MEUN	Mobile end user nodes
MEVICO	Mobile Networks Evolution for Individual Communications Experience
MIMO	Multiple input and multiple output
MIP	Mobile IP
MIPv6	Mobile IPv6
MLB	Mobility load balancing
MM	Mobility management
MME	Mobility management entity
MNO	Mobile network operator
MO	Mobile operator
MOS	Mean Opinion Score
MPG	Mobile Personal Grid
MPLS	Multiprotocol label switching
MSC	Mobile switching center
MSOP	Mobile virtual network operator service optimization
MSS	Mobile switching systems
MTC	Machine-type communications
MTU	Maximum transmission unit
MVNA	Mobile virtual network aggregator
MVNE	Mobile virtual network enabler
MVNO	Mobile virtual network operator
MVO	Mobile virtual operator
NaaS	Network as a service
NAS	Nonaccess stratum
NAT	Network address translation
NBI	Northbound interface

NBS	Name-based sockets
NFV	Network function virtualization
NFVI	Network function virtual infrastructure
NFVO	Network function virtualization orchestrator
NGN	Next-generation network
NIDS	Network Intrusion Detection Systems
NMS	Network monitoring system
NNSF	NAS Node Selection Function
NOMA	Nonorthogonal multiple access
NOS	Network operating system
NRM	Network resource manager
NSC	Network service chaining
NSD	Network Service Descriptor
NSN	Nokia Solutions and Networks
NSS	Network Service Subsystem
NVP	Network virtualization platform
OAM	Operation and management
OF	OpenFlow
OFDM	Orthogonal frequency division multiplexing
OIF	Optical Internet Forum
ONF	Open Network Foundation
OpenSig	Open signaling
OpEx	Operational expenditure
OS	Operating system
OS3E	Open Science, Scholarship, and Services Exchange
OSS	Optimal subscriber services
OTT	Over the top
OVS	Open virtual switch
OWR	Open wireless architecture
P2P	Peer to Peer
P4P	Provider Portal for P2P Applications
PaaS	Platform as a service
PC	Personal computers
PCC	Policy Control and Charging
PCEF	Policy Control Enforcement Function
PCRF	Policy and Charging Rules Function
PDCP	Packet Data Convergence Protocol
PDN-GW	Packet Data Network Gateway
PDP	Packet Data Protocol
PDSN	Packet Data Serving Node
PFB	Per-flow behavior
P-GW	Packet Data Network Gateway
PID	Provider-defined identifier
PIN	Place in the network
PKI	Public key infrastructure
PMIPv6	Proxy Mobile IPv6

PO	Primary operator
PoC	Push over cellular
POP	Points of presence
POS	Packet over SONET
POTS	Plain old telephone systems
PPP	Point-to-Point Protocol
PR	Path Record
PSTN	Public switched telephone network
PTT	Push to talk
QCI	QoS class identifier
QEN	Quality enforcement
QMON	Quality monitoring
QoE	Quality of experience
QoS	Quality of service
QRULE	Quality rules
RAN	Radio access network
RAP	Radio access point
RBAC	Role-based access control
REST	Representational State Transfer
RF	Radio frequency
RFC	Request for Comments
RLOC	Routing locators
RLSE	Regional-level spectral efficiency
RMON	Remote monitoring
RNAP	Resource Negotiation and Pricing
RNC	Radio network controller
ROI	Return of investment
RPC	Remote Procedure Calls
RRC	Radio Resource Control
RRM	Radio Resource Management
RTP	Real-Time Transport Protocol
RTT	Round-trip time
S/P-GW	Serving/Packet Data Network
SaaS	Security as a service
SaaS	Software as a service
SAE	System architecture evolution
SatCom	Satellite communication
SBC	Session border controller
SBI	Southbound interface
SCTP	Stream Control Transmission Protocol
SD	Standard definition
SDM	Software-defined monitoring
SDM CTRL	Software-defined monitoring controllers
SDMN	Software-defined mobile networks
SDN	Software-defined networks
SDNRG	Software-Defined Networking Research Group

SDP	Session Description Protocol
SDR	Software-defined radio
SDS	Software-defined security
SDWN	Software-defined wireless networks
SENSS	Software-defined security service
SG	Study groups
SGSN	Serving GPRS Support Node
S-GW	Serving Gateways
SHA	Secure Hash Algorithm
SID	Service identifier
SIEM	Security Information and Event Management
SIGMONA	SDN Concept in Generalized Mobile Network Architectures
SILUMOD	Simulation Language for User Mobility Models
SIM	Subscriber identity module
SINR	Signal-to-interference-plus-noise ratio
SIP	Session Initiation Protocol
SLA	Service-level agreement
SMS	Short Message Services
SNMP	Simple Network Monitoring Protocol
SNR	Signal-to-noise ratio
SO	Secondary operator
SOA	Service-oriented architecture
SON	Self-organizing networks
SONET	Synchronous Optical Networking
SP	Service providers
SPR	Subscription Profile Repository
SP-SDN	Service provider SDN
SRAM	Static random access memory
SSID	Service set identifier
SSL	Secure Sockets Layer
SSS	Standard subscriber services
STUN	Simple traversal of UDP over NAT
SUMA	Software-defined unified monitoring agent
SuVMF	Software-defined unified virtual monitoring function for SDN-based networks
TAI	Tracking area ID
TCAM	Ternary content-addressable memory
TCP	Transmission Control Protocol
TDF	Traffic Detection Function
TDM	Time-division multiplexed
TDMA	Time-division multiple access
TEID	Tunnel endpoint identifier
TFT	Traffic flow template
TLS	Transport Layer Security
ToR	The Onion Router
ToS	Theft of Service
TR	Tunnel routers

TV	Television
UDP	User Datagram Protocol
UDR	User Data Repository
UE	User equipment
UE-AMBR	UE aggregate maximum bit rate
UGC	User-generated content
UL	Uplink
UMTS	Universal Mobile Telecommunications Systems
URI	Uniform resource identifier
URL	Uniform resource locator
USIM	Universal Subscriber Identity Module
USS	User-Supportive Subsystem
VDI	Virtual desktop infrastructure
vEPC	Virtual evolved packet core
VHF	Very high frequency
VIM	Virtualized infrastructure manager
VIRMANEL	Virtual MANET Lab
VLAN	Virtual local area network
VLD	Virtual Link Descriptor
VLR	Visitor Location Register
VM	Virtual machines
VN	Virtual networks
VNE	Virtualized network element
VNF	Virtualized network function
VNFD	Virtual Network Function Descriptor
VNFD	VNF Descriptor
VNFFGD	Virtual Network Function Forwarding Graph Descriptor
VNFM	VNF manager
VNI	Visual Networking Index
VNO	Virtual network operator
VO	Virtual operator
VOD	Video on demand
VoIP	Voice over IP
VPMNO	Virtual private mobile network operator
VPN	Virtual private networks
VPNS	Virtual private network systems
VS	Virtual switches
WAP	Wireless Application Protocol
WCDMA	Wideband code division multiple access
Wi-Fi	Wireless fidelity
WiMAX	Worldwide Interoperability for Microwave Access
WLAN	Wireless local area network
WPAN	Wireless personal area network
WWAN	Wireless wide area network
XML	Extensible Markup Language

Part I

Introduction

1

Overview

Madhusanka Liyanage,[1] Mika Ylianttila,[2] and Andrei Gurtov[3]

[1] Centre for Wireless Communication, University of Oulu, Oulu, Finland
[2] Centre for Internet Excellence, University of Oulu, Oulu, Finland
[3] Helsinki Institute for Information Technology HIIT, Aalto University, Espoo, Finland

Future mobile network architectures need to evolve to cope with future demand for high bandwidth, a large and evolving set of services with new specific requirements, high-level security, low energy consumption, and optimal spectrum utilization. Specifically, the increasing number of mobile users and services will result in the increasing capacity requirements for the mobile network. On the other hand, it is expected that mobile data traffic will grow faster than the fixed Internet during the upcoming years. Thus, accommodating this expected traffic growth is an imminent requirement of future mobile networks.

In order to keep up with the traffic growth, mobile networks have not only to go through architecture processes to optimize the current resources but also to add new components/ technologies that increase the capacity. However, mobile backhaul networks contain remarkably complex and inflexible devices. Although the interfaces of a cellular network are globally standardized, still most of these devices are vendor specific. Thus, mobile operators do not have flexibility to "mix and match" capabilities from different vendors. In another aspect, the standardization process for mobile networks is a long-lasting process. Although operators find promising concepts, they need to wait years to implement them in their networks. This might bury lots of interesting opportunities due to the lack of support.

On these grounds, software defined networking (SDN) is one of the promising technologies that are expected to solve these limitations in current mobile networks. SDN provides the required improvements in flexibility, scalability, and performance to adapt the mobile network to keep up with the expected growth. Software defined mobile networking (SDMN) is directing the current mobile network toward a flow-centric model that employs inexpensive hardware and a logically centralized controller. SDN enables the separation of the data forwarding plane from the control planes (CP). The SDN-enabled switches, routers, and gateways are controlled through an SDN

Software Defined Mobile Networks (SDMN): Beyond LTE Network Architecture, First Edition.
Edited by Madhusanka Liyanage, Andrei Gurtov, and Mika Ylianttila.
© 2015 John Wiley & Sons, Ltd. Published 2015 by John Wiley & Sons, Ltd.

controller/network operating system (NOS) and are seen as virtual resources. The CP of the mobile networking elements can be deployed onto an operator cloud for computing.

In this paradigm, each operator has the flexibility to develop his own networking concepts, optimize his network, and address specific needs of his subscribers. Furthermore, software-programmable network switches in SDMN use modern agile programming methodologies. These software methodologies can be developed, enhanced, and upgraded at much shorter cycles than the development of today's state-of-the-art mobile backhaul network devices.

The acquisition of virtualization into Long-Term Evolution (LTE) mobile networks brings the economic advantage in two ways. First, SDMN requires inexpensive hardware such as commodity servers and switches instead of expensive mobile backhaul gateway devices. Second, the introduction of SDN technology to mobile networks allows entering new actors in the mobile network ecosystem such as independent software vendors (ISV), cloud providers, and Internet service providers (ISP) that will change the business model of mobile networks.

Thus, the concept of SDMN would change the network architecture of the current LTE 3rd Generation Partnership Project (3GPP) networks. SDN will also open up new opportunities for traffic, resource, and mobility management as well as impose new challenges on network security. Many academic and industrial researchers are working on the deployment of SDMNs. We believe that ideas stemming from design and experiments with SDMN provide indispensable knowledge for anybody interested in next-generation mobile networks.

The overview chapter starts with Section 1.1, which contains a discussion on the limitation of the present mobile network. The SDMN architecture and its components are presented in Section 1.2. The key benefits of SDMN architectures are described in Section 1.3. Section 1.4 contains the conclusion.

1.1 Present Mobile Networks and Their Limitations

The mobile communication was introduced in the 1980s. The first generation of mobile networks supports only the voice call services and the connectivity speed up to 56 kbps. However, the mobile network technology achieved a tremendous development during the last four decades. Today's mobile networks support various network services such as amended mobile Web access, Internet Protocol (IP) telephony, gaming services, high-definition mobile television (TV), videoconferencing, 3D television, cloud computing, and high-speed broadband connectivity up to several Gbps [1].

With inbuilt mobility support, these services fuel the attraction toward the mobile broad-band networks instead of wired Internet. It is expecting that the mobile data traffic will be exceeding the wired data traffic in the near future. On the other hand, mobile networks have to provide carrier-grade high-quality services for their subscribers even while copping with these traffic demands.

It is challenging to satisfy all these requirements by using present-day mobile network architecture. Present-day mobile networks are facing various limitations, and they can be categorized as below [2, 3]:

- *Scalability limitations* – The rapid increment of mobile traffic usage is projected due to new bandwidth-hungry mobile services such as online streaming, video calls, and high-definition mobile TV. The existing static overprovisioned mobile networks are inflexible and costly to scale to keep up with the increasing traffic demand.

- *Complex network management* – Significant expertise and platform resources are required to manage the present mobile network. In most cases, backhaul devices are lacking of common control interfaces. Therefore, straightforward tasks such as configuration or policy enforcement also require a significant amount of effort.
- *Manual network configuration* – Most of the network management systems are manually intensive, and trained operators are required [4–7] to achieve even moderate security [8]. However, these manual configurations are prone to misconfiguration errors. Also, it is expensive and taking a long time to troubleshoot such errors. According to the Yankee Group report [4], 62% of network downtime in multivendor networks happened due to human errors. Furthermore, 80% of IT budgets are spent on maintenance and operations of the network.
- *Complex and expensive network devices* – Some of the mobile backhaul devices have to handle extensive amount of work. For instance, Packet Data Network Gateway (PDN GW) is responsible for many important data plane (DP) functions such as traffic monitoring, billing, quality-of-service (QoS) management access control, and parental controls in LTE networks. Thus, the devices are complex and expensive.
- *Higher cost* – Mobile operators do not have flexibility to "mix and match" capabilities from different vendors' devices. Therefore, they cannot build their network by using the cheap equipment from different vendors. It directly increases the CAPEX of the network. On the other hand, the manual configuration and inflexibility increase the OPEX of the network.
- *Inflexibility* – The standardization process for mobile networks is a long-lasting process. It requires many months or years to introduce new services. Furthermore, the implementation of new service also takes weeks or months due to the manually intensive service activation, delivery, and assurance.

Apart from these key issues, future mobile networks will face critical network congestion issues. Regardless of the limited radio bandwidth, the demand for mobile data is increasing rapidly. Therefore, mobile network operators have to use smaller cells to accommodate the traffic growth, which ultimately increases the number of base stations in the network. It is predicted that the global number of cellular sites will reach up to 4 million by the end of 2015. It was only 2.7 million by the end of 2010 [1]. Therefore, mobile backhaul networks will face congestion in a manner similar to data center networks due to the increment of mobile broadband traffic and the number of base stations as a solution.

1.2 Software Defined Mobile Network

The adaptation of SDN and virtualization concepts to the mobile network domain will solve the previously mentioned issues. The SDN concepts not only solve these issues but also improve the flexibility, scalability, and performance of a telecommunication network. SDN is originally designed for fixed networks. However, mobile networks have different requirements than fixed networks such as mobility management, precious traffic transportation, efficient protection of the air interface, higher QoS, the heavy use of tunneling in packet transport, and more. Therefore, SDMN concept is proposed as an extension of SDN paradigm to support mobile network-specific functionality. Furthermore, SDMN has a greater degree of service awareness and optimum use of network resources than original SDN concepts.

The least telecommunication architectures such as the 3GPP Evolved Packet Core (EPC) explained the advantages of the separation of the CP from the DP. EPC supported this separation to some extent. However, SDN enables the complete separation of the CP from the data forwarding plane. Furthermore, standardization organizations such as the Internet Engineering Task Force (IETF) and the European Telecommunications Standards Institute (ETSI) are interested to utilize network function virtualization (NFV) concepts in telecommunication networks. The SDN concepts help to adapt the NFV functions as well. Basically, the NFV concepts facilitate on-demand provision and online scale-up for mobile networks.

The SDMN architecture is now directing the current mobile network toward a flow-centric model that employs inexpensive hardware and a logically centralized controller. SDMN is basically an approach to networking in which the CP is decoupled from telecom-specific hardware and given to a software application called the controller. The SDN-enabled switches, routers, and gateways are controlled through the SDN controller and NOS. The CP of the mobile networking elements can deploy in an operator cloud as virtual components. Furthermore, the modern agile programming methodologies can be used to program and upgrade the performance of software-programmable network switches in SDMNs. These software methodologies can be developed, enhanced, and upgraded at much shorter cycles than the development of today's state-of-the-art mobile backhaul network devices. In this paradigm, each operator has the flexibility to develop his own networking concepts to address specific needs of his subscribers and optimize his network to achieve better performance. During the past few years, many academic and industrial researchers are working on the deployment of SDMNs. The integration of SDN in mobile networks is proposed in various papers [3, 9–11].

Figure 1.1 illustrates the basic architecture of SDMNs [2, 12].

Basically, SDMN splits the CP and the DP of the mobile network. It allows centralizing all the controlling functionalities. The DP now consists of low-end switches and links among them.

The SDMN architecture can be divided into three layers [2, 12, 13]:

1. **DP layer**
   The DP layer is also known as the infrastructure layer. It consists of the network elements such as switches and other devices. These switches support packet switching and forwarding functions. Base stations are connected to DP switches at the border. However, the SDMN architecture is transparent to the existing radio technologies. Similarly, border switches at the core network are connected to the Internet to off-load the mobile subscriber traffic.
2. **Network controller**
   The logically centralized controller provides the consolidated control functionality of the DP switches. The control protocol (e.g., OpenFlow [14], Beacon [15], Maestro [16], and DevoFlow [17]) is used by the controller to communicate with the DP elements. Basically, the controller uses the control protocol to install flow rules in each DP switch to route the traffic along the mobile network DP. The boundary between the network controller and the DP layer is traversed by the southbound application programming interface (API).
   The NOS is run on top of the controller to support the control functions.
3. **Application layer**
   The application layer consists of all the controlling and business applications of the mobile network. The traditional mobile network elements such as Policy and Charging Rules Function (PCRF), Home Subscriber Server (HSS), Mobility Management Entity (MME), and Authentication, Authorization, and Accounting (AAA) are now software

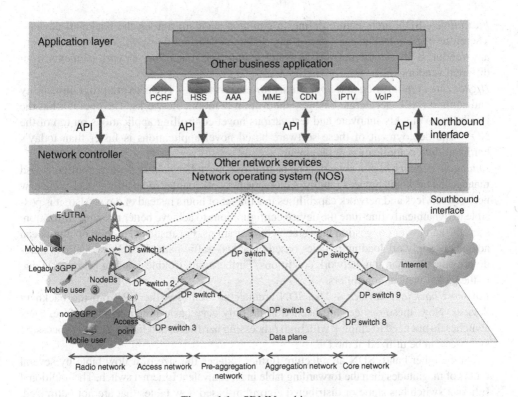

Figure 1.1 SDMN architecture.

applications that are running on top of NOS. The boundary between the application layer and the network controller is traversed by the northbound API.

Control elements perform the traditional functionalities and assist NOS to handle mobile network functionalities such as mobility management, resource management, and traffic transportation.

Thus, the adaptation of SDN changes the network architecture of the current mobile networks. Moreover, SDN will also open up new opportunities in various sections in the mobile network. Especially, it provides various benefits for traffic, resource, and mobility management as well as imposes new challenges on network security.

1.3 Key Benefits of SDMN

The adaption of SDN concepts offers various benefits for the entire mobile networks including wireless access segments, mobile backhaul networks, and core networks. Here, we present the key benefits of SDMNs [2, 11–13]:

- *Logically centralized controlling* – A centralized controller can take control decision based on the global view of the network. These decisions are more accurate, optimum, and efficient than the existing autonomous system-based mechanisms.

- *Flexibility* – SDN architecture defines a common standard among the backhaul devices. Therefore, the controller can control any SDN-enabled mobile network component from any vendor. It allows the network operator to mix and match the network elements from different vendors.
- *Higher rate of innovation and opportunity for new services* – The network programmability and common APIs accelerate business innovation in mobile networks. The operator has the flexibility to quickly innovate and test various novel controlling applications on top of the NOS. The deployment of these software-based novel applications is faster than today's hardware-based application deployment.
- *Automatic network management* – The centralized controller help to deploy, configure, and manage the backhaul devices rapidly. Automatic network management allows deploying new network services and network capabilities in a matter of hours instead of days. Also, it is possible to dynamically fine-tune the device configurations to achieve better resource utilization, security, and lower congestion than static configurations. For instance, mobile operators can adaptively apply off-loading policies based on actual traffic patterns. Today's static policies do not adapt to changing network conditions. Furthermore, troubleshooting is very fast due to the global view at controllers.
- *Low-cost backhaul devices* – The SDN architecture removes the CP from the backhaul devices. Now, these devices are needed to do only very basic functions. Therefore, SDN switches do not need to employ with high processing hardware, and low-cost, low processing switches can be utilized at the DP.
  On the other hand, SDN architecture further reduces the size of a flow table by several orders of magnitudes than the forwarding table in an equivalent Ethernet switch. The traditional Ethernet switch has static or distributed algorithm-based flow tables that are not optimized. Therefore, even small wiring closet switches typically contain a million entries. However, the flow-based traffic routing and centralized controlling will optimize the flow rules in the switches, and these rules can be dynamically revoked or added. Therefore, SDN switches now have much smaller flow tables that need to keep track of flows in progress. In most of the cases, the flow tables can be small enough to be held in a tiny fraction of a switching chip. Therefore, SDN switches now contain low-capacity memory, and the cost of the devices is drastically decreasing. Even for a campus-level switch, where perhaps tens of thousands of flows could be ongoing, it can still use on-chip memory that saves cost and power [18].
- *More granular network control* – The flow-based control model in SDN architecture allows applying the flow control policies at a very granular level such as the session, user, device, and application levels. Also, it is possible to dynamically change these control policies based on observed network behaviors. For instance, the operator is able to provide priority for high-revenue-producing corporate customers than lower-yielding consumers.
- *Heterogeneous network support and interoperability* – The flow-based traffic transport model in SDN is well suited to provide end-to-end communications across heterogeneous network technologies, such as Global System for Mobile Communications (GSM), 3G, 4G, Wi-Fi, code division multiple access (CDMA), and more. Also, it provides compatibility for future 5G-like network technologies.
- *Efficient segmentation* – SDN architecture supports efficient network segmentation. The software-based segmentations can be used to provide services for extremely popular mobile virtual network operator (MVNO) services. For instance, FlowVisor and language-level isolations can be used here.

- *Efficient access control network* – The centralized controlling allows deploying efficient intercell interference management algorithms. It allows taking efficient and optimal resource management decisions and improves the utilization of scarce radio-frequency (RF) spectrum. In addition, computational-intensive processing can be off-loaded to cloud devices by reducing the costs and increasing the scalability.
- *Path optimization* – The network controller can optimize the end-to-end path by considering the global view of the network. In a mobile environment, fast and efficient path optimization mechanisms are important as they support millions of mobile subscribers who change their locations rapidly. The centralized path optimization procedures are more efficient, faster, and optimum than the existing distributed path optimization mechanisms.
- *On-demand provision and online scale-up* – The SDN concepts enable adaptation of network virtualization. The virtualization of network devices offers the on-demand provisioning of resources when needed and scaling up of resources whatever demands are requested.

1.4   Conclusion

The growing traffic demand and new bandwidth-hungry mobile services increase the strain on present mobile networks. Thus, many academic and industrial researches are currently underway exploring different architectural paths for future mobile networks. The development of beyond LTE architecture is presently in its early research stages. Researchers are seeking new architectural paths that not only increase the network capacity but also solve the existing limitations in present LTE architecture.

In common terms, SDN is considered as "the radical new idea in networking." It has provided various benefits for fixed and wired network. Thus, SDN is considered as one of the promising technologies that can solve limitations in current mobile networks. SDN provides the required improvements in flexibility, scalability, and performance to adapt the mobile network to keep up with the expected growth. Therefore, SDN will likely play a crucial role in the design phase of beyond LTE mobile networks. A deep understanding of this emerging SDMN concept is necessary to address the various challenges of the future SDN-enabled mobile networks. The book will provide a comprehensive knowledge for researchers who are interested in SDMN concepts. Moreover, it covers all the technical areas such as virtualized transport and network management, resource and mobility management, mobile network security, and technoeconomic modeling concepts.

References

[1] M. Liyanage, M. Ylianttila, and A. Gurtov. A case study on security issues in LTE backhaul and core networks. In B. Issac (ed.), *Case Studies in Secure Computing—Achievements and Trends*. Taylor & Francis, Boca Raton, FL, 2013.
[2] K. Pentikousis, Y. Wang, and W. Hu. Mobileflow: Toward software-defined mobile networks. IEEE Communications Magazine, 51(7):44–53, 2013.
[3] L. E. Li, Z. M. Mao, and J. Rexford. Toward software-defined cellular networks. European Workshop on Software Defined Networking (EWSDN). IEEE, Darmstadt, Germany, 2012.
[4] Z. Kerravala. *Configuration management delivers business resiliency*. The Yankee Group, Boston, MA, 2002.
[5] T. Roscoe, S. Hand, R. Isaacs, R. Mortier, and P. Jardetzky. Predicate routing: Enabling controlled networking. SIGCOMM Computer Communication Review, 33(1):65–70, 2003.

[6] A. Wool. The use and usability of direction-based filtering in firewalls. Computers & Security, 26(6):459–468, 2004.

[7] G. Xie, J. Zhan, D. A. Maltz, H. Zhang, A. Greenberg, and G. Hjalmtysson. Routing design in operational networks: A look from the inside. In Proceedings of the SIGCOMM, Portland, OR, September 2004.

[8] A. Wool. A quantitative study of firewall configuration errors. IEEE Computer, 37(6):62–67, 2004.

[9] A. Gudipati, D. Perry, L. E. Li, and S. Katti. SoftRAN: Software defined radio access network. In Proceedings of the Second ACM SIGCOMM Workshop on Hot Topics in Software Defined Networking. ACM, Chicago, IL, 2013.

[10] X. Jin, L. E. Li, L. Vanbever, and J. Rexford. CellSDN: Software-defined cellular core networks. Opennet Summit, Santa Clara, CA, 2013.

[11] X. Jin, L. E. Li, L. Vanbever, and J. Rexford. SoftCell: Taking control of cellular core networks. arXiv preprint arXiv:1305.3568, 2013.

[12] K. Christos, S. Ahlawat, C. Ashton, M. Cohn, S. Manning, and S. Nathan. OpenFlow™-enabled mobile and wireless networks, White Paper. Open Network Foundation, Palo Alto, CA, 2013.

[13] Y. Liu, A. Y. Ding, and S. Tarkoma. Software-Defined Networking in Mobile Access Networks. University of Helsinki, Helsinki, 2013.

[14] N. McKeown, T. Anderson, H. Balakrishnan, G. Parulkar, L. Peterson, J. Rexford, S. Shenker, and J. Turner. OpenFlow: Enabling innovation in campus networks. ACM SIGCOMM Computer Communication Review, 38(2):69–74, 2008.

[15] OpenFlowHub. BEACON. http://www.openflowhub.org/display/Beacon (accessed January 17, 2015).

[16] Z. Cai, A. L. Cox, and T. E. Ng. Maestro: A system for scalable OpenFlow control. In Rice University Technical Report, 2010. Available at: http://www.cs.rice.edu/~eugeneng/papers/TR10-11.pdf (accessed January 17, 2015).

[17] J. C. Mogul, J. Tourrilhes, P. Yalagandula, P. Sharma, A. R. Curtis, and S. Banerjee. DevoFlow: Cost-effective flow management for high performance enterprise networks. In Proceedings of the ninth ACM Workshop on Hot Topics in Networks (HotNets). ACM, New York, 2010.

[18] M. Casado, M. J. Freedman, J. Pettit, J. Luo, N. McKeown, and S. Shenker. Ethane: Taking control of the enterprise. ACM SIGCOMM Computer Communication Review 37(4):1–12, 2007.

2

Mobile Network History

Brian Brown, Rob Gonzalez, and Brian Stanford
Cisco Systems, Herndon, VA, USA

2.1 Overview

Few things have changed the way we work, live, and play more than the evolution of mobile networking. In 1990, landlines were the big moneymaker for the service providers (SPs). Little did they know that a massive disruption in the marketplace was going to develop and in the middle would be mobile networking. Since the acceleration of this change through the decades, kids and teenagers today have never seen a pay phone, regularly carry a small computer (smartphone) in their pocket, and are connected close to 24 h a day. As technology continues to evolve, new areas of development are coming to the forefront of the discussion. Virtualization, orchestration, and scalability are now big concerns as more and more applications, data sources, and users around the world discover the possibilities. Network connectivity is critical to provide mobile users the services and experience they are looking for today (Fig. 2.1).

As the demand for mobile services has changed from the beginning of the commercially available GSM in the early 1990s to the evolution to a packet-based architecture in GPRS, to a more robust service in UMTS, and to the more mature designs in LTE and beyond, the supporting core architecture has also changed. In this chapter, we will discuss the history and evolution of the mobile network. The initial demands and drivers for a mobile network have also changed over the years. While voice was the primary service during the initial architectures, data and video have overwhelmingly dwarfed voice. What technology changes pushed the network to places unheard of in the early parts of the mobility movement? Why did the industry start where it did, and why did it end up where it has? How have the network owners changed over the years, and are the same services driving the push for progress? How have user expectations changed along with these technology changes?

Software Defined Mobile Networks (SDMN): Beyond LTE Network Architecture, First Edition.
Edited by Madhusanka Liyanage, Andrei Gurtov, and Mika Ylianttila.
© 2015 John Wiley & Sons, Ltd. Published 2015 by John Wiley & Sons, Ltd.

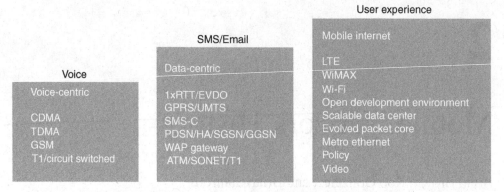

Figure 2.1   Evolution of the mobile industry.

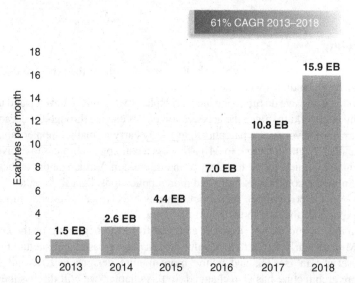

Figure 2.2   Global mobile traffic per month. Source: Cisco Systems (2014) Cisco Visual Networking Index (VNI).

2.2   The Evolution of the Mobile Network

At the end of 2013, global mobile data traffic reached 1.5 exabytes per month [1] and is expected to reach 15.9 exabytes per month by 2018 (Fig. 2.2).

The network has come a long way from the days of the 56K leased lines and T1s along with the large, clunky mobile phone in the early 1990s. Today, as the mobile endpoints become more and more numerous, the amount of data required to serve these endpoints becomes an issue along with managing the network access of these mobile devices. In 2013, over half a billion (526 million) mobile devices and connections were added. The Internet of Things is contributing to the increase in mobile traffic led by the transition to smarter mobile devices, emergence of wearable devices, and the increase in machine-to-machine (M2M) connections.

Today, the new data center architecture touts virtualization, orchestration, and scalability. Are these new concepts for the mobile network and where did it all start?

2.2.1 Sharing Resources

Back in the 1980s, because voice was the main driver for all SP business, the fledgling data network, including the mobile data network, was modeled after the existing voice network. Voice was carried over dedicated time-division multiplexed (TDM) lines designed to tightly pack in the newly digitized voice traffic into 64-kilobit channels. If a user in Atlanta, Georgia, needed to communicate with a user in San Francisco, California, a dedicated voice channel was established in order to support the 64-kilobit voice call. This dedicated channel could be used in any way for the duration of the call but may be wasted if no one was speaking. The bandwidth was reserved for this connection until the end users disconnected. Users had become accustomed to high-quality voice services due to the high availability (HA) of the legacy voice networks. Today, users are more tolerant of voice anomalies like choppy voice and dropped calls as they have migrated to more data usage.

Today, the ideas of sharing compute and storage resources are not new concepts. They are an old concept applied to a more current technical architecture. The mobile SPs in the early 1990s used the existing technology of the day to share what resources they had between users in order to efficiently deliver voice services. In the United States, a provider could multiplex 28 T1s into a single T3 and further multiplex many T3s into larger optical circuits. As newer multiplexing technologies were introduced, the result of sharing those resources was that the cost of a single T1 continued to drop. Asynchronous Transfer Mode (ATM) was also used to share high bandwidth connections among many customers. These shared large-capacity trunks were used to pass thousands of customer circuits between two endpoints on a network. ATM provided the higher bandwidth and quality of service (QOS) needed for evolving networks that were transporting voice, video, and data. Additionally, once the infrastructure was installed, new customers were quickly and easily added relative to before.

As the network transitioned to more of a packet-based architecture and voice and data were converged onto the same network, sharing bandwidth now improved even more, enabling the ability to share more resources. Circuits were no longer used per user but could now carry multiple users and applications, further moving the architecture to a shared space. Where before, in the circuit-switched world, a dedicated circuit had to be established and maintained during a call, the new packet-based network could share bandwidth on a per-packet level. As we migrated to the packet-based network, other technology contributed to the efficiency of the overall architecture. For example, silence suppression on voice trunks was one feature that would not transmit packets if an end user were not speaking, thereby increasing the sharable bandwidth in the network.

The network infrastructure was designed with HA in mind, so SPs had to reserve bandwidth to account for network node or link failures. The addition of redundant paths and redundant nodes along with many protocols has increased the complexity of the network infrastructure. SPs operating a highly complex network have higher operational cost, and they tend to have less network availability. The sharing of network resources that reduced the cost of providing the converged infrastructure for the SPs has now put a higher priority on HA and QOS due to the impact of network outages. As user demands grow, SPs must find a cost-effective way to add more users to the existing infrastructure along with the ability to deploy new services. Along with these technology changes and user demands, SPs are finding it harder to maintain the same level of services while expanding their networks.

In addition to sharing transport resources, the mobile packet core was also in transition. Where before, dedicated resources had to be installed and maintained for every specific function of the mobile packet core, now these functions can be shared on the same computing platform.

2.2.2 Orchestration

Wikipedia defines computing orchestration as "the automated arrangement, coordination, and management of complex computer systems, middleware, and services." This term is now widely used when talking about the cloud network. In layman's terms, orchestration is the higher-level coordination of hardware and software components, stitched together to support a particular service or application, and the active management and monitoring of that network. Again, this is not a new concept; however, technology today enables companies to automate the process in ways unheard of several years ago. How far back can we go to witness the beginning of the orchestration concept in action in the fixed or mobile communication system?

In the early days of telephony, a phone call was initiated by picking up the receiver and telling the operator the number that you wanted to connect. The operator then physically connected a wire to the next hop of the circuit. Since most calls were local, the operator connected the two callers together. When the call was completed, the operator had to disconnect the circuit. Could you imagine having to tell an operator each time you wanted to go to a new Web site? Fortunately, we have automated many of the processes that used to be done manually. This is the journey of orchestration.

In the early days of commercial mobile services, mobile SPs held a monopoly on the network, the devices connected to the network, and the applications that ran over the network. Development of new applications took months, if not years, to create. When a new application was ready to be introduced into the network, the orchestration of that implementation was affected by several factors.

Computing technology was still being developed, and all documentation was still largely being delivered using paper copies of service orders and directives. This communication started at the headquarters location and was delivered to any and all field locations with tasks to complete. As each field location completed its tasks, it communicated this back to the headquarters. After the field locations were completed, the service was then tested. Once testing was completed, the application could be put into service. This process was thorough, but not fast. Communication was slow, network visibility was limited, and the motivation to quickly implement the service was lacking. Competition was not yet a driver for improvements in this process.

Today, with the proliferation of cheap memory, large database capabilities, fast network communications, virtualized computing, and standard connections, the scenario above can be automated and completed in a fraction of the time. The concept is still the same. A central figure dictates what needs to be done at each location in the network. Once the completion has been acknowledged, the service or application can now be put into service.

2.2.3   Scalability

As mobility services grew in popularity, the SPs needed to address scalability issues. These issues included expanding points of presence (POPs) to facilitate additional equipment to terminate more users and additional circuits for core capacity along with determining where to connect these new core links. The early models used to design voice networks continued to be used as mobile voice services grew. Traditional point-to-point circuits connecting early services (2G/3G) to simple hub-and-spoke architectures were the norm. As data services grew, the design model had to evolve to provide cost-effective bandwidth in the network—evolution to a mixed architecture of hub-and-spoke architecture for S1 interfaces and meshed architecture for X2 interfaces. The outcome of the changes with network design provided ample bandwidth in the core, so QOS was not addressed in this area. The access into the network from the edge was all that was needed to address QOS. As data usage grew, providing the customer with a good experience had to be addressed. Due to the high cost of core bandwidth, it quickly became cost prohibitive to keep adding enough capacity to handle peak data load in the core. Network modeling evolved to include a more complex QOS strategy, which addressed the increased cost of adding additional bandwidth. As SPs have upgraded their core network technology from TDM, ATM, POS, and Ethernet for transport, additional technology has evolved to provide path selection along with guarantees to the user traffic carried on SP networks. SPs want to ensure that signaling traffic takes the shortest path in order to preserve both voice calls and data sessions as users are moving, which require the voice call or data session to switching from one base station to another. Scalability design usually involves a trade-off of maximum size of the network elements (the number of links, nodes, and protocols) and minimum availability. SPs need to ensure that the network will remain stable in order to provide the best user experience but also need to expand the network. This will continue to be one of the main challenges with the evolving mobility networks.

2.3   Limitations and Challenges in Current Mobile Networks

With the introduction of the smartphone and the consumer's ability to change the operating system and choose their own applications, customers achieved true control of their personal devices for the first time. Prior to this, the control consumers wielded was limited to the decision of which phone to buy only. After making that fateful decision, they were constrained to use the application offered on that phone. If customers wanted a navigation application, they were forced to pay whatever fee the phone provider wanted to charge for that application.

Happily, this is no longer the case. Now, the consumer can pay a small one-time fee for an application or use one of thousands of free applications. Choices in devices also increased with not only phones but tablets, computers, cellular modems, and personal hotspots to share cellular connections. The smartphone is no longer a mere toy for the personal sphere; it is an essential business tool, and there are many choices depending on the needs and personal preferences of the consumer. But with this growth in choice comes a range of additional pressures on SPs who must support their customer's evolving tastes and behaviors. Smartphones require more data and are expected to be 96% of the global mobile traffic by 2018 [1] (Fig. 2.3).

Customers now use the network in different ways and not just for phone calls and SMS messaging. We are sending email, surfing the Web, accessing private corporate networks, editing

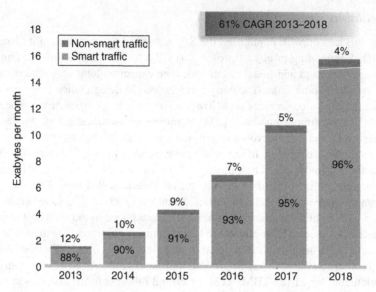

Figure 2.3 Effect of smart mobile devices and connections growth on traffic. Smart traffic is generated by devices with advanced computing capabilities and a 3G connection (or higher). Source: Cisco Systems (2014) Cisco Visual Networking Index (VNI).

documents, listening to music, watching video, and even watching live TV. Some of us hold our digital lives in these devices with everything from the personal and work spheres coexisting: family photos and video, the family calendar, corporate email, and confidential documents of all stripes. The user is also more mobile than ever, and we expect a similar experience on our smart-phones to that of sitting at our desk in the office. With mobile devices, the business user can have even more now with the use of virtual desktop infrastructure (VDI) that allows them to access a virtual intestine of their desktop with secure access to the corporate network and applications. Compounding pressures exerted by user expectations for performance and flexibility are yet other phenomena in the new landscape of the technology world.

Over the top (OTT) refers to video, music, and other services provided over the Internet rather than being provided by the mobile SP. OTT providers such as Amazon Prime, Netflix, Hulu, iTunes, and numerous other content applications are using large amounts of data and are not under the provider's control. OTT traffic accounts for the majority of mobile Internet traffic today as it is continuing to grow. Some of these services compete directly with the mobile provider and affect the potential revenue as well as drive the continual need for more bandwidth.

The old world of primarily voice and SMS applications and for which much of the current network apparatus was built requires far less bandwidth compared with the new video-intensive applications that are far more "bursty" in nature. This type of traffic is bandwidth intensive depending on what the user is doing and is latency sensitive. If the latency is too large or variable, the user experience will be reduced and can possibly drive the user to find other means of working including switching providers. SPs must find a way to adapt to the new wave of users but can't neglect older services either.

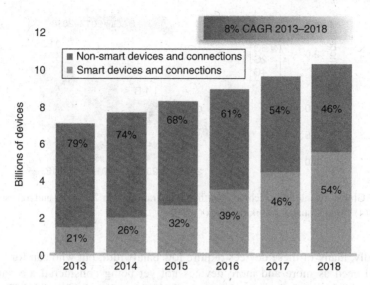

Figure 2.4 Global growth of smart mobile devices and connections. Smart devices are those having advanced computing capabilities with a minimum of 3G connectivity. Source: Cisco Systems (2014) Cisco Visual Networking Index (VNI).

With the rapid evolution of the mobile phone systems, mobile providers need to maintain multiple networks to support all of their 2G, 3G, and LTE services. Though a large percentage of users are migrating to new LTE-enabled smartphones [1], there still exists a significant block of older technology consumers that demands the care and feeding of multiple networks (Fig. 2.4).

While some reuse of components can be accomplished, the demands of the "new network" require the insertion or overlay of a number of new devices and technologies. This leads to a proliferation of support systems (e.g., OSS/BSS/NMS) required to keep these systems functional.

Providers also must balance the need to have capacity to cover current term needs with the costs of maintaining unused bandwidth. The data usage increase has led to carriers needing to increase backhaul capacity with various high-speed connections including metro Ethernet and dark fiber. The time to add capacity can be days to months and varies by type of connections as well as whether they own or lease the circuits. Adding backhaul circuits can also require multiple truck rolls to install, provision, and test new circuits. While mobile carriers are needing to upgrade their networks to handle increased traffic, they are also seeing a reduction in the price of voice minutes, declining text messaging, and loss of revenue to OTT services. At the same time customers are demanding more data bandwidth and lower costs. The challenges continue to increase for the SP.

Looming on the horizon are the anticipated growth and corresponding demands on the underlying networks and systems expected from the advent of the Internet of Things. An anticipated explosion in connectivity driven by everything from sensors on cars and other inanimate objects to health information from body sensors could overwhelm the existing infrastructure. Mobile-to-mobile connections will grow from 341 million in 2013 to over 2 billion by 2018 [1] (Fig. 2.5).

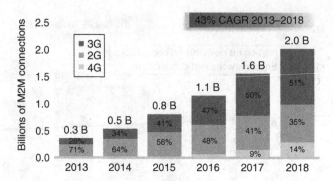

Figure 2.5 Global machine-to-machine growth and migration from 2G to 3G and 4G. Source: Cisco Systems (2014) Cisco Visual Networking Index (VNI).

Admittedly, many of these devices require low bandwidth, but who can foresee the full growth and need as more and more devices not yet being considered are added to the constellation of devices being added each day? History tells us that no one truly can, but we only know that there will be surprises.

2.4   Requirement in Future Mobile Networks

Understanding the challenges of today and looking at the trends of the future, it is clear that the way mobile networks are managed and built needs to change. There is the need for better orchestration as well as in-depth analytics to ensure the successful operations of mobile networks in the future. Monthly global mobile data traffic will surpass 15 exabytes per month by 2018 from close to 10 billion mobile devices and connections [1].

Mobile providers will have to adapt to the growing number of devices and traffic demands. To do this, they need to be able to provision, manage, and optimize the mobile network end to end. Using common tools and standardized application programming interfaces (API) to communicate with devices will remove a great deal of network complexity. This allows providers to quickly and efficiently apply new or updated service policies to their evolving network's traffic engineering needs at any point in the network.

An increasing number of mobile users will connect from a fixed location. This fixed nature places an increasing demand in areas of concentrated populations. Users will demand a seamless transition between fixed and mobile connections. With increasing mobile data rates available to users, mobile providers will become a wireline broadband competitor. LTE networks growth may grow faster as mobile users continue to demand similar service from both fixed and mobile networks. To assist with more fixed users as well as to deal with areas of dense population of mobile users, mobile providers are looking at small cells. Small cells have limited range but use low power and can be deployed in congested areas such as stadiums, city downtown areas, auditoriums, and neighborhoods to off-load mobile data traffic as providers deal with the large growth in traffic. There are forecasts for millions of small cells in the near future and adding to the complexity of the network that needs to be operated. Mobile off-load traffic is seen to increase from 1.2 exabytes/month in 2013 to 17.3 exabytes/month by 2018 [1] (Fig. 2.6).

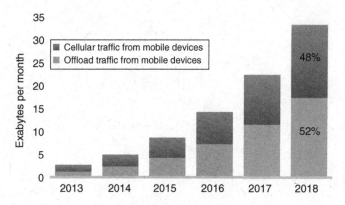

Figure 2.6 Fifty-two percent of total mobile data traffic will be off-loaded by 2018. Source: Cisco Systems (2014) Cisco Visual Networking Index (VNI).

The need to extract data from the network is becoming a must for mobile providers—from analyzing and even predicting usage patterns to location-based services. With location-based services, the mobile provider can work with stores to provide customers with coupons as they enter the store or provide offers for new services including Wi-Fi access for repeat customers. For this to be effective, mobile providers must be able to quickly extract the network inelegance, process the data to determine if the store or merchant has any offers, and then present the end user with the offer. This will require not only the ability to collect the data but also the computing resources to identify the needed information and act of that information.

Networks are already complex to manage, but with the addition of possibility of millions of small cells in the near future, this becomes even more challenging and impossible to optimize manually. Extracting information from the network will also allow mobile providers to apply business intelligence to optimize the network.

Without a standard API to collect information from the network and a robust orchestration system to automate the changes, the network will not be able to respond to the traffic demands. The standards will also reduce the case of vendor lock-in due to proprietary features or proprietary management systems. Operators need to be able to have a high-level control point where they can quickly apply changes to the network as a whole and manage dense deployments.

Without this simplicity, operators will continue to struggle with very large complex networks that make deployment more time consuming at a time when networks need to be more flexible to meet the ever-changing demands of end users.

Reference

[1] Cisco Systems (2014) Cisco Visual Networking Index (VNI). http://www.cisco.com/c/en/us/solutions/service-provider/visual-networking-index-vni/index.html (accessed February 17, 2015).

3

Software Defined Networking Concepts

Xenofon Foukas,[1,2] Mahesh K. Marina,[1] and Kimon Kontovasilis[2]
[1] *The University of Edinburgh, Edinburgh, UK*
[2] *NCSR "Demokritos", Athens, Greece*

3.1 Introduction

Software defined networking (SDN) is an idea that has recently reignited the interest of network researchers for programmable networks and shifted the attention of the networking community to this topic by promising to make the process of designing and managing networks more innovative and simplified compared to the well-established but inflexible current approach.

Designing and managing computer networks can become a very daunting task due to the high level of complexity involved. The tight coupling between a network's control plane (where the decisions of handling traffic are made) and data plane (where the actual forwarding of traffic takes place) gives rise to various challenges related to its management and evolution. Network operators need to manually transform high-level policies into low-level configuration commands, a process that for complex networks can be really challenging and error prone. Introducing new functionality to the network, like intrusion detection systems (IDS) and load balancers, usually requires tampering with the network's infrastructure and has a direct impact on its logic, while deploying new protocols can be a slow process demanding years of standardization and testing to ensure interoperability among the implementations provided by various vendors.

The idea of programmable networks has been proposed as a means to remedy this situation by promoting innovation in network management and the deployment of network services through programmability of the underlying network entities using some sort of an open network API. This leads to flexible networks able to operate according to the user's needs in a direct analogy to how programming languages are being used to reprogram computers in order to perform a number of tasks without the need for continuous modification of the underlying hardware platform.

Software Defined Mobile Networks (SDMN): Beyond LTE Network Architecture, First Edition.
Edited by Madhusanka Liyanage, Andrei Gurtov, and Mika Ylianttila.
© 2015 John Wiley & Sons, Ltd. Published 2015 by John Wiley & Sons, Ltd.

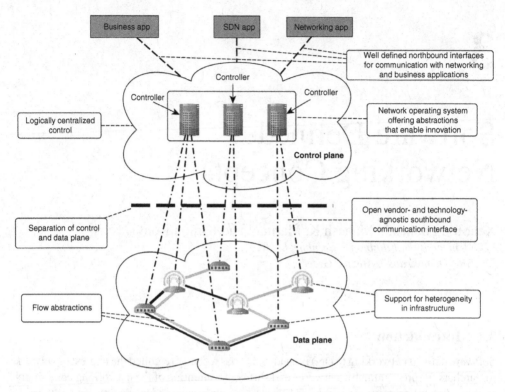

Figure 3.1   SDN in a nutshell—key ideas underlying the SDN paradigm.

SDN is a relatively new paradigm of a programmable network that changes the way that networks are designed and managed by introducing an abstraction that decouples the control from the data plane, as illustrated in Figure 3.1. In this approach, a software control program, referred to as the *controller*, has an overview of the whole network and is responsible for the decision making, while the hardware (routers, switches, etc.) is simply responsible for forwarding packets into their destination as per the controller's instructions, typically a set of packet handling rules.

The separation of the logically centralized control from the underlying data plane has quickly become the focus of vivid research interest in the networking community since it greatly simplifies network management and evolution in a number of ways. New protocols and applications can be tested and deployed over the network without affecting unrelated network traffic; additional infrastructure can be introduced without much hassle; and middleboxes can be easily integrated into the software control, allowing new potential solutions to be proposed for problems that have long been in the spotlight, like managing the highly complex core of cellular networks.

This chapter is a general overview of SDN for readers who have just been exposed to the SDN paradigm as well as for those requiring a survey of its past, present, and future. Through the discussion and the examples presented in this chapter, the reader should be able to comprehend why and how SDN shifts paradigms with respect to the design and management of networks and to understand the potential benefits that it has to offer to a number of interested parties like network operators and researchers.

The chapter begins by presenting a comprehensive history of programmable networks and their evolution to what we nowadays call SDN. Although the SDN hype is fairly recent, many of its underlying ideas are not new and have simply evolved over the past decades. Therefore, reviewing the history of programmable networks will provide to the reader a better understanding of the motivations and alternative solutions proposed over time, which helped to shape the modern SDN approach.

The next part of this chapter focuses on the building blocks of SDN, discussing the concept of the *controller* and giving an overview of the state of the art by presenting different design and implementation approaches. It also clarifies how the communication of the data and control plane could be achieved through a well-defined API by giving an overview of various emerging SDN programming languages. Moreover, it attempts to highlight the differences of SDN to other related but distinct technologies like network virtualization. Additionally, some representative examples of existing SDN applications are discussed, allowing the reader to appraise the benefits of exploiting SDN to create powerful applications.

The final part of the chapter discusses the impact of SDN to both the industry and the academic community by presenting the various working groups and research communities that have been formed over time describing their motivations and goals. This in turn demonstrates where the current research interest concentrates, which SDN-related ideas have been met with widespread acceptance, and what are the trends that will potentially drive future research in this field.

3.2   SDN History and Evolution

While the term *programmable* is used to generalize the concept of the simplified network management and reconfiguration, it is important to understand that in reality it encapsulates a wide number of ideas proposed over time, each having a different focus (e.g., control or data plane programmability) and different means of achieving their goals. This section reviews the history of programmable networks right from its early stages, when the need for network programmability first emerged, up to the present with the dominant paradigm of SDN. Along these lines, the key ideas that formed SDN will be discussed along with other alternatives that were proposed and affected SDN's evolution but that were not met with the same widespread success.

3.2.1   Early History of Programmable Networks

As already mentioned, the concept of programmable networks dates its origins back in the mid-1990s, right when the Internet was starting to experience widespread success. Until that moment, the usage of computer networks was limited to a small number of services like email and file transfers. The fast growth of the Internet outside of research facilities led to the formation of large networks, turning the interest of researchers and developers in deploying and experimenting with new ideas for network services. However, it quickly became apparent that a major obstacle toward this direction was the high complexity of managing the network infrastructure. Network devices were used as black boxes designed to support specific protocols essential for the operation of the network, without even guaranteeing vendor interoperability. Therefore, modifying the control logic of such devices was not an option, severely restricting network evolution. To remedy this situation, various efforts focused on finding novel solutions for creating more open, extensible, and programmable networks.

Two of the most significant early ideas proposing ways of separating the control software from the underlying hardware and providing open interfaces for management and control were of the open signaling (OpenSig) [1] working group and from the active networking [2] initiative.

3.2.1.1   OpenSig

The OpenSig working group appeared in 1995 and focused on applying the concept of programmability in ATM networks. The main idea was the separation of the control and data plane of networks, with the signaling between the planes performed through an open interface. As a result, it would be possible to control and program ATM switches remotely, essentially turning the whole network into a distributed platform, greatly simplifying the process of deploying new services.

The ideas advocated by the OpenSig community for OpenSig interfaces acted as motivation for further research. Toward this direction, the Tempest framework [3], based on the OpenSig philosophy, allowed multiple switch controllers to manage multiple partitions of the switch simultaneously and consequently to run multiple control architectures over the same physical ATM network. This approach gave more freedom to network operators, as they were no longer forced to define a single unified control architecture satisfying the control requirements of all future network services.

Another project aimed at designing the necessary infrastructure for the control of ATM networks was Devolved Control of ATM Networks (DCAN) [4]. The main idea was that the control and management functions of the ATM network switches should be stripped from the devices and should be assigned to external dedicated workstations. DCAN presumed that the control and management operations of multiservice networks were inherently distributed due to the need of allocating resources across a network path in order to provide quality of service (QoS) guarantees. The communication between the management entity and the network was performed using a minimalistic protocol, much like what modern SDN protocols like OpenFlow do, adding any additional management functionality like the synchronization of streams in the management domain. The DCAN project was officially concluded in mid-1998.

3.2.1.2   Active Networking

The active networking initiative appeared in the mid-1990s and was mainly supported by DARPA [5, 6]. Like OpenSig, its main goal was the creation of programmable networks that would promote network innovations. The main idea behind active networking is that resources of network nodes are exposed through a network API, allowing network operators to actively control the nodes as they desire by executing arbitrary code. Therefore, contrary to the static functionality offered by OpenSig networks, active networking allowed the rapid deployment of customized services and the dynamic configuration of networks at run-time.

The general architecture of active networks defines a three-layer stack on active nodes. At the bottom layer sits an operating system (NodeOS) multiplexing the node's communication, memory, and computational resources among the packet flows traversing the node. Various projects proposing different implementations of the NodeOS exist, with some prominent examples being the NodeOS project [7] and Bowman [8]. At the next layer exist one or more execution environments providing a model for writing active networking applications, including ANTS [9] and PLAN [10]. Finally, at the top layer are the active applications themselves, that is, the code developed by network operators.

Two programming models fall within the work of the active networking community [6, 11]: the *capsule model*, in which the code to be executed is included in regular data packets, and the *programmable router/switch model*, in which the code to be executed at network nodes is established through out-of-band mechanisms. Out of the two, the capsule model came to be the most innovative and most closely associated with active networking [6]. The reason is that it offered a radically different approach to network management, providing a simple method of installing new data plane functionality across network paths. However, both models had a significant impact and left an important legacy, since many of the concepts met in SDN (separation of the control and data plane, network APIs, etc.) come directly from the efforts of the active networking community.

3.2.2 Evolution of Programmable Networks to SDN

3.2.2.1 Shortcomings and Contributions of Previous Approaches

Although the key concepts expressed by these early approaches envisioned programmable networks that would allow innovation and would create open networking environments, none of the proposed technologies was met with widespread success. One of the main reasons for this failure was the lack of compelling problems that these approaches managed to solve [5, 6]. While the performance of various applications like content distribution and network management appeared to benefit from the idea of network programmability, there was no real pressing need that would turn the shift to the new paradigm into a necessity, leading to the commercialization of these early ideas.

Another reason for which active networking and OpenSig did not become mainstream was their focus on the wrong user group. Until then, the programmability of network devices could be performed only by programmers working for the vendors developing them. The new paradigm advocated as one of its advantages the flexibility it would give to end users to program the network, even though in reality the use case of end user programmers was really rare [6]. This clearly had a negative impact on the view that the research community and most importantly the industry had for programmable networks, as it overshadowed their strong points, understating their value for those that could really benefit like ISPs and network operators.

Furthermore, the focus of many early programmable network approaches was in promoting data instead of control plane programmability. For instance, active networking envisioned the exposure and manipulation of resources in network devices (packet queues, processing, storage, etc.) through an open API but did not provide any abstraction for logical control. In addition, while one of the basic ideas behind programmable networks was the decoupling of the control from the data plane, most proposed solutions made no clear distinction between the two [5]. These two facts hindered any attempts for innovation in the control plane, which arguably presents more opportunities than the data plane for discovering compelling use cases.

A final reason for the failure of early programmable networks was that they focused on proposing innovative architectures, programming models, and platforms, paying little or no attention to practical issues like the performance and the security they offered [6]. While such features are not significant key concepts of network programmability, they are important factors when it comes to the point of commercializing this idea. Therefore, even though programmable networks had many theoretical advantages, the industry was not eager to adopt such solutions unless pressing performance and security issues were resolved.

Clearly, the aforementioned shortcomings of early programmable network attempts were the stumbling blocks to their widespread success. However, these attempts were really significant, since they defined for the first time key concepts that reformed the way that networks are perceived and identified new research areas of high potential. Even their shortcomings were of high significance, since they revealed many deficiencies that should be addressed if the new paradigm was to be successful one day. All in all, these early attempts were the cornerstones that shaped the way to the more promising and now widely accepted paradigm of SDN.

3.2.2.2 Shift to the SDN Paradigm

The first years of the 2000s saw major changes in the field of networking. New technologies like ADSL emerged, providing high-speed Internet access to consumers. At that moment, it was easier than ever before for an average consumer to afford an Internet connection that could be used for all sorts of activities, from email and teleconference services to large file exchanges and multimedia. This mass adoption of high-speed Internet and of all the new services that accompanied it had cataclysmic effects for networks, which saw their size and scope increase along with traffic volumes. Industrial stakeholders like ISPs and network operators started emphasizing on network reliability, performance, and QoS and required better approaches in performing important network configuration and management functions like routing, which at the time were primitive at best. Additionally, new trends in the storage and management of information like the appearance of cloud computing and the creation of large data centers made apparent the need for virtualized environments, accompanied by network virtualization as a means to support their automated provisioning, automation, and orchestration.

All these problems constituted compelling use cases that programmable networks promised to solve and shifted the attention of the networking community and the industry to this topic once more. This shift was strengthened by the improvement of servers that became substantially better than the control processors in routers, simplifying the task of moving the control functions outside network devices [6]. A result of this technological shift was the emergence of new improved network programmability attempts, with the most prominent example being SDN.

The main reason for the apparent success of SDN is that it managed to build on the strong points of early programmable network attempts while at the same time succeeded in addressing their shortcomings. Naturally, this shift from early programmable networks to SDN did not occur at once, but, as we shall now see, went through a series of intermediate steps.

As already mentioned, one of the major drawbacks of early programmable networking attempts was the lack of a clear distinction between the control and data plane of network devices. The Internet Engineering Task Force (IETF) Forwarding and Control Element Separation (ForCES) [12] working group tried to address this by redefining the internal architecture of network devices through the separation of the control from the data plane. In ForCES, two logical entities could be distinguished: the forwarding element (FE), which operated in the data plane and was responsible for per-packet processing and handling, and the control element (CE), which was responsible for the logic of network devices, that is, for the implementation of management protocols, for control protocol processing, etc. A standardized interconnection protocol lay between the two elements enforcing the forwarding behavior to the FE as directed by the CE. The idea behind ForCES was that by allowing the forwarding and control planes to evolve separately and by providing a standard means of interconnection, it was possible to develop different types of FEs (general purpose or specialized) that could be combined with third-party control, allowing greater flexibility for innovation.

Another approach targeting the clean separation of the CE and FE of network devices was the 4D project [13]. Like ForCES, 4D emphasized the importance of separating the decision logic from the low-level network elements. However, in contrast to previous approaches, the 4D project envisioned an architecture based on four planes: a decision plane responsible for creating a network configuration, a dissemination plane responsible for delivering information related to the view of the network to the decision plane, a discovery plane allowing network devices to discover their immediate neighbors, and a data plane responsible for forwarding traffic. One experimental system based on the 4D architecture was Tesseract [14], which enabled the direct control of a network under the constraint of a single administrative domain. The ideas expressed in the 4D project acted as direct inspiration for many projects related to the controller component of SDNs, since it gave the notion of a logically centralized control of the network.

A final project worth mentioning during the pre-SDN era is SANE/Ethane [15, 16]. Ethane was a joint attempt made in 2007 by researchers in the universities of Stanford and Berkeley to create a new network architecture for the enterprise. Ethane adopted the main ideas expressed in 4D for a centralized control architecture, expanding it to incorporate security. The researchers behind Ethane argued that security could be integrated to network management, as both require some sort of policy, the ability to observe network traffic, and a means to control connectivity. Ethane achieved this by coupling very simple flow-based Ethernet switches with a centralized controller responsible for managing the admittance and routing of flows by communicating with the switches through a secure channel. A compelling feature of Ethane was that its flow-based switches could be incrementally deployed alongside conventional Ethernet switches and without any modification to end hosts required, allowing the widespread adoption of the architecture. Ethane was implemented in both software and hardware and was deployed at the campus of Stanford University for a period of a few months. The Ethane project was very significant, as the experiences gained by its design, implementation, and deployment laid the foundation for what would soon thereafter become SDN. In particular, Ethane is considered the immediate predecessor of OpenFlow, since the simple flow-based switches it introduced formed the basis of the original OpenFlow API.

3.2.2.3 The Emergence of SDN

In the second half of the 2000s, funding agencies and researchers started showing interest in the idea of network experimentation at scale [6]. This interest was mainly motivated by the need to deploy new protocols and services, targeting better performance and QoS in large enterprise networks and the Internet, and was further strengthened by the success of experimental infrastructures like PlanetLab [17] and by the emergence of various initiatives like the US National Science Foundation's Global Environment for Network Innovations (GENI).

Until then, large-scale experimentation was not an easy task to perform; researchers were mostly limited in using simulation environments for evaluation, which, despite their value, could not always capture all the important network-related parameters in the same manner as a realistic testbed would.

One important requirement of such infrastructure-based efforts was the need for network programmability, which would simplify network management and network service deployment and would allow multiple experiments to be run simultaneously at the same infrastructure, each using a different set of forwarding rules. Motivated by this idea, a group of researchers at Stanford created the Clean Slate Program. In the context of this project, which had as a mission to "reinvent the Internet," the OpenFlow protocol was proposed as a means for researchers to run experimental protocols in everyday networking environments. Similarly to previous approaches like ForCES, OpenFlow followed the principle of decoupling the control and forwarding plane and standardized the information exchanges between the two using a simple communication protocol. The solution proposed by OpenFlow, which provided architectural support for programming the network, led to the creation of the term SDN to encapsulate all the networks following similar architectural principles. The fundamental idea behind SDNs compared to the conventional networking paradigm is the creation of horizontally integrated systems through the separation of the control and the data plane while providing an increasingly sophisticated set of abstractions.

Looking back at all the milestones and important programmable network projects presented in this section, we can conclude that the road to SDN was indeed a long one with various ideas being proposed, tested, and evaluated, driving research in this field even further. SDN was not so much of a new idea, as it was the promising result of the distilled knowledge and experience obtained through many of the ideas presented in this section. What SDN managed to do differently compared to these ideas is that it integrated the most important network programmability concepts into an architecture that emerged at the right time and had compelling use cases for a great number of interested parties. Even though it remains to be seen whether SDN will be the next major paradigm shift in networking, the promise it demonstrates is undeniably very high.

3.3   SDN Paradigm and Applications

In this section, we focus on the key ideas underlying the SDN paradigm, the most recent instance in the evolution of programmable networks. In order to better understand the SDN concepts and to comprehend the benefits that this paradigm promises to deliver, we need to examine it both macro- and microscopically. For this, we begin this section by presenting a general overview of its architecture before going into an in-depth analysis of its building blocks.

3.3.1   Overview of SDN Building Blocks

As already mentioned, the SDN approach allows the management of network services through the abstraction of lower-level functionality. Instead of dealing with low-level details of network devices regarding the way that packets and flows are managed, network administrators now only need to use the abstractions available in the SDN architecture. The way that this is achieved is by decoupling the control plane from the data plane following the layered architecture illustrated in Figure 3.1.

At the bottom layer, we can observe the data plane, where the network infrastructure (switches, routers, wireless access points, etc.) lies. In the context of SDN, these devices have been stripped of all control logic (e.g., routing algorithms like BGP) simply implementing a set of forwarding operations for manipulating network data packets and flows, providing an abstract open interface for the communication with the upper layers. In the SDN terminology, these devices are commonly referred to as network *switches*.

Moving to the next layer, we can observe the control plane, where an entity referred to as the *controller* lies. This entity encapsulates the networking logic and is responsible for providing a programmatic interface to the network, which is used to implement new functionality and perform various management tasks. Unlike previous approaches like ForCES, the control plane of SDN is ripped entirely from the network device and is considered to be logically centralized, while physically it can be either centralized or decentralized residing in one or more servers, which control the network infrastructure as a whole.

An important aspect that distinguishes SDN from previous programmable network attempts is that it has introduced the notion of the network operating system abstraction [18]. Recall that previous efforts like active networking proposed some sort of node operating system (e.g., NodeOS) for controlling the underlying hardware. A network operating system offers a more general abstraction of network state in switches, revealing a simplified interface for controlling the network. This abstraction assumes a logically centralized control model, in which the applications view the network as a single system. In other words, the network operating system acts as an intermediate layer responsible for maintaining a consistent view of network state, which is then exploited by control logic to provide various networking services for topological discovery, routing, management of mobility, and statistics.

At the top of the SDN stack lies the application layer, which includes all the applications that exploit the services provided by the controller in order to perform network-related tasks, like load balancing, network virtualization, etc. One of the most important features of SDN is the openness it provides to third-party developers through the abstractions it defines for the easy development and deployment of new applications in various networked environments from data centers and WANs to wireless and cellular networks. Moreover, the SDN architecture eliminates the need for dedicated middleboxes like firewalls and IDS in the network topology, as it is now possible for their functionality to be implemented in the form of software applications that monitor and modify the network state through the network operating system services. Obviously, the existence of this layer adds great value to SDN, since it gives rise to a wide range of opportunities for innovation, making SDN a compelling solution both for researchers and the industry.

Finally, the communication of the controller to the data plane and the application layer can be achieved through well-defined interfaces (APIs). We can distinguish two main APIs in the SDN architecture: (i) a *southbound* API for the communication between the controller and the network infrastructure and (ii) a *northbound* API defining an interface between the network applications and the controller. This is similar to the way communication is achieved among the hardware, the operating system, and the user space in most computer systems.

Having seen the general overview of the SDN architecture, it is now time for an in-depth discussion of each of the building blocks just presented. Some examples of SDN applications will be discussed in the next section.

3.3.2   SDN Switches

In the conventional networking paradigm, the network infrastructure is considered the most integral part of the network. Each network device encapsulates all the functionality that would be required for the operation of the network. For instance, a router needs to provide the proper hardware like a ternary content-addressable memory (TCAM) for quickly forwarding packets, as well as sophisticated software for executing distributed routing protocols like BGP. Similarly, a wireless access point needs to have the proper hardware for wireless connectivity as well as software for forwarding packets, enforcing access control, etc. However, dynamically changing the behavior of network devices is not a trivial task due to their closed nature.

The three-layered SDN architecture presented in Section 3.3.1 changes this by decoupling the control from the forwarding operations, simplifying the management of network devices. As already mentioned, all forwarding devices retain the hardware that is responsible for storing the forwarding tables (e.g., application-specific integrated circuits (ASICs) with a TCAM) but are stripped of their logic. The controller dictates to the switches how packets should be forwarded by installing new forwarding rules through an abstract interface. Each time a packet arrives to a switch, its forwarding table is consulted and the packet is forwarded accordingly.

Even though in the earlier overview of SDN a clean three-layered architecture was presented, it remains unclear what the boundaries between the control and the data plane should be. For example, active queue management (AQM) and scheduling configuration are operations that are still considered part of the data plane even in the case of SDN switches. However, there is no inherent problem preventing these functions from becoming part of the control plane by introducing some sort of abstraction allowing the control of low-level behavior in switching devices. Such an approach could turn out to be beneficial, since it would simplify the deployment of new more efficient schemes for low-level switch operations [19].

On the other hand, while moving all control operations to a logically centralized controller has the advantage of easier network management, it can also raise scalability issues if physical implementation of the controller is also centralized. Therefore, it might be beneficial to retain some of the logic in the switches. For instance, in the case of DevoFlow [20], which is a modification of the OpenFlow model, the packet flows are distinguished into two categories: small ("mice") flows handled directly by the switches and large ("elephant") flows requiring the intervention of the controller. Similarly, in the DIFANE [21] controller, intermediate switches are used for storing the necessary rules, and the controller is relegated to the simple task of partitioning the rules over the switches.

Another issue of SDN switches is that the forwarding rules used in the case of SDN are more complex than those of conventional networks, using wildcards for forwarding packets, considering multiple fields of the packet like source and destination addresses, ports, application, etc. As a result, the switching hardware cannot easily cope with the management of packets and flows. In order for the forwarding operation to be fast, ASICs using TCAM are required. Unfortunately, such specialized hardware is expensive and power consuming, and as a result, only a limited number of forwarding entries for flow-based forwarding schemes can be supported in each switch, hindering network scalability. A way to cope with this would be to introduce an assisting CPU to the switch or somewhere nearby to perform not only control plane but also data plane functionalities, for example, let the CPU forward the "mice" flows [22], or to introduce new architectures that would be more expressive and would allow more actions related to packet processing to be performed [23].

The issue of hardware limitations is not only restricted to fixed networks but is extended to the wireless and mobile domains as well. The wireless data plane needs to be redesigned in order to offer more useful abstractions similarly to what happened with the data plane of fixed networks. While the data plane abstractions offered by protocols like OpenFlow support the idea of decoupling the control from the data plane, they cannot be extended to the wireless and mobile field unless the underlying hardware (e.g., switches in backhaul cellular networks and wireless access points) starts providing equally sophisticated and useful abstractions [5].

Regardless of the way that SDN switches are implemented, it should be made clear that in order for the new paradigm to gain popularity, backward compatibility is a very important factor. While pure SDN switches that completely lack integrated control exist, it is the hybrid approach (i.e., support of SDN along with traditional operation and protocols) that would probably be the most successful at these early steps of SDN [11]. The reason is that while the features of SDN present a compelling solution for many realistic scenarios, the infrastructure in most enterprise networks still follows the conventional approach. Therefore, an intermediate hybrid network form would probably ease the transition to SDN.

3.3.3   SDN Controllers

As already mentioned, one of the core ideas of the SDN philosophy is the existence of a network operating system placed between the network infrastructure and the application layer. This network operating system is responsible for coordinating and managing the resources of the whole network and for revealing an abstract unified view of all components to the applications executed on top of it. This idea is analogous to the one followed in a typical computer system, where the operating system lies between the hardware and the user space and is responsible for managing the hardware resources and providing common services for user programs. Similarly, network administrators and developers are now presented with a homogeneous environment easier to program and configure much like a typical computer program developer would.

The logically centralized control and the generalized network abstraction it offers make the SDN model applicable to a wider range of applications and heterogeneous network technologies compared to the conventional networking paradigm. For instance, consider a heterogeneous environment composed of a fixed and a wireless network comprised of a large number of related network devices (routers, switches, wireless access points, middleboxes, etc.). In the traditional networking paradigm, each network device would require individual low-level configuration by the network administrator in order to operate properly. Moreover, since each device targets a different networking technology, it would have its own specific management and configuration requirements, meaning that extra effort would be required by the administrator to make the whole network operate as intended. On the other hand, with the logically centralized control of SDN, the administrator would not have to worry about low-level details. Instead, the network management would be performed by defining a proper high-level policy, leaving the network operating system responsible for communicating with and configuring the operation of network devices.

Having discussed the general concepts behind the SDN controller, the following subsections take a closer look at specific design decisions and implementation choices made at this core component that can prove to be critical for the overall performance and scalability of the network.

3.3.3.1   Centralization of Control in SDN

As already discussed, the SDN architecture specifies that the network infrastructure is logically controlled by a central entity responsible for management and policy enforcement.
However, it should be made clear that logically centralized control does not necessarily also
imply physical centralization.

There have been various proposals for physically centralized controllers, like NOX [18] and
Maestro [24]. A physically centralized control design simplifies the controller implementation. All switches are controlled by the same physical entity, meaning that the network is not
subject to consistency-related issues, with all the applications seeing the same network state
(which comes from the same controller). Despite its advantages, this approach suffers from
the same weakness that all centralized systems do, that is, the controller acts as a single point
of failure for the whole network. A way to overcome this is by connecting multiple controllers
to a switch, allowing a backup controller to take over in the event of a failure. In this case, all
controllers need to have a consistent view of the network; otherwise, applications might fail to
operate properly. Moreover, the centralized approach can raise scalability concerns, since all
network devices need to be managed by the same entity.

One approach that further generalizes the idea of using multiple controllers over the network is to maintain a logically centralized but physically decentralized control plane. In this
case, each controller is responsible for managing only one part of the network, but all controllers communicate and maintain a common network view. Therefore, applications view the
controller as a single entity, while in reality control operations are performed by a distributed
system. The advantage of this approach, apart from not having a single point of failure anymore, is the increase in performance and scalability, since only a part of the network needs to
be managed by each individual controller component. Some well-known controllers that
belong to this category are Onix [25] and HyperFlow [26]. One potential downside of decentralized control is once more related to the consistency of the network state among controller
components. Since the state of the network is distributed, it is possible that applications served
by different controllers might have a different view of the network, which might make them
operate improperly.

A hybrid solution that tries to encompass both scalability and consistency is to use two
layers of controllers like the Kandoo [27] controller does. The bottom layer is composed of a
group of controllers that do not have knowledge of the whole network state. These controllers
only run control operations that require knowing the state of a single switch (local information
only). On the other hand, the top layer is a logically centralized controller responsible for
performing network-wide operations that require knowledge of the whole network state. The
idea is that local operations can be performed faster this way and do not incur any additional
load to the high-level central controller, effectively increasing the scalability of the network.

Apart from the ideas related to the level of physical centralization of controllers, there have
been other proposed solutions related to their logical decentralization. The idea of logical
decentralization comes directly from the early era of programmable networks and from the
Tempest project. Recall that the Tempest architecture allowed multiple virtual ATM networks to
operate on top of the same set of physical switches. Similarly, there have been proposals for
SDN proxy controllers like FlowVisor [28] that allow multiple controllers to share the same
forwarding plane. The motivation for this idea was to enable the simultaneous deployment of
experimental and enterprise networks over the same infrastructure without affecting one another.

Before concluding our discussion on the degree of centralization with SDN controllers, it is important to examine the concerns that can be raised regarding their performance and applicability over large networking environments.

One of the most frequent concerns raised by SDN skeptics is the ability of SDN networks to scale and be responsive in cases of high network load. This concern comes mainly from the fact that in the new paradigm control moves out of network devices and goes in a single entity responsible for managing the whole network traffic. Motivated by this concern, performance studies of SDN controller implementations [29] have revealed that even physically centralized controllers can perform really well, having very low response times. For instance, it has been shown that even primitive single-threaded controllers like NOX can handle an average workload of up to 200 thousand new flows per second with a maximum latency of 600 ms for networks composed of up to 256 switches. Newer multithreaded controller implementations have been shown to perform significantly better. For instance, NOX-MT [30] can handle 1.6 million new flows per second in a 256-switch network with an average response time of 2 ms in a commodity eight-core machine of 2 GHz CPUs. Newer controller designs targeting large industrial servers promise to improve the performance even further. For instance, the McNettle [31] controller claims to be able to serve networks of up to 5000 switches using a single controller of 46 cores with a throughput of over 14 million flows per second and latency under 10 ms.

Another important performance concern raised in the case of a physically decentralized control plane is the way that controllers are placed within the network, as the network performance can be greatly affected by the number and the physical location of controllers, as well as by the algorithms used for their coordination. In order to address this, various solutions have been proposed, from viewing the placement of controllers as an optimization problem [32] to establishing connections of this problem to the fields of local algorithms and distributed computing for developing efficient controller coordination protocols [33].

A final concern raised in the case of physically distributed SDN controllers is related to the consistency of the network state maintained at each controller when performing policy updates due to concurrency issues that might occur by the error-prone, distributed nature of the logical controller. The solutions of such a problem can be similar to those of transactional databases, with the controller being extended with a transactional interface defining semantics for either completely committing a policy update or aborting [34].

3.3.3.2 Management of Traffic

Another very important design issue of SDN controllers is related to the way that traffic is managed. The decisions about traffic management can have a direct impact on the performance of the network, especially in cases of large networks composed of many switches and with high traffic loads. We can divide the problems related to traffic management into two categories: control granularity and policy enforcement.

Control Granularity

The control granularity applied over network traffic refers to how fine or coarse grained the controller inspection operations should be in relation to the packets traversing the network [11]. In conventional networks, each packet arriving at a switch is examined individually, and a routing decision is made as to where the packet should be forwarded depending on the information it carries (e.g., destination address). While this approach generally works for conventional networks, the same cannot be said for SDN. In this case, the per-packet approach becomes infeasible to implement across any sizeable network, since all packets would have to pass through the controller that would need to construct a route for each one of them individually.

Due to the performance issues raised by the per-packet approach, most SDN controllers follow a flow-based approach, where each packet is assigned to some flow according to a specific property (e.g., the packet's source and destination address and the application it is related with). The controller sets up a new flow by examining the first packet arriving for that flow and configuring the switches accordingly. In order to further off-load the controller, an extra coarse-grained approach would be to enforce control based on an aggregation flow match instead of using individual flows.

The main trade-off when examining the level of granularity is the load in the controller versus the QoS offered to network applications. The more fine grained the control, the higher the QoS. In the per-packet approach, the controller can always make the best decisions for routing each individual packet, therefore leading to improved QoS. On the other end, enforcing control over an aggregation of flows means that the controller decisions for forwarding packets do not fully adapt to the state of the network. In this case, packets might be forwarded through a suboptimal route, leading to degraded QoS.

Policy Enforcement

The second issue in the management of traffic is related to the way that network policies are applied by the controller over network devices [11]. One approach, followed by systems like Ethane, is to have a *reactive* control model, where the switching device consults the controller every time a decision for a new flow needs to be made. In this case, the policy for each flow is established to the switches only when an actual demand arises, making network management more flexible. A potential downside of this approach is the degradation of performance, due to the time required for the first packet of the flow to go to the controller for inspection. This performance drop could be significant, especially in cases of controllers that are physically located far away from the switch.

An alternative policy enforcement approach would be to use a *proactive* control model. In this case, the controller populates the flow tables ahead of time for any traffic that could go through the switches and then pushes the rules to all the switches of the network. Using this approach, a switch no longer has to request directions by the controller to set up a new flow and instead can perform a simple lookup at the table already stored in the TCAM of the device. The advantage of proactive control is that it eliminates the latency induced by consulting the controller for every flow,

3.3.4 SDN Programming Interfaces

As already mentioned, the communication of the controller with the other layers is achieved through a southbound API for the controller–switch interactions and through a northbound API for the controller–application interactions. In this section, we briefly discuss the main concepts and issues related to SDN programming by separately examining each point of communication.

3.3.4.1 Southbound Communication

The southbound communication is very important for the manipulation of the behavior of SDN switches by the controller. It is the way that SDN attempts to "program" the network. The most prominent example of a standardized southbound API is OpenFlow [35]. Most projects related to SDN assume that the communication of the controller with the switches is OpenFlow based, and therefore, it is important to make a detailed presentation of the OpenFlow approach. However, it should be made clear that OpenFlow is just one (rather popular) out of many possible implementations of controller–switch interactions. Other alternatives, for example, DevoFlow [20], also exist, attempting to solve performance issues that OpenFlow faces.

Overview of OpenFlow
Following the SDN principle of decoupling the control and data planes, OpenFlow provides a standardized way of managing traffic in switches and of exchanging information between the switches and the controller, as Figure 3.2 illustrates. The OpenFlow switch is composed of two logical components. The first component contains one or more flow tables responsible for maintaining the information required by the switch in order to forward packets. The second component is an OpenFlow client, which is essentially a simple API allowing the communication of the switch with the controller.

The flow tables consist of flow entries, each of which defines a set of rules determining how the packets belonging to that particular flow will be managed by the switch (i.e., how they will be processed and forwarded). Each entry in the flow table has three fields: (i) a packet header defining the flow, (ii) an action determining how the packet should be processed, and (iii) statistics, which keep track of information like the number of packets and bytes of each flow and the time since a packet of the flow was last forwarded.

Once a packet arrives at the OpenFlow switch, its header is examined, and the packet is matched to the flow that has the most similar packet header field. If a matching flow is found, the action defined in the action field is performed. These actions include the forwarding of the

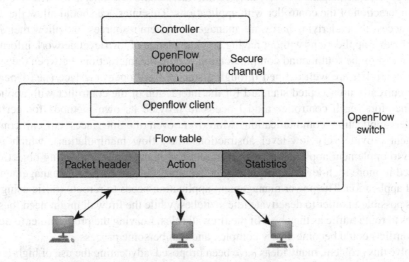

Figure 3.2 Design of an OpenFlow switch and communication with the controller.

packet to a particular port in order to be routed through the network, the forwarding of the packet in order to be examined by the controller, or the rejection of the packet. If the packet cannot be matched to any flow, it is treated according to the action defined in a table-miss flow entry.

The exchange of information between the switch and the controller happens by sending messages through a secure channel in a standardized way defined by the OpenFlow protocol. This way, the controller can manipulate the flows found in the flow table of the switch (i.e., add, update, or delete a flow entry) either proactively or reactively as discussed in the basic controller principles. Since the controller is able to communicate with the switch using the OpenFlow protocol, there is no longer a need for network operators to interact directly with the switch.

A particularly compelling feature of OpenFlow is that the packet header field can be a wild-card, meaning that the matching to the header of the packet does not have to be exact. The idea behind this approach is that various network devices like routers, switches, and middleboxes have a similar forwarding behavior, differing only in terms of which header fields they use for matching and the actions they perform. OpenFlow allows the use of any subset of these header fields for applying rules on traffic flows, meaning that it conceptually unifies many different types of network devices. For instance, a router could be emulated by a flow entry using a packet header performing a match only on the IP address, while a firewall would be emulated through a packet header field containing additional information like the source and destination IP addresses and port numbers as well as the transport protocol employed.

3.3.4.2   Northbound API

As already discussed, one of the basic ideas advocated in the SDN paradigm is the existence of a network operating system, lying between the network infrastructure and the high-level services and applications, similarly to how a computer operating system lies between the hardware and the user space. Assuming such a centralized coordination entity and based on the basic operating system principles, a clearly defined interface should also exist in the SDN architecture for the interaction of the controller with applications. This interface should allow the applications to access the underlying hardware, manage the system resources, and allow their interaction with other applications without having any knowledge of low-level network information.

In contrast to the southbound communication, where the interactions between the switches and the controller are well defined through a standardized open interface (i.e., OpenFlow), there is currently no accepted standard for the interaction of the controller with applications [11]. Therefore, each controller model needs to provide its own methods for performing controller–application communication. Moreover, even the interfaces current controllers implement provide· very low-level abstractions (i.e., flow manipulation), which make it difficult to implement applications with different and many times conflicting objectives that are based in more high-level concepts. As an example, consider a power management and a firewall application. The power management application needs to reroute traffic using as few links as possible in order to deactivate idle switches, while the firewall might need these extra switches to route traffic as they best fit the firewall rules. Leaving the programmer to deal with these conflicts could become a very complex and cumbersome process.

To solve this problem, many ideas have been proposed, advocating the use of high-level network programming languages responsible for translating policies to low-level flow constraints, which in turn will be used by the controller to manage the SDN switches. These network programming languages can also be seen as an intermediate layer in the SDN architecture, placed between the application layer and the controller in a similar manner as to how high-level programming languages like C++ and Python exist on top of the assembly language for hiding the complex low-level details of the assembly language from the programmer. Some examples of such high-level network programming languages include Frenetic [36] and Pyretic [37].

3.3.5 SDN Application Domains

In order to demonstrate the applicability of SDN in a wide range of networking domains, we briefly present two characteristic examples in which SDN could prove to be beneficial: data centers and cellular networks. Of course, the list of SDN applications is not only limited to these domains but is also extended in many others, from enterprise networks, WLANs, and heterogeneous networks to optical networks and the Internet of Things [5, 11].

3.3.5.1 Data Center Networks

One of the most important requirements for data center networks is to find ways to scale in order to support hundreds of thousands of servers and millions of virtual machines. However, achieving such scalability can be a challenging task from a network perspective. First of all, the size of forwarding tables increases along with the number of servers, leading to a requirement for more sophisticated and expensive forwarding devices. Moreover, traffic management and policy enforcement can become very important and critical issues, since data centers are expected to continuously achieve high levels of performance.

In traditional data centers, the aforementioned requirements are typically met through the careful design and configuration of the underlying network. This operation is in most cases performed manually by defining the preferred routes for traffic and by placing middleboxes at strategic choke points on the physical network. Obviously, this approach contradicts the requirement for scalability, since manual configuration can become a very challenging and error-prone task, especially as the size of the network grows. Additionally, it becomes increasingly difficult to make the data center operate at its full capacity, since it cannot dynamically adapt to the application requirements.

The advantages that SDN offers to network management come to fill these gaps. By decoupling the control from the data plane, forwarding devices become much simpler and therefore cheaper. At the same time, all control logic is delegated to one logically centralized entity. This allows the dynamic management of flows, the load balancing of traffic, and the allocation of resources in a manner that best adjusts the operation of the data center to the needs of running applications, which in turn leads to increased performance [38]. Finally, placing middleboxes in the network is no longer required, since policy enforcement can now be achieved through the controller entity.

3.3.5.2 Cellular Networks

The market of cellular mobile networks is perhaps one of the most profitable in telecommunications. The rapid increase in the number of cellular devices (e.g., smartphones and tablets) during the past decade has pushed the existing cellular networks to their limits. Recently, there has been significant interest in integrating the SDN principles in current cellular architectures like the 3G Universal Mobile Telecommunications System (UMTS) and the 4G Long-Term Evolution (LTE) [39].

One of the main disadvantages of current cellular network architectures is that the core of the network has a centralized data flow, with all traffic passing through specialized equipment, which packs multiple network functions from routing to access control and billing (e.g., packet gateway in LTE), leading to an increase of the infrastructural cost due to the complexity of the devices and raising serious scalability concerns. Moreover, cell sizes of the access network tend to get smaller in order to cover the demands of the ever-increasing traffic and the limited wireless spectrum for accessing the network. However, this leads to increased interference among neighboring base stations and to the fluctuation of load from one base station to another due to user mobility, rendering the static allocation of resources no longer adequate.

Applying the SDN principles to cellular networks promises to solve some of these deficiencies. First of all, decoupling the control from the data plane and introducing a centralized controller that has a complete view of the whole network allow network equipment to become simpler and therefore reduce the overall infrastructural cost. Moreover, operations like routing, real-time monitoring, mobility management, access control, and policy enforcement can be assigned to different cooperating controllers, making the network more flexible and easier to manage. Furthermore, using a centralized controller acting as an abstract base station simplifies the operations of load and interference management, no longer requiring the direct communication and coordination of base stations. Instead, the controller makes the decisions for the whole network and simply instructs the data plane (i.e., the base stations) on how to operate. One final advantage is that the use of SDN eases the introduction of virtual operators to the telecommunication market, leading to increased competitiveness. By virtualizing the underlying switching equipment, all providers become responsible for managing the flows of their own subscribers through their own controllers, without the requirement to pay large sums for obtaining their own infrastructure.

3.3.6 Relation of SDN to Network Virtualization and Network Function Virtualization

Two very popular technologies closely related to SDN are network virtualization and network function virtualization (NFV). In this subsection, we briefly attempt to clarify their relationship to SDN, since these technologies tend to become the cause of confusion especially for those recently introduced to the concept of SDN.

Network virtualization is the separation of the network topology from the underlying physical infrastructure. Through virtualization, it is possible to have multiple "virtual" networks deployed over the same physical equipment, with each of them having a much simpler topology compared to that of the physical network. This abstraction allows network operators to construct networks as they see fit without having to tamper with the underlying infrastructure, which can turn out to be a difficult or even impossible process. For instance, through network virtualization, it becomes possible to have a virtual local area network (VLAN) of hosts spanning multiple physical networks or to have multiple VLANs on top of a single physical subnet.

The idea behind network virtualization of decoupling the network from the underlying physical infrastructure bears resemblance to that advocated by SDN for decoupling the control from the data plane and therefore naturally becomes a source of confusion. The truth is that none of the two technologies is dependent on the other. The existence of SDN does not readily imply network virtualization. Similarly, SDN is not necessarily a prerequisite for achieving network virtualization. On the contrary, it is possible to deploy a network virtualization solution over an SDN network, while at the same time an SDN network could be deployed in a virtualized environment.

Since its appearance, SDN has closely coexisted with network virtualization, which acted as one of the first and perhaps the most important use cases of SDN. The reason is that the architectural flexibility offered by SDN acted as an enabler for network virtualization. In other words, network virtualization can be seen as a solution focusing on a particular problem, while SDN is one (perhaps the best at this moment) architecture for achieving this. However, as already stressed, network virtualization needs to be seen independently from SDN. In fact, it has been argued by many that network virtualization could turn out to be even bigger technological innovation than SDN [6].

Another technology that is closely related but different from SDN is NFV [40]. NFV is a carrier-driven initiative with a goal to transform the way that operators architect networks by employing virtualization-related technologies to virtualize network functions such as intrusion detection, caching, domain name service (DNS), and network address translation (NAT) so that they can run in software. Through the introduction of virtualization, it is possible to run these functions over generic industry-standard high-volume servers, switches, and storage devices instead of using proprietary purpose-built network devices. This approach reduces operational and deployment costs, since operators no longer need to rely on expensive proprietary hardware solutions. Finally, flexibility in network management increases as it is possible to quickly modify or introduce new services to address changing demands.

The decoupling of network functions from the underlying hardware is closely related to the decoupling of the control from the data plane advocated by SDN, and therefore, the distinction of the two technologies can be a bit vague. It is important to understand that even though closely related, SDN and NFV refer to different domains. NFV is complementary to SDN but does not depend on it, and vice versa. For instance, the control functions of SDN could be implemented as virtual functions based on the NFV technology. On the other hand, an NFV orchestration system could control the forwarding behavior of physical switches through SDN. However, neither technology is a requirement for the operation of other, but both could benefit from the advantages each can offer.

3.4   Impact of SDN to Research and Industry

Having seen the basic concepts of SDN and some important applications of this approach, it is now time to briefly discuss the impact of SDN to the research community and the industry. While the focus of each interested party might be different, from designing novel solutions exploiting the benefits of SDN to developing SDN-enabled products ready to be deployed in commercial environments, their involvement in the evolution of SDN helps in shaping the future of this technology. Seeing what the motivation and the focus of current SDN-related attempts will provide us with indications of what will potentially drive future research in this field.

3.4.1 Overview of Standardization Activities and SDN Summits

Recently, several standardization organizations have started focusing on SDN, each working in providing standardized solutions for a different part of the SDN space. The benefits of such efforts are very significant, since standardization is the first step toward the wide adoption of a technology.

The most relevant standardization organization for SDN is considered the Open Networking Foundation (ONF) [41], which is a nonprofit industry consortium founded in 2011. It has more than 100 company members including telecom operators, network and service providers, equipment vendors, and networking and virtualization software suppliers. Its vision is to make SDN the new norm for networks by transforming the networking industry to a software industry through the open SDN standards. To achieve this, it attempts to standardize and commercialize SDN and its underlying technologies, with its main accomplishment the standardization of the OpenFlow protocol, which is also the first SDN standard. ONF has a number of working groups working in different aspects of SDN from forwarding abstractions, extensibility, configuration, and management to educating the community on the SDN value proposition.

The IETF, which is a major driving force in developing and promoting Internet standards, also has a number of working groups focusing on SDN in a broader scope than just OpenFlow. The Software Defined Networking Research Group (SDNRG) [42] focuses on identifying solutions related to the scalability and applicability of the SDN model as well as for developing abstractions and programming languages useful in the context of SDN. Finally, it attempts to identify SDN use cases and future research challenges. On a different approach, the Interface to the Routing System (I2RS) [43] working group is developing an SDN strategy, counter to the OpenFlow approach, in which traditional distributed routing protocols can run on network hardware to provide information to a centrally located manager. Other SDN-related IETF working groups include application-layer traffic optimization (ALTO) [44] using SDN and CDNI [45] studying how SDN can be used for content delivery network (CDN) interconnection.

Some study groups (SGs) of ITU's Telecommunication Standardization Sector (ITU-T) [46] are also looking on SDN for public telecommunication networks. For instance, Study Group 13 (SG13) is focusing on a framework of telecom SDN and on defining requirements of formal specification and verification methods for SDN. Study Group 11 (SG11) is developing requirements and architectures on SDN signaling, while Study Group 15 (SG15) has started discussions on transport SDN.

Other standardization organizations that have also been interested in applying the SDN principles include the Optical Internetworking Forum (OIF) [47], the Broadband Forum (BBF) [48], and the Metro Ethernet Forum (MEF) [49]. OIF is responsible for promoting the development and deployment of interoperable optical networking systems, and it supports a working group to define the requirements for a transport network SDN architecture. BBF is a forum for fixed line broadband access and core networks, working on a cloud-based gateway that could be implemented using SDN concepts. Finally, MEF has as its goal to develop, promote, and certify technical specifications for carrier Ethernet services. One of its directions is to investigate whether MEF services could fit within an ONF SDN framework.

Apart from the work performed on standardizing SDN solutions, there exist a number of summits for sharing and exploring new ideas and key developments produced in the SDN research community. The Open Networking Summit (ONS) is perhaps the most important

SDN event having as a mission "to help the SDN revolution succeed by producing high-impact SDN events." Other SDN-related venues have also started emerging like the SDN & NFV Summit for solutions on network virtualization, the SDN & OpenFlow World Congress, the SIGCOMM Workshop on Hot Topics in Software Defined Networking (HotSDN), and the IEEE/IFIP International Workshop on SDN Management and Orchestration (SDNMO).

3.4.2   SDN in the Industry

The advantages that SDN offers compared to traditional networking have also made the industry focus on SDN either for using it as a means to simplify management and improve services in their own private networks or for developing and providing commercial SDN solutions.

Perhaps one of the most characteristic examples for the adoption of SDN in production networks is Google, which entered in the world of SDN with its B4 network [50] developed for connecting its data centers worldwide. The main reason for moving to the SDN paradigm, as explained by Google engineers, was the very fast growth of Google's back-end network. While computational power and storage become cheaper as scale increases, the same cannot be said for the network. By applying SDN principles, the company was able to choose the networking hardware according to the features it required, while it managed to develop innovative software solutions. Moreover, the centralized network control made the network more efficient and fault tolerant providing a more flexible and innovative environment, while at the same time it led to a reduction of operational expenses. More recently, Google revealed Andromeda [51], a software defined network underlying its cloud, which is aimed at enabling Google's services to scale better, cheaper, and faster. Other major companies in the field of networking and cloud services like Facebook and Amazon are also planning on building their next-generation network infrastructure based on the SDN principles.

Networking companies have also started showing interest in developing commercial SDN solutions. This interest is not limited in developing specific products like OpenFlow switches and network operating systems; rather, there is a trend for creating complete SDN ecosystems targeting different types of customers. For instance, companies like Cisco, HP, and Alcatel have entered the SDN market, presenting their own complete solutions intended for enterprises and cloud service providers, while telecommunication companies like Huawei are designing solutions for the next generation of telecom networks, with a specific interest in LTE and LTE-Advanced networks. In 2012, VMware acquired an SDN start-up called Nicira in order to integrate its network virtualization platform (NVP) to NSX, VMware's own network virtualization and security platform for software defined data centers. The list of major companies providing SDN solutions constantly grows, with many others like Broadcom, Oracle, NTT, Juniper, and Big Switch Networks recognizing the benefits of SDN and proposing their own solutions.

3.4.3   Future of SDN

Going back to the beginning of this discussion and looking at all the intermediate steps that led to modern software defined networks, it is tricky to predict what lies in the future. Previous attempts for redesigning the network architecture have shown that very promising technologies can fail due to lack of the proper conditions, while success depends on a number of factors from finding compelling use cases for the emerging technology to managing its adoption not only by the research community but by the industry as well. The way that SDN deals with these matters makes it a very promising candidate for being the next major disruption in the networking field. The benefits of applying the SDN principles in different types of networks, the unification of heterogeneous environments, and the wide number of applications that this paradigm offers demonstrate its very high potential to become a major driving force commercially in the very near future especially for cloud service providers, network operators, and mobile carriers. It remains to be seen whether these predictions will be confirmed and to what extent SDN will deliver its promises.

References

[1] A. T. Campbell, I. Katzela, K. Miki, and J. Vicente. "Open signaling for ATM, internet and mobile networks (OPENSIG'98)." ACM SIGCOMM Computer Communication Review 29.1 (1999): 97–108.

[2] D. L. Tennenhouse, J. Smith, D. Sincoskie, D. J. Wetherall, and G. J. Minden. "A survey of active network research." IEEE Communications Magazine 35.1 (1997): 80–86.

[3] J. E. Van der Merwe, S. Rooney, I. Leslie, and S. Crosby. "The tempest-a practical framework for network programmability." IEEE Network 12.3 (1998): 20–28.

[4] "Devolved control of ATM networks." Available from http://www.cl.cam.ac.uk/research/srg/netos/old-projects/dcan/ (accessed January 19, 2015).

[5] J. Qadir, N. Ahmed, and N. Ahad. "Building programmable wireless networks: An architectural survey." arXiv preprint arXiv:1310.0251 (2013).

[6] N. Feamster, J. Rexford, and E. Zegura. "The road to SDN." ACM Queue 11.12 (2013): 20–40.

[7] N. Shalaby, Y. Gottlieb, M. Wawrzoniak, and L. Peterson. "Snow on silk: A nodeOS in the Linux kernel." *Active Networks*. Springer, Berlin (2002): 1–19.

[8] S. Merugu, S. Bhattacharjee, E. Zegura, and K. Calvert. "Bowman: A node OS for active networks." INFOCOM 2000. Proceedings of the Nineteenth Annual Joint Conference of the IEEE Computer and Communications Societies; Tel Aviv 3 (2000): 1127–1136.

[9] D. J. Wetherall, J. V. Guttag, and D. L. Tennenhouse. "ANTS: A toolkit for building and dynamically deploying network protocols." Open Architectures and Network Programming, 1998 IEEE: pp. 117, 129; April 3–4, 1998. doi: 10.1109/OPNARC.1998.662048.

[10] M. Hicks, P. Kakkar, J. T. Moore, C. A. Gunter, and S. Nettles. "PLAN: A packet language for active networks." ACM SIGPLAN Notices 34.1 (1998): 86–93.

[11] B. Nunes, M. Mendonca, X. Nguyen, K. Obraczka, and T. Turletti. "A survey of software-defined networking: Past, present, and future of programmable networks." Communications Surveys & Tutorials, IEEE, 16 (3) (2014): 1617–1634, third quarter.

[12] L. Yang, R. Dantu, T. Anderson, and R. Gopal. "Forwarding and control element separation (ForCES) framework." RFC 3746, (2004). Available at https://tools.ietf.org/html/rfc3746 (accessed February 17, 2015).

[13] A. Greenberg, G. Hjalmtysson, D. A. Maltz, A. Myers, J. Rexford, G. Xie, H. Yan, J. Zhan, and H. Zhang. "A clean slate 4D approach to network control and management." ACM SIGCOMM Computer Communication Review 35.5 (2005): 41–54.

[14] H. Yan, D. A. Maltz, T. S. Eugene Ng, H. Gogineni, H. Zhang, and Z. Cai. "Tesseract: A 4D network control plane." 4th USENIX Symposium on Networked Systems Design & Implementation 7; Cambridge, MA (2007): 369–382.

[15] M. Casado, T. Garfinkel, A. Akella, M. J. Freedman, D. Boneh, N. McKeown, and S. Shenker. "SANE: A protection architecture for enterprise networks." 15th USENIX Security Symposium; Vancouver, BC, Canada (2006): 137–151.

[16] M. Casado, M. J. Freedman, J. Pettit, J. Luo, N. McKeown, and S. Shenker. "Ethane: Taking control of the enterprise." ACM SIGCOMM Computer Communication Review 37.4 (2007): 1–12.

[17] B. Chun, D. Culler, T. Roscoe, A. Bavier, L. Peterson, M. Wawrzoniak, and M. Bowman. "Planetlab: An overlay testbed for broad-coverage services." ACM SIGCOMM Computer Communication Review 33.3 (2003): 3–12.

[18] N. Gude, T. Koponen, J. Pettit, B. Pfaff, M. Casado, N. McKeown, and S. Shenker. "NOX: Towards an operating system for networks." ACM SIGCOMM Computer Communication Review 38.3 (2008): 105–110.

[19] A. Sivaraman, K. Winstein, S. Subramanian, and H. Balakrishnan. "No silver bullet: Extending SDN to the data plane." Proceedings of the Twelfth ACM Workshop on Hot Topics in Networks 19; Maryland (2013): 1–7.

[20] A. R. Curtis, J. C. Mogul, J. Tourrilhes, P. Yalagandula, P. Sharma, and S. Banerjee. "Devoflow: Scaling flow management for high-performance networks." ACM SIGCOMM Computer Communication Review 41.4 (2011): 254–265.

[21] M. Yu, J. Rexford, M. J. Freedman, and J. Wang. "Scalable flow-based networking with DIFANE." ACM SIGCOMM Computer Communication Review 40.4 (2010): 351–362.

[22] G. Lu, R. Miao, Y. Xiong, and C. Guo. "Using cpu as a traffic co-processing unit in commodity switches." Proceedings of the First Workshop on Hot Topics in Software Defined Networks; Helsinki, Finland (2012): 31–36.

[23] P. Bosshart, G. Gibb, H.-S. Kim, G. Varghese, N. McKeown, M. Izzard, F. Mujica, and M. Horowitz. "Forwarding metamorphosis: Fast programmable match-action processing in hardware for SDN." SIGCOMM Computer Communication Review 43.4 (2013): 99–110.

[24] Z. Cai, A. L. Cox, and T. E. Ng. "Maestro: A system for scalable OpenFlow control." Technical Report TR10-08. Texas: Rice University (2010).

[25] T. Koponen, M. Casado, N. Gude, J. Stribling, L. Poutievski, M. Zhu, R. Ramanathan, Y. Iwata, H. Inoue, T. Hama, and S. Shenker. "Onix: A distributed control platform for large-scale production networks." 9th USENIX Symposium on Operating Systems Design and Implementation, OSDI 10; Vancouver, BC, Canada (2010): 1–6.

[26] A. Tootoonchian and Y. Ganjali. "Hyperflow: A distributed control plane for openflow." Proceedings of the 2010 Internet Network Management Conference on Research on Enterprise Networking. San Jose, CA: USENIX Association (2010): 3–8.

[27] S. H. Yeganeh and Y. Ganjali. "Kandoo: A framework for efficient and scalable offloading of control applications." Proceedings of the First Workshop on Hot Topics in Software Defined Networks. Helsinki, Finland: ACM (2012): 19–24.

[28] R. Sherwood, G. Gibb, K.-K. Yap, G. Appenzeller, M. Casado, N. McKeown, and G. Parulkar. "Flowvisor: A network virtualization layer." Technical Report, OpenFlow Switch Consortium (2009). Available from http://archive.openflow.org/downloads/technicalreports/openflow-tr-2009-1-flowvisor.pdf (accessed February 17, 2015).

[29] A. Shalimov, D. Zuikov, D. Zimarina, V. Pashkov, and R. Smeliansky. "Advanced study of SDN/OpenFlow controllers." Proceedings of the 9th Central & Eastern European Software Engineering Conference in Russia. Moscow: ACM (2013).

[30] A. Tootoonchian, S. Gorbunov, Y. Ganjali, M. Casado, and R. Sherwood. "On controller performance in software-defined networks." USENIX Workshop on Hot Topics in Management of Internet, Cloud, and Enterprise Networks and Services (Hot-ICE) 54; San Jose, CA (2012).

[31] A. Voellmy and J. Wang. "Scalable software defined network controllers." Proceedings of the ACM SIGCOMM 2012 Conference on Applications, Technologies, Architectures, and Protocols for Computer Communication. Helsinki, Finland: ACM (2012): 289–290.

[32] B. Heller, R. Sherwood, and N. McKeown. "The controller placement problem." Proceedings of the First Workshop on Hot Topics in Software Defined Networks. Helsinki, Finland: ACM (2012): 7–12.

[33] S. Schmid and J. Suomela. "Exploiting locality in distributed sdn control." Proceedings of the Second ACM SIGCOMM Workshop on Hot Topics in Software Defined Networking. Hong Kong, China: ACM (2013): 121–126.

[34] M. Canini, P. Kuznetsov, D. Levin, and S. Schmid. "Software transactional networking: Concurrent and consistent policy composition." Proceedings of the Second ACM SIGCOMM Workshop on Hot Topics in Software Defined Networking. Hong Kong, China: ACM (2013): 1–6.

[35] N. McKeown, T. Anderson, H. Balakrishnan, G. Parulkar, L. Peterson, J. Rexford, S. Shenker, and J. Turner. "OpenFlow: enabling innovation in campus networks." ACM SIGCOMM Computer Communication Review 38.2 (2008): 69–74.

[36] N. Foster, R. Harrison, M. J. Freedman, C. Monsanto, J. Rexford, A. Story, and D. Walker. "Frenetic: A network programming language." ACM SIGPLAN Notices 46.9 (2011): 279–291.

[37] C. Monsanto, J. Reich, N. Foster, J. Rexford, and D. Walker. "Composing software-defined networks." 10th USENIX Symposium on Networked Systems Design & Implementation; Lombard, IL (2013): 1–13.

[38] M. Al-Fares, S. Radhakrishnan, B. Raghavan, N. Huang, and A. Vahdat. "Hedera: Dynamic flow scheduling for data center networks." 7th USENIX Symposium on Networked Systems Design & Implementation 10; San Jose, CA (2010): 19–34.

[39] L. E. Li, Z. M. Mao, and J. Rexford. "Toward software-defined cellular networks." European Workshop on Software Defined Networking (EWSDN). Darmstadt, Germany: IEEE (2012): 7–12.

[40] C. Cui, H. Deng, D. Telekom, U. Michel, and H. Damker. "Network functions virtualisation." Available from http://portal.etsi.org/NFV/NFV_White_Paper.pdf (accessed 19 January 2015).

[41] Open Networking Foundation (ONF). Available from https://www.opennetworking.org (accessed 19 January 2015).

[42] IRTF. "Software-defined networking research group (SDNRG)." Available from https://irtf.org/sdnrg (accessed 19 January 2015).

[43] IETF. "Interface to the routing system (i2rs)." Available from http://datatracker.ietf.org/wg/i2rs/ (accessed 19 January 2015).

[44] IETF. "ALTO and software defined networking (SDN)." Available from http://www.ietf.org/proceedings/84/slides/slides-84-alto-5 (accessed 19 January 2015).

[45] IETF. "CDNI request routing with SDN." Available from http://www.ietf.org/proceedings/84/slides/slides-84-cdni-1.pdf (accessed 19 January 2015).

[46] "ITU Telecommunication Standardization Sector." Available from http://www.itu.int/en/ITU-T (accessed 19 January 2015).

[47] Optical Internetworking Forum (OIF). Available from http://www.oiforum.com (accessed 19 January 2015).

[48] "Broadband Forum and SDN." Available from http://www.broadband-forum.org/technical/technicalwip.php (accessed 19 January 2015).

[49] "MEF—Metro Ethernet Forum." Available from http://metroethernetforum.org (accessed 19 January 2015).

[50] S. Jain, A. Kumar, S. Mandal, J. Ong, L. Poutievski, A. Singh, S. Venkata, J. Wanderer, J. Zhou, M. Zhu, J. Zolla, U. Hölzle, S. Stuart, and A. Vahdat. "B4: Experience with a globally-deployed software defined WAN." Proceedings of the ACM SIGCOMM 2013 Conference on SIGCOMM. Hong Kong, China: ACM (2013): 3–14.

[51] A. Vahdat. "Enter the Andromeda zone—Google Cloud Platform's latest networking stack." Available from http://googlecloudplatform.blogspot.gr/2014/04/enter-andromeda-zone-google-cloud-platforms-latest-networking-stack.html (accessed 19 January 2015).

4

Wireless Software Defined Networking

Claude Chaudet[1] and Yoram Haddad[2]

[1] *Telecom ParisTech, Institut Telecom, Paris, France*
[2] *Jerusalem College of Technology, Jerusalem, Israel*

4.1 Introduction

The recent years have witnessed the advent of a digital, mobile, and connected society in which digital devices are used for countless applications. People use wireless devices such as computers, smartphones, or tablets for communicating, producing, accessing, and sharing knowledge; shopping; interacting with public services; taking and storing photographs: finding directions: listening to music: watching videos: studying; or playing. More usages are expected to appear in the next years as the terminal size and cost decrease. The availability and the performance of wireless communications have obviously played a major role in this evolution.

Compared to the digital world of the beginning of the year 2000, people now often possess multiple terminals (smartphone, tablet, computer) and select one of the terminals based on the situation. A professional can work on a presentation from his office computer, make a few modifications on his tablet on the plane, and rehearse it on his smartphone on the metro while getting to a meeting where he will use his laptop to display it. Someone may start watching a movie at home on his TV and finish it in the bus while traveling to work. These scenarios, which are a reality today, have an important impact on the networks.

First, data is more and more often synchronized among terminals, thanks to cloud services. However, this permanent synchronization process consumes wireless channel bandwidth and creates important data flows that remind peer-to-peer applications. If the average size of the exchanged files is smaller, the number of exchanges, compared to the wireless link capacity, is important. Besides, the deployment of self-hosted cloud solutions induces similar user-to-user traffic patterns, except that few peer-to-peer applications were hosted by moving devices.

Software Defined Mobile Networks (SDMN): Beyond LTE Network Architecture, First Edition.
Edited by Madhusanka Liyanage, Andrei Gurtov, and Mika Ylianttila.

Handovers are therefore more and more frequent, and concern multimedia data that can tolerate losses better than delays, as well as file transfer that can tolerate delays as long as the TCP connections remain. Finally, when looking at the scenario involving terminal switching while watching a streaming movie, content adaptation may be required.

To fully satisfy the user, wireless infrastructure networks and wireless LANs (WLANs) should therefore find efficient ways to manage horizontal and vertical handover content adaptation while preserving their scarce bandwidth and reducing as much as possible their exploitation costs. Numerous research initiatives are driving the technology toward these objectives. Femtocells and power control improve spatial efficiency; traffic off-loading and multihoming allow to share bandwidth between technologies; in-network data replication and information-centric networks (ICN) help content distribution; power savings and green networking improve the operational costs; etc. Some key challenges remain, though: how to implement these techniques and make sure they interact properly.

Operators therefore need a way to manage their wireless network as a whole efficiently and in an evolutionary manner. Besides, several optimization possibilities would also require coordination and cooperation between operators, users, and infrastructures, which is difficult to achieve in practice. This advocates for a paradigm shift in the way wireless networks are managed. A network should be able to increase or decrease its capacity on demand, providing the desired quality of service (QoS) to users when possible and avoiding disturbances to close networks when possible. This requires a bandwidth management entity capable of examining not only the demand and the user service-level agreements (SLAs) but also the state of the different access points (APs) or the wireless channel status and to take appropriate decisions. This management entity shall build its vision of the network by gathering data from various network devices and perform global optimization to ultimately suggest connection options to each user and for each application. This paradigm is referred to as *network virtualization* or *software defined networking* (SDN).

As exposed in the other chapters of this book, the SDN concept emerged as a way to foster innovation, allowing experimenters to use a production network without noticeable impact, and as an elegant way to provide better QoS or to improve network reliability. OpenFlow [1] was introduced in 2011 following these concepts and has since then been pushed by the industry, leading to the creation of the Open Networking Foundation (ONF), a nonprofit organization that manages what has now become a standard. OpenFlow basically consists in separating the control and forwarding planes by letting the interconnection equipment match packets based on a cross-layer set of 12 fields (the 12-tuple) and by deporting all the intelligence to central entities called controllers. It is these controllers that decide the policy that applies in case of successful/ unsuccessful matching. Multiple controllers can coexist and manage multiple and independent visions of the network called *slices* in the OpenFlow terminology. SDN in the OpenFlow vision therefore consists in three functions: flow-based forwarding that requires packet matching against a flow table, status reporting from each interconnection device, and slicing that requires the capability for each interconnection element to isolate traffic. An OpenFlow architecture is composed of the interconnection devices that only perform matching and forwarding, the controllers that decide and publish policies regarding flows handling, and an entity, called *FlowVisor*, that presents a sliced vision of the physical infrastructure to the controllers.

We refer the reader to other chapters of this book for details on this aspect and will focus, in this present chapter, on the wireless extension of SDN that can be called *wireless SDN* or *software defined wireless networks* (SDWN). Section 4.2 elaborates around the concept of wireless SDN and the extension to OpenFlow for wireless. We then present related works and some existing projects in Section 4.3. Section 4.4 exposes the opportunities that a successful wireless SDN implementation could bring, while Section 4.5 lists some of the key challenges to overcome.

4.2 SDN for Wireless

The software defined network concept has initially been imagined with data centers and fixed networks in mind. However, the wireless world could greatly benefit from such a framework. Indeed, most wireless technologies have to face limited resources. Cellular networks have to deal with an ever-increasing mobile traffic that is handled with difficulty by the radio access networks. Mobile operators try to imagine solutions to off-load data to WLANs in urban areas, which can theoretically accommodate more elastic flows thanks to random access. However, wireless LANs also become congested, as the unlicensed frequency bands are utilized by multiple technologies whose traffic also increases. Moreover, the deployment of WLAN APs in urban areas is denser and denser and uncoordinated, making interference mitigation difficult.

Besides, the mobility of users also has an impact on the available resources. Users should be able to keep their communications and connections open while traveling, which requires the operator to predict user mobility and to reserve a part of the resources for users passing from cell to cell. Several handover procedures are possible, but this process often requires dedicating a part of the resources to mobility management.

These problems are all related to the scarce capacity offered by wireless channels compared to the demand of the applications, which is often scaled on the wired connection performance. Classical strategy to solve this issue, in the past, consisted in improving the spectral efficiency by working on modulation or coding. However, this process has a limit, and it is now common to read that we are reaching Shannon capacity limit. The most optimistic predictions place today's technology within 20% of this limit, and the ultimate efforts to improve signal over interference plus noise ratio are expected to be very difficult.

However, recent reports show, using frequency band scanning, that only 2% of the wireless spectrum between 30 MHz and 3 GHz is effectively used in some areas of the world.[1] However, the frequency bands are allocated by national regulators and are granted with caution to applications in order to keep spectrum free for the future and to preserve space for potential tactical communications. If software defined radio (SDR) (aka cognitive radio) provides ways to make different classes of users cohabit on a single frequency band, hardware is not mature enough to make this procedure seamless. Moreover, changing regularly channels and consequently the offered bandwidth requires a strong service adaptation layer.

The fact is that this large frequency band already hosts dozens of wireless transmission technologies. Most of them are unidirectional (FM radio, broadcast television, etc.) and are not suited for data communications. However, WLAN technologies (e.g., IEEE 802.11), cellular broadband networks (e.g., UMTS, LTE, WiMAX), or even satellite networks, for example, can all be utilized by connected data services, even though their performance levels are heterogeneous. However, using these technologies conjointly or successively requires to be able to perform handovers between technologies and between operators,

---

[1]http://www.sharedspectrum.com/papers/spectrum-reports/

which both pose serious technical issues. Operators already respond to the lack of bandwidth on certain networks by off-loading traffic to other networks. For instance, in the access networks, several mobile operators have a solution to off-load traffic from the 3G network to a partner Wi-Fi network when a user is in range [2], which nevertheless causes disconnection and performance issues for mobile users, as handover is not properly handled especially regarding security. Authentication and encryption keys often need to be reestablished, which causes delays that are often incompatible with high mobility. Besides, off-loading needs to be carefully managed, as it could easily cause overload. The most emblematic examples of such situations come from other domains. Electrical networks use off-loading extensively, and the lack of coordination between operators has been one of the causes of the escalade in electrical failure of November 4, 2006, that led to a blackout in occidental Europe [3]. If the consequences in telecommunication networks will never reach this level, undesired saturation can appear.

A third solution to the scarcity of the wireless spectrum consists in enhancing the spatial efficiency of wireless communications. Bringing the users closer to their serving base station would allow to reduce both mobile terminal and base station transmission powers, therefore generating less interference on close cells. Femtocells in cellular networks work this way, even though their goal is more to improve coverage than to reduce interferences. Nevertheless, a generalization of this principle brings its load of issues. If each user deploys a miniature plug-and-play base station, there will be no possibility for the operator to control its position, and global planning strategies are therefore impossible. There is a slight chance that wireless spectrum-related problems will appear in some areas instead of being solved, especially when different operators work on the same band of frequencies (e.g., Wi-Fi, UMTS).

These examples show that if solutions exist to alleviate issues related to the spectrum scarcity, the coordination is necessary to manage the radio resource globally, even across operators, and to adapt the network behavior to the user traffic. And that is precisely where software defined networks can help. Operators could use the feedback features from the APs or base stations as well as the slicing facilities. Moreover, as the free bands of the wireless spectrum become overcrowded, decoupling the traffic operators that provide users with connectivity from the infrastructure operators that could run different or heterogeneous technologies would certainly alleviate the problem.

In September 2013, the ONF published a brief paper entitled "OpenFlow-Enabled Mobile and Wireless Networks" [4]. This document describes several outcomes of wireless SDN, focusing on radio access network performance optimization. They list several key challenges for implementing wireless SDN before focusing on two major issues: wireless channel resource management through intercell interference reduction and mobile traffic management through roaming and off-loading. If this shows a part of the potential of wireless SDN, this document also reminds that the wireless channel has some specificity that makes the SDN concept tough to implement. That's why this marriage has not been extensively studied yet. However, the set of problems it brings is also what makes it interesting to study.

First of all, the wireless medium is a fundamentally shared medium. The few free frequency bands such as the ISM bands are shared by multiple clients and multiple technologies. Concerning slicing, if some solutions have been proposed to reduce the interference between technologies (e.g., Bluetooth and IEEE 802.11), the narrowness of the available bandwidth makes the number of independent channels too low to efficiently implement slicing when considering that close APs may interfere. In reserved frequency bands such as cellular networks, similar problems arise when the number of mobile virtual network operator (MVNO) increases.

Concerning status report, the problem is even more difficult. First, the wireless channel state changes very frequently, especially in an indoor scenario. Fading and shadowing, for example, can easily make a link disappear suddenly, provoking frequent updates on link state that need to be considered in the routing protocols. A controller therefore needs to evaluate more than the simple channel or device load; it also needs to acquire information on the link stability, for example. This variation is partially due to the variations in the physical environment (doors closing, people passing, etc.), but also to the presence of close APs that have their own traffic pattern and that do not necessarily belong to the same operator. Discovering these potential interferers is challenging.

Finally, given the variability of the channel conditions, the status report from the AP is likely to generate a lot of control traffic that could pass over wireless links whose bandwidth is limited.

4.2.1 Implementations: OpenRoads and OpenRadio

OpenRoads is the adaptation of OpenFlow to wireless networks. It relies on OpenFlow to separate control path and data path through the FlowVisor open API. As for the wired case with OpenFlow, in OpenRoads, the network OS constitutes the interface between the infrastructure and the applications that observe and control the network.

The OpenRoads project built a demonstration platform composed of Wi-Fi and WiMAX APs that has been used for academic courses. Successful student projects demonstrate the potential of the approach through the development of applications or of an n-casting mobility manager. Basically, this consist in multiplexing at the receiver the same traffic coming from multiple base stations at the same time and from different networks such as Wi-Fi and WiMAX. Packet duplication facilitates the handover between technologies and increases the QoS. Some other mobility manager implementations succeed to improve the handover process with a reported reduced packet loss rate [5] and represent no more than a dozen lines of code [6]. It is worth to mention that there exist lots of wireless platforms over the academic world. The specificity of wireless SDN labs is that they provide wireless platforms but with a wired backbone control [7].

If these projects were only early developments, full utilization of all the features of OpenRoads is expected in the midterm, when the infrastructure includes programmable radio hardware platform. These flexible radio interfaces will be controlled through an API that allows external selection of various physical parameters, modulation, and coding scheme, for example. This control over the entire protocol stack is difficult to achieve, but it is necessary to allow full flexibility at the data plane level. Even with such capabilities, it is still not trivial to implement the rule-action abstraction in wireless world.

In the current version, slicing in OpenRoads is implemented by creating virtual interfaces on the same AP and assigning different service set identifier (SSID) to each interface. Each SSID can be considered as a separated slice and may be managed by a different controller. Even though the controller can apply different policies to different users, it is limited by the physical constraints and the hardware capabilities of today's APs. All slices should use the same channel and power settings, for example, which limits isolation. This demonstrates that it is fundamentally different to run SDN on regular wired switches than on wireless APs.

More recently, Bansal et al. [8] describe OpenRadio, an implementation of the separation between a decision and a processing plane that are similar to the control and forwarding plane of OpenFlow. They demonstrate the effectiveness of the concept by implementing Wi-Fi and LTE protocols over generic DSPs. OpenRadio, which is one of the SDN projects of the Open Networking Research Center (ONRC) at Stanford University, targets dense wireless networks and the development of a wireless controller, as well as the provision of slicing functionalities.

4.2.2 SDR versus SDN

SDR, also called cognitive radio, designates a set of techniques designed to cope with the shortage of available wireless channels. SDR makes a limited use of hardware components (for digital/analog conversion) and runs most of the tasks in software (filtering, coding, modulation, etc.). This software-based architecture is exactly in the line of the software defined network philosophy, as it allows remote definition of the channel parameters. Hence, SDR technologies appear at least as a fundamental building block for wireless SDN.

However, if SDR can clearly help in implementing wireless SDN, it is not a complete solution. First of all, SDR technologies generally try to fill in the so-called white space, that is, the frequency bands that are not used. However, as soon as a channel is used, even partly, it is considered as busy by the SDR and classified out of the usable bands. In addition, the focus of SDR is on the physical layer, but SDN require to consider the full protocol stack.

When looking from a higher point of view, SDN should allow to balance load between different operators, which not only requires to adapt the physical parameters of the base stations to facilitate the user handoff but also to exchange information between the operators to make the service providers aware of the resource blocks (in the LTE terminology) they were granted. SDR can therefore provide the capability to define dynamically and remotely the physical layer parameters, but does not represent a complete solution.

4.3 Related Works

Although the concept of SDN is recent, the developments around the various platforms and implementation issues already lead to a few publications. Some articles investigate the benefits and challenges behind the deployment of SDN in wireless networks of embedded devices like personal area networks or body area networks [9], which require device energy management and in-network aggregation. We stand here more from the infrastructure's point of view, even though terminal cooperation should be highly beneficial.

Our intent in this section is not to provide here a complete state of the art, but to select some works that illustrate some of the challenges and opportunities mentioned in Sections 4.4 and 4.5.

Concerning wireless SDN, Dely et al. [10] propose an interesting implementation of a wireless mesh network using SDN. In this work, each radio interface is split into two virtual interfaces, one for data transmission and one for the control packets. These virtual interfaces use two different SSIDs. At the controller level, a monitoring and control server is used in parallel of the controller, whose role is to maintain a topology database used by the controller to compute the optimal data paths. The implementation provides performance results that demonstrate that some serious issues exist regarding rule activation time (i.e., the time required to set up a new rule for a new flow by a remote controller), rule processing time (which can be important if lots of rules have to be parsed before the matching to succeed for an incoming flow), and the famous scalability issue related to the volume of traffic generated by the control plane. Yang et al. [11] describe briefly the architecture of a radio access network based on SDN that separates the network in four hierarchical levels: one that represents each operator, one that virtualizes infrastructure devices, one that manages the wireless spectrum, and the last one that constitutes the wireless SDN.

Some works deal with the performance of OpenFlow in the wired case. For instance, see Ref. [12], which focuses on the interaction between a router and the controller and evaluates the mean sojourn time of a packet in the system, taking into account the probability that no rule matches the flow the packet belongs to. It is shown that the sojourn time depends mainly on the processing speed of the controller used where the measurements show that it lies between 220 and 245 μs. It also evaluates the probability that a packet is dropped due to a limited buffer space at the controller. Bianco et al. [13] focus on the data plane and compare network efficiency with and without OpenFlow regarding throughput and packet latency. For instance, it is shown that OpenFlow experiences a performance (latency and throughput) drop of 11% of the packets compared to regular layer 3 routing, when small packets (64 bytes) are considered. However, almost equivalent performance is measured when packet size is slightly increased (96 bytes or more). As for the OpenRoads wireless extension of OpenFlow, Yap et al. [7] showed that the increased load generated by the communications between the devices (switches, etc.) and their controller is not important and stands for <0.05% of all traffic volume transferred in the reported experiences.

4.4 Wireless SDN Opportunities

The potential of SDN and the OpenFlow approaches has already been demonstrated into numerous examples concerning the network infrastructure. This paragraph lists a few technical opportunities that are specific to wireless, either by addressing wireless-specific problems (interferences, fast channel quality evolutions, etc.) or by using the specificities of the wireless medium (e.g., the broadcast channel) to improve generic tasks.

4.4.1 Multinetwork Planning

WLAN standards, as well as most unlicensed technologies, have a limited number of independent channels. Wi-Fi, for example, has only three nonoverlapping channels in the 2.4 GHz band and eight in the 5.2 GHz band. New technologies at the physical layer, such as ultrawide band, could increase this number, but there will always be a limit to the number of terminals that can operate without any interference in a given geographic area. Today, such interference problems appear at the building level, between APs of neighbor users, and tomorrow, close personal or body area networks could also interfere in a similar manner. Figure 4.1 represents the channels attributed to different Wi-Fi APs by their owners. Terminals located at the intersection of the two red or green circles are likely to experience poor network performance as the emissions of the two corresponding APs will interfere and cause collisions. Figure 4.2 represents a more efficient allocation in which close cells operate on different frequencies.

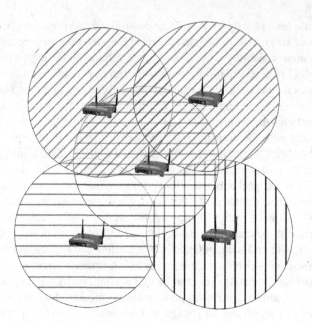

Figure 4.1   Uncoordinated Wi-Fi channel allocation.

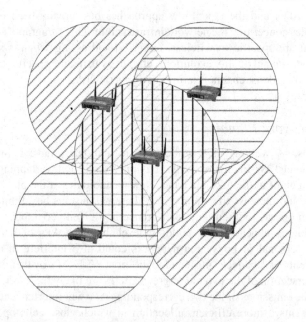

Figure 4.2   Interference-free Wi-Fi channel allocation.

SDN could provide help when it comes to such wireless network planning. It is indeed possible to create zone-specific and operator-independent controllers, which can even be distributed processes, that would be capable of aggregating statistics coming from the APs and could decide of channel allocations and transmission power of the access points in order to minimize interdependencies and interferences, similarly to the Control and Provisioning of Wireless Access Points (CAPWAP) protocol [14]. CAPWAP, which has been implemented in some Wi-Fi AP models, could integrate smoothly in global multitechnology SDN, with CAPWAP serving as a monitoring and control interface for Wi-Fi and compatible APs.

Nevertheless, realizing such a controller is not easy for several reasons. First, the algorithmic problem to solve can be solvable or not. Frequency allocation is a graph vertex k-coloring problem, which may or may not be feasible, depending on the number of independent channels. Brooks' theorem proves that at most $\Delta + 1$ channels are required to provide independent access in a graph whose maximum degree is Δ, but the density of the terminals within a single transmission range will most likely reach a value superior to the number of available channels. Power control could help, as it reduces the transmission range and hence the density, but it requires solving a plane or 3D packing problem, and the disconnection probability increases as the transmission ranges decrease.

In any case, such a power and frequency allocation problem requires full collaborations from the APs, which are the only ones capable of monitoring the different channel occupancy levels. A single nonparticipating access point could break the solution, and game-theoretic algorithms should be imagined to solve these issues.

4.4.2   Handovers and Off-Loading

Today's devices are highly portable, ergonomic enough to allow users to interact with online services while moving, and they are also equipped with multiple wireless interfaces. A state-of-the-art smartphone can potentially be connected to the Internet through a cellular access (LTE, UMTS, etc.), Wi-Fi, and even Bluetooth LE if a compatible gateway is around. A device could therefore select the best network AP(s) to access remote services based on a combination of criteria that include link quality (throughput, SINR, etc.), stability, billing, QoS capabilities, as well as the operators' preferences regarding, for example, off-loading. As the operators providing each technology access are not necessarily the same, the terminal is likely to be multihomed and to realize conjointly three types of handovers.

Classical "horizontal" handovers across APs are common in cellular networks and are well handled by operators through mobility prediction and resource soft reservation. As Wi-Fi localization is relatively precise today in urban areas, a cellular operator may expect a slight improvement in mobility prediction. One can imagine to implement local mobility management-dedicated controllers that aggregate statistic on signal strength or disconnection events coming from various technologies and send back prereservation orders to the appropriate neighbor cells.

Vertical handover across technologies happens when a cellular operator favors off-loading to Wi-Fi networks. Off-loading today suffers from the short range of Wi-Fi APs, which makes the connection very episodic, and from the long authentication delay. Even if EAP-SIM 802.1X method allows sharing the authentication token across technologies, the process still takes too much time, and a user traveling by car in a city often experiences connection and disconnection events that are not properly handled by the terminal. In this case, SDN could help by providing cross-technology soft handover, transmitting simultaneously the same data across both networks until the connection stability is confirmed. SDN could also realize the authentication procedure without involving the terminal.

Cross-operator handovers may happen when a mobile user temporarily or permanently leaves the coverage area of an operator he used to start a connection, for example, leaving the home Wi-Fi network when going outside, leaving the cellular operator when taking an elevator, or even crossing a frontier and roaming to a new cellular operator. Today, the user expects its TCP connection to break; however, SDN could help, in this case, to preserve it by automating the implementation of a Mobile IP-like scenario. In these situations, using SDN could help to reduce the handoff latency by facilitating the initial Mobile IP handshake. Local user location and travel monitoring can help predict handoffs and trigger prereservation of resources in the upcoming operator network, while a roaming-dedicated controller can prepare the control packets necessary to change Care-of Address and register to the new foreign agent.

Going a bit further, a multihomed terminal could easily decide dynamically how to use its multiple connections based on QoS considerations. Specific wireless technologies can be bound for a specific application. For instance, Wi-Fi is suitable for applications that require high throughput but can tolerate a variable delay. Cellular 3G technology provides a more consistent bandwidth and a good coverage. If SDN are able to have a good vision of which networks and technologies are available in different areas, they could "intelligently" allocate specific resource of a specific wireless technology for a specific user QoS requirement [15]. Traffic could also, in particular cases, be split among technologies, as illustrated in Figure 4.3. Let us take the example of a videoconferencing chat. The images of the participants require a high throughput, but the speech is the most important flow, as its clarity has a real impact on the quality of experience. One could imagine transferring the video data over a WLAN, while the speech arrives through a link that supports resource reservation such as a cellular network.

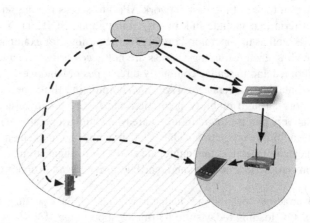

Figure 4.3 Example of a terminal receiving two parts of a data flow from heterogeneous networks: the switch decides to forward a part of the traffic through a more reliable cellular network rather than using the congested Wi-Fi channel.

Nevertheless, the vision from the infrastructure is imprecise and may not be accurate enough to take a clever decision for a given terminal. In this case, SDN could bring to the user information on the available networks, their stability, their coverage, and so on in order to feed a local decision process.

4.4.3   Dead Zone Coverage

One interesting case in cellular coverage is the dead zone. This is critical in some rural zones or large forest or also occurs when there is a disaster. Hasan et al. [16], for example, examine the scenario in which multiple operators share the same network infrastructure, and not only the antennas, to cover rural areas.

In some cases, a user located in a dead zone is not covered by an AP but is still within the range of another user. Both users can communicate via ad hoc networks, but the "uncovered" user will not be able to be reached by the AP since the AP does not know that the user is in fact reachable through ad hoc. Thanks to the concept of wireless SDN (with respect to what we mention in Section 4.5.2), we can use the topology discovery capabilities at the controller to help backhauling users in some dead zone through ad hoc network via neighboring users. Of course, this implies some reports by the end user to its covering AP to help it build an accurate view of the topology of the users covered. But since the controller gathers in the same place the network topology information, it won't be difficult to run an algorithm that identifies the possible holes and in consequence establishes "virtual links" to the devices physically disconnected from the AP. Therefore, merging ad hoc network and mesh network with wireless SDN exhibits very large amount of opportunities that needs to be further investigated.

4.4.4   Security

The monitoring capacity of OpenFlow and OpenRoads can provide a clear vision of a network status to an entity in charge of detecting intrusions or abnormal behavior. The network load and the packet distributions per protocol can be compared to statistics, and a process could decide if the current traffic matches the expected values for this date and time. Suspicious situations may indicate an intrusion or the presence of inside computers participating to a botnet, for example, control traffic can also be examined. A high ARP activity may raise a warning when the network topology and traffic did not change. Besides generic intrusion detection based on signatures, more specific situations can be addressed:

- The status reports can be used to detect misbehaving users or routers in a collaborative network (e.g., mesh network or public Wi-Fi network) in which users are expected to route traffic for mobile clients in exchange for a similar service. The owner of a participating AP could insert a traffic shaper behind its shared AP to preserve its uplink bandwidth. Comparing the different statistics coming from close APs and for upstream and downstream interconnection devices, this kind of behavior could be detected.
- Similarly, in a QoS-enabled wireless network, an emitter always classifying its frames into the highest priority class to gain access to the medium more often than its share could be detected by close APs and terminals (provided that terminals also participate into statistics collection; see following text).

- User locations and movements can be tracked by looking at the MAC addresses uploaded by the APs of a LAN [7]. These statistics could be correlated with larger-scale movements obtained from a 3G system (which may cause privacy issues, as mentioned in the following), which could help detect physical intrusion attempts and orient a sensor network.

4.4.5   CDN and Caching

Wireless SDN could use content-centric paradigm where the base station could gather and store some data so that it can be delivered in a timely manner to delay sensitive application. The controller could identify the users' needs and according to this associate the users with the closest base stations that hold the required data. An interesting use case could be people working in finance in the Wall Street district in New York. It is likely that a large number of the users require the same data from some marketplace all the time and with very short delays. Instead of forwarding the query of each user to the marketplace server, it could instead gather and store the data in the base stations and deliver it in downlink as soon as requested.

4.5   Wireless SDN Challenges

The previous section detailed some potential benefits of SDN for the general wireless scenario. However, the wireless medium specificities also bring its load of challenges that need to be addressed.

4.5.1   Slice Isolation

As mentioned earlier in Section 4.2, defining slices in a wireless network is not easy as link or channel isolation cannot be guaranteed. Among a single infrastructure, it is possible to plan frequency usage and to control on which frequency each AP operates to minimize or avoid interferences. Depending on the number of available channels, it can be possible to create independent slices if the operator has a full control over the channel space. However, when it comes to open technologies, Wi-Fi, for example, a given operator does not control the AP locations, which are placed by the customers in their homes, and does not control the close AP frequencies either, especially when multiple operators are present in the area. Therefore, the creation of a wireless link that is isolated and that does not disrupt close networks cannot be guaranteed unless the channel space is under control.

4.5.2   Topology Discovery and Topology-Related Problems

One of the required functions of SDN concerns the upward communication of network status. However, the status of a wireless network is not easy to measure, especially when multiple interferences can impact the quality of transmissions. Ideally, an AP could want to identify close APs and determine various parameters such as their channel(s), their output power, or their traffic patterns. This process relates to topology discovery in multihop networks and could benefit from the classical techniques used in these networks (periodically transmit neighbor discovery packets that include list of 1-hop neighbors to automatically discover 2-hop neighborhood). However, it has to be performed carefully:

• First, there may exist extended versions of the classical hidden/exposed terminal situations in which a mobile user is in range of two APs that do not see each other. As the two APs are not mutually aware of their existence, their transmissions are not correlated, and the middle user suffers from collisions on its downstream traffic, getting a poor QoS. This means that these two APs are not as independent as they are able to detect, and the only possible solution to this kind of issues is to implicate the terminal using uplink feedback or explicit control packets.
• Second, discovering topology may require more than simply identifying the APs and their operation frequencies. Two close access point operating on the same channel could indicate either a problematic situation if the two APs interfere or could correspond to a desired situation, in which one extends the range of the second one. If channel usage reports are used to derive policies, these situations should be identified.

4.5.3 Resource Evaluation and Reporting

Orthogonally to topology discovery, an infrastructure, for status report purposes, should be able to evaluate its available resources (e.g., channel capacity). However, this is far from obvious. The problem of identifying potential interferers has already been mentioned when discussing topology discovery. It is even more serious when it comes to resource evaluation because even if it is not necessary to identify the different interferers, their timely pattern needs to be imagined. If it is possible to measure and report the current network status, predicting its evolution even in the next few seconds is almost impossible. It means that the usage of these statistics is limited and that this information shall not be trusted to implement strong guarantees.

4.5.4 User and Operator Preferences

Slicing and status report are necessary to implement SDN and, as seen earlier, are difficult to implement in the wireless domain. However, there are also other issues to solve that are more related to the user experience than to the proper network controller operation. The network can define access point physical parameters and behavior in order to minimize interferences or to enhance the global network performance. Nevertheless, these objectives may be blind to the individual user preferences or can be counterproductive.

The question of how to specify and take into account user preferences is also expected to arise in the wireless context, as wireless technologies enable mobility and mobility implies that a given user will be successively or simultaneously connected to various networks with different pricing policies and different levels of QoS and trust. Today's solution has some logical yet simple policies (prefer Ethernet over Wi-Fi as it has a better performance, prefer Wi-Fi over UMTS because it off-loads the operator's network, etc.). The questions of how to implement a personal mobility manager that makes the user able to specify its own preferences in a simple way and how to mix these preferences with global- or operator-level objectives are important.

4.5.5 Nontechnical Aspects (Governance, Regulation, Etc.)

4.5.5.1 Interactions between Operators

A desired efficient network would be one where users could migrate freely between the infrastructures of different providers. The service provider to whom the users subscribe could realize the payment to the infrastructure holders. For this purpose, we need a clear distinction between the network infrastructure and the service delivered. But this openness gives rise to nonobvious economic and regulatory issues.

Let us bring the example of the electrical market that, in Europe, has seen the separation from the electrical produced companies and the electrical transporter companies. There are numerous advantages over this model, the first one being allowing a more open concurrency without requiring new providers to deploy their own power transport and distribution lines. Even though associations of transport providers exist (UCTE in Europe, ETRANS in Switzerland), the existence of multiple such associations and their limited communication impact the vision operators have of other infrastructures, even though they are interdependent. This lack of a shared real-time vision of the network status has been one of the major causes of the 2003 Italian–Swiss electrical blackout [17]. Obviously, the consequences of a separate vision are not that critical in IP networks. To avoid similar situations in IP networks (which would indeed have less severe consequences), there is a need for redefining peering and transit contracts, as the operators connecting the users shall not own the infrastructure and consequently not have a real-time vision of the network status.

4.5.5.2 Service and Forwarding Provides Interactions

As for every service where there is a decoupling between the service provided and the infrastructure needed to provide this service, we face also in our case the major question which is: Can the service provider be also an infrastructure provider, and vice versa? Historically, when the service is only at its beginning, the same company that built the infrastructure also provides the service. This is understandable since the deployment of infrastructure is very expensive. But with time, governments seek more competitiveness, and the market is open to new competitors that are generally service providers that lease infrastructure from original operators. To some extent, we can cite the MVNO case in the cellular field, but this should be generalized to all wireless and wired technologies. One of the issues that arise from this challenge is the question of fairness. For instance, assume that we allow a service provider to be also an infrastructure provider, and then we will have some problem of objectivity of the measurements that are required for the users to be able to compare the service quality of different providers and in different networks. This is only one out of the many questions that show how complex are the regulatory decisions regarding the separation between authorities.

4.5.5.3 Privacy-Related Issues

As mentioned before, SDN relies a lot on status reports from the access points and even from the users. These status reports can easily include user identities or MAC addresses, which can be used to predict mobility but also to finely track users, causing some serious privacy issues.

Such tracking issues already exist in cellular networks; nevertheless, the cellular operators are clearly identified, and the tracking resolution is far lower than what Wi-Fi could achieve.

Moreover, besides the users' tracking issues, the published statistics could be exploited for malicious activities. Let us imagine, for example, that a user publishes statistics on his home network usage to a neighborhood-level controller to regulate the area wireless network. A malicious user gaining access to the controller could use these statistics to determine when the user is home and when he is away.

4.6   Conclusion

In this chapter, we examined the potential benefits and the challenges behind the adaptation of an OpenFlow-like software defined network paradigm to wireless networks. The presence of the controller that can be a mixture of a centralized entity and a collection of distributed small controllers brings multiple advantages when it comes to managing the radio resource and the user mobility. As this controller gathers data from various measurement points and potentially from various technologies, it is able to take informed decisions on all the radio parameters and to optimize, even using machine learning, the network operation. This controller can even exchange information with other operators' controller entities, which would allow collaboration without giving full access to confidential data.

However, implementing wireless SDN also poses some intrinsic challenges, and even though some implementations have already been demonstrated, the demonstration scenarios remain modest or controlled today. The task appears difficult but not impossible and the aim of this chapter is to point out hard points that need to be solved.

References

[1] McKeown N, Anderson T, Balakrishnan H, Parulkar G, Peterson L, Rexford J, Shenker S, Turner J. OpenFlow: Enabling innovation in campus networks. ACM SIGCOMM Computer Communication Review. 2008;38(2), 69–74.

[2] Lee K, Rhee I, Lee J, Chong S, Yi Y. Mobile data offloading: How much can WiFi deliver? In: Proceedings of ACM CoNEXT 2010. Philadelphia, USA; 2010.

[3] Union for the Co-ordination of Transmission of Electricity (UCTE). Final Report on the European System Disturbance on 4 November 2006. Union for the Co-ordination of Transmission of Electricity; 2006.

[4] Open Networking Foundation. OpenFlow-Enabled Mobile and Wireless Networks; 2013. ONF Solution Brief.

[5] Yap KK, Sherwood R, Kobayashi M, Huang TY, Chan M, Handigol N, McKeown N, Parulkar G. Blueprint for introducing innovation into wireless mobile networks. In: Proceedings of the Second ACM SIGCOMM Workshop on Virtualized Infrastructure Systems and Architectures (VISA'10). New Delhi, India; 2010, 20–32.

[6] Yap KK, Kobayashi M, Sherwood R, Huang TY, Chan M, Handigol N, McKeown N. OpenRoads: Empowering research in mobile networks. ACM SIGCOMM Computer Communication Review. 2010;40(1), 125–126.

[7] Yap KK, Kobayashi M, Underhill D, Seetharaman S, Kazemian P, McKeown N. The Stanford OpenRoads deployment. In: Proceedings of the 4th ACM International Workshop on Experimental Evaluation and Characterization (WINTECH'09). Beijing, China; 2009.

[8] Bansal M, Mehlman J, Katti S, Levis P. OpenRadio: A programmable wireless dataplane. In: Proceedings of the First Workshop on Hot Topics in Software Defined Networks (HotSDN'12). Helsinki, Finland; 2012.

[9] Costanzo S, Galluccio L, Morabito G, Palazzo S. Software defined wireless networks: Unbridling SDNs. In: Proceedings of the 2012 European Workshop on Software Defined Networking. Darmstadt, Germany; 2012.

[10] Dely P, Kassler A, Bayer N. OpenFlow for wireless mesh networks. In: Proceedings of 20th International Conference on Computer Communications and Networks (ICCCN). Maui, HI, USA; 2011.

[11] Yang M, Li Y, Jin D, Su L, Ma S, Zeng L. OpenRAN: A software-defined RAN architecture via virtualization. In: Proceedings of the ACM SIGCOMM 2013 Conference. Hong Kong, China; 2013.

[12] Jarschel M, Oechsner S, Schlosser D, Pries R, Goll S, Tran-Gia P. Modeling and performance evaluation of an OpenFlow architecture. In: Proceedings of the 23rd International Teletraffic Congress (ITC); 2011, 1–7.

[13] Bianco A, Birke R, Giraudo L, Palacin M. OpenFlow switching: Data plane performance. In: Proceedings of the IEEE International Conference on Communications (ICC). Cape Town, South Africa; 2010.

[14] Calhoun P, Montemurro M, Stanley D. Control And Provisioning of Wireless Access Points (CAPWAP) Protocol Specification; 2009. RFC 5415.

[15] Yap KK, Katti S, Parulkar G, McKeown N. Delivering capacity for the mobile internet by stitching together networks. In: Proceedings of the 2010 ACM Workshop on Wireless of the Students, by the Students, for the Students (S3'10). Chicago, IL, USA; 2010.

[16] Hasan S, Ben-David Y, Scott C, Brewer E, Shenker S. Enhancing rural connectivity with software defined networks. In: Proceedings of the 3rd ACM Symposium on Computing for Development (ACM DEV'13). Bangalore, India; 2013.

[17] Johnson CW. Analysing the causes of the Italian and Swiss blackout, 28th September 2003. In: Proceedings of the 12th Australian Workshop on Safety Critical Systems and Software and Safety-Related Programmable Systems. Adelaide, Australia; 2007.

5

Leveraging SDN for the 5G Networks
Trends, Prospects, and Challenges

Akram Hakiri and Pascal Berthou
Univ de Toulouse, LAAS-CNRS, Toulouse, France

5.1 Introduction

Mobile and wireless connectivities have made tremendous growth during the last decade. Today 3G and 4G mobile systems are becoming an essential part of the terrestrial back-haul to provide connectivity through IP core network (i.e., evolved packet core (EPC)). They also focus toward providing seamless connection to cellular networks such as 3G, LTE, WLAN, and Bluetooth. The fifth generation (5G) is being seen as a user-centric concept instead of operator centric as in 3G or service centric as in 4G. Mobile terminals will be able to combine multiple flows coming from different technologies. Multimode mobile terminals have been seen toward the 4G cellular network. They aim to provide single user terminal that can cooperate in different wireless networks and overcome the design problem of power consumption and cost old mobile terminals. The Open Wireless Architecture (OWR) [1] is targeted to support multiple existing wireless air interfaces as well as future wireless communication standard in an open architecture platform. Nevertheless, the growing demand and the diverse patterns of mobile traffic place an increasing strain on cellular networks. To cater to the large volumes of traffic delivered by the new services and applications, the future 5G wireless/mobile broadband [2] network will provide the fundamental infrastructure for billions of new devices with less predictable traffic patterns that will join the network.

The 5G wireless network should enable the development and exploitation of massive capacity and massive connectivity of complex and powerful heterogeneous infrastructures. Accordingly, the network should be capable of handling the complex context of operations to support the increasingly diverse set of new and yet unforeseen services, users, and applications (i.e., including

Software Defined Mobile Networks (SDMN): Beyond LTE Network Architecture, First Edition.
Edited by Madhusanka Liyanage, Andrei Gurtov, and Mika Ylianttila.
© 2015 John Wiley & Sons, Ltd. Published 2015 by John Wiley & Sons, Ltd.

smart cities, mobile industrial automation, vehicle connectivity, machine-to-machine (M2M) modules, video surveillance, etc.), all with extremely diverging requirements, which will push mobile network performance and capabilities to their extremes. Additionally, it should provide flexible and scalable use of all available noncontiguous spectrums (e.g., further LTE enhancements to support small cells (nonorthogonal multiple access (NOMA) [3] and future radio access (FRA)) for wildly different network deployment scenarios in an energy-efficient and secure manner).

To address these key challenges, there is a need to enhance the future networks through intelligence, to proceed to successful deployment and realization of a powerful wireless world. Principles of virtual network management and operation, network function virtualization (NFV), and software defined networking (SDN) [4] are redefining the network architecture to support the new requirements of a new ecosystem in the future. SDN techniques have been seen as promising enablers for this vision of carrier cloud, which will likely play a crucial role in the design of 5G wireless networks. Accordingly, the future SDN-enabled 5G communications have to properly address key challenges and requirements driven by multiple societies, users, and operators, which would give them greater freedom to balance operational parameters, such as network resilience, service performance, and quality of experience (QoE).

5.2 Evolution of the Wireless Communication toward the 5G

Figure 5.1 depicts an overview of the wireless world toward the 5G. The figure gives a multidimensional overview of significant design challenges that 5G technology will face to simultaneously meet the future services, such as to achieve cost-effective resource provisioning and ecosystem, built with novel technologies such as SDN and network virtualization.

5.2.1 Evolution of the Wireless World

The mobile communication system has evolved through the first generation (1G), to the second generation (2G) and third generation (3G) through the 4G or Long-Term Evolution-Advanced (LTE-A) of mobile/cellular communications, with the typical service improvement and cost efficiency for each generation. For example, 1G (i.e., advanced mobile phone system (AMPS)) and 2G (i.e., GSM and GPRS) were designed for circuit-switched voice application. 3G (i.e., UMTS) and 4G (i.e., LTE-A) were developed for packet-switched services including multimedia, wideband data, and mobile Internet services. Meanwhile, there has been the introduction of other local, metropolitan, and wide-area wireless/cellular technologies such as microcells, femtocell, picocell, small cells, etc.

The other evolution, which has emerged in the past decade, aimed to exploit heterogeneous wireless communication comprised the wireless access infrastructure in both licensed and unlicensed parts of the wireless spectrum. It was intended to interconnect cellular system to wireless access networks (i.e., WLAN, WiMAX, etc.) to improve the service delivery and application provisioning end to end. Thus, the network was composed of different mixed types of infrastructures, forming heterogeneous networks (HetNets) [5]. Driven by both technical and economic incentives, the proliferation of HetNets had offered opportunities to satisfy users and applications in terms of their capabilities to support the new services.

Another important direction, which is expected to characterize beyond the 4G and 5G wireless networks, concerns the deployment of the application-driven networks.

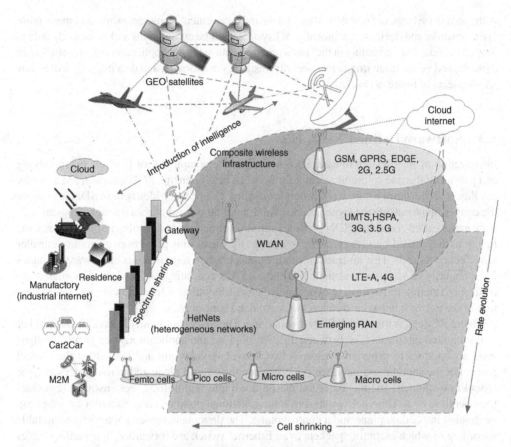

Figure 5.1 A view of the wireless world.

Application-driven networks consist of interconnecting end-user devices, M2M modules, and several machines, sensors, and actuators, the so-called Internet of Things (IoT) with billions of objects connected to the Internet for supplying big data applications. In parallel, in recent years the introduction and deployment of cloud-based concepts have emerged as an important solution offering enterprises a potentially cost-effective business model. For example, mobile users can use cloud-connected devices through public and private Mobile Personal Grid (MPG). Given the dynamic needs and supply of the network resource with rich resources available in the cloud, mobile users can benefit from resource virtualization to accommodate the different requirements as mobile devices move around within the mobile cloud. Furthermore, the integration of satellites in the future 5G networks reveals many challenges to support flexible, programmable, and secure infrastructure. The intersection of the cloud, satellites, big data, M2M, and 5G will bring about an exciting new automated future.

5G networks will not be based on routing and switching technologies anymore. They will be open, more flexible, able to support HetNets, and able to evolve more easily than the traditional networks. They will be able to provide convergent network communication across multitechnology networks (e.g., packet and optical networks) and provide open communication system to cooperate with satellite systems, cellular networks, clouds and data centers, home gateways, and many more open networks and devices. Additionally, 5G systems will be autonomous and sufficiently able to adapt their behavior depending in the user's requirements to handle application-driven networks in dynamic and versatile environments. Security, resiliency, robustness, and data integrity will be key requirements of future networks.

5.3   Software Defined Networks

Introduction of intelligence toward 5G can address the complexity of HetNets by specifying and providing flexible solutions to cater for network heterogeneity. SDN has emerged as a new intelligent architecture for network programmability. The primary idea behind SDN is to move the control plane outside the switches and enable external control of data through a logical software entity called controller. SDN provides simple abstractions to describe the components, the functions they provide, and the protocol to manage the forwarding plane from a remote controller via a secure channel. This abstraction captures the common requirements of forwarding tables for a majority of switches and their flow tables. This centralized up-to-date view makes the controller suitable to perform network management functions while allowing easy modification of the network behavior through the centralized control plane.

Figure 5.2 depicts the overall SDN architecture. The SDN community has adopted a number of northbound interfaces (i.e., between the control plane and applications) that provide higher-level abstractions to program various network-level services and applications at the control plane. For the southbound interface (i.e., between the control plane and network devices), the OpenFlow standard [6] has emerged as the dominant technology. For instance, consider the operation of an Ethernet switch. From the functional point of view, Ethernet switches can be divided into a data plane and a control plane. The data plane represents a forwarding table according to which incoming packets to an Ethernet switch are forwarded. Forwarding tables consist of entries that indicate which output port the received Ethernet frames should be sent. Populating of forwarding table with these entries is the task of the control plane. The control plane is a set of actions exerted on the received Ethernet frames to decide their destination ports. In order to quickly perform frame processing, these actions are implemented in hardware together with the forwarding table.

SDN makes it possible to manage the entire network through intelligent orchestration and provisioning systems. Thus, it allows on-demand resource allocation, self-service provisioning, and truly virtualized networking and secures cloud services. Thus, the static network can evolve into an extensible vendor-independent service delivery platform capable of responding rapidly to changing business, end-user, and market needs, which greatly simplifies the network design and operation. Consequently, the devices themselves no longer need to understand and process thousands of protocol standards but merely accept instructions from the SDN controllers.

The value of SDN in 5G wireless networks lies specifically in its ability to provide new capabilities like network virtualization, automating and creating new services on top of the virtualized resources, in secure and trusted networks. Also, SDN enables the separation of the control logic from vendor-specific hardware to open and vendor-neutral software controllers. Thus, it enables implementing routing and data processing functions of wireless infrastructure into software packages in general-purpose computer or even in the cloud.

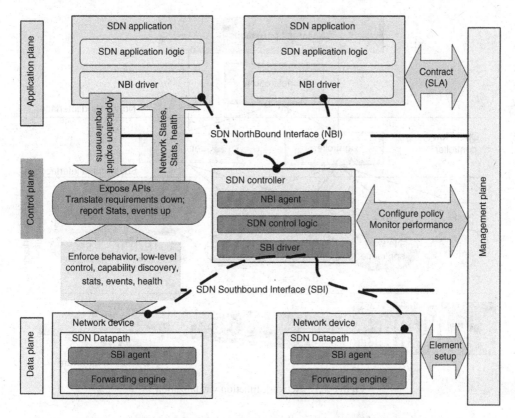

Figure 5.2 Reference architecture of software defined networking.

5.4 NFV

One of the most interesting complementary technology of SDN that has the potential to dramatically impact the future 5G networking and how to refactor the architecture of legacy networks is virtualizing as many network functions as possible, the so-called NFV. The aim of NFV is to virtualize (known also as network softwarization) a set of network functions by deploying them into software packages, which can assembled and chained to create the same services provided legacy networks. It is possible, for example, to deploy a virtualized session border controller (SBC) [7] in order to protect the network infrastructure more easily than installing the conventional complex and expensive network equipments. The concept of NFV is inherited from the classical server virtualization that could by installing multiple virtual machines run different operating systems, software, and processes.

Traditionally, network operators had always preferred the use of dedicated highly available black box network equipments to deploy their networks. However, this old approach inevitably leads to long time to market (CAPEX) and requires a competitive staff (OPEX) to deploy and run them. As depicted in Figure 5.3, the NFV technology aims to build an end-to-end infrastructure and enable the consolidation of many HetNet devices by moving network functions from dedicated hardware onto general-purpose computing/storage platforms such as servers. The network

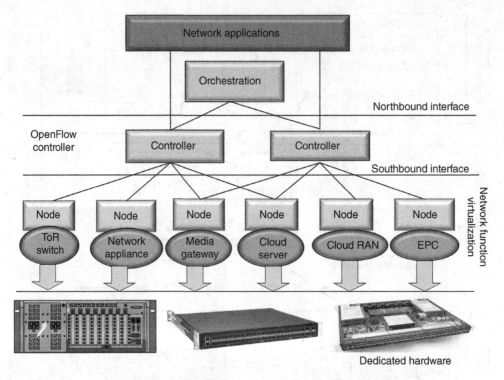

Figure 5.3   Network function virtualization.

functions are implemented in software packages that can be deployed in virtualized infrastructure, which will allow for new flexibilities in operating and managing mobile networks.

Another important topic in 5G carrier-grade mobile networks that may be improved by implementing NFV in cloud infrastructures is resilience. Implementing network functions in data centers allows transparent migration between either virtual machines or real machines. Furthermore, implementing mobile network functions in data centers will enable more flexibility in terms of resource management, assignment, and scaling. This impact the development of ecosystems and energy efficiency of networks, as overprovisioning can be avoided by only using the necessary amount of resources.

NFV is currently discussed in the context of virtualizing the core network as well as centralizing the baseband processing within radio access networks (RAN). Examples of mobile network virtualization are used for Cloud-RAN (C-RAN). C-RAN can use virtualized software modules running in different virtual machines. Additionally, enhancing NFV with SDN may off-load the centralized location within network nodes that require high-performance connections between radio access point (RAP) and data centers. Decentralizing these connections with SDN will enable managing HetNet nodes (i.e., picocell, macrocell, etc.) and heterogeneous backhaul connectivity such as fiber, wireless, etc.

Another concept that received a lot of attention with the evolution of SDN and NFV is the network service chaining (NSC) [8]. The NSC aimed to help carrier-grade networks to provide continuous delivery of services based on dynamic network function orchestration and automated deployment mechanisms to improve operational efficiency. Because SDN moves the management functions out of the hardware and places them in controller software running in general-purpose server and NFV moves network functions out of hardware and puts them onto software, too, building service chaining no longer requires hardware, so there is no need for overprovisioning since additional servers can be added when needed.

One example of an increasingly complex network platform is the 3GPP EPC, which requires multiple functions (e.g., network address translation (NAT), service access policing for VPN, video platforms and VoIP, infrastructure firewall protection, etc.) typically installed in independent boxes. Carrier-grade networks should define statically provisioned service chains for customer traffic crosses several middle boxes. In the future 5G networks, carrier-grade networks will not use monolithic, closed, and mainframe-like boxes to provide a single service. SDN- and NFV-driven service chaining can improve flexible allocation, orchestration, and management of cross-layer (L2–L7) network functions and services and provide the substrate for dynamic network service chains.

5.5 Information-Centric Networking

The information-centric networking (ICN) is a novel network architecture that is receiving a lot of attention in the 5G networks. ICN consists of new communication model that revolve around the production, consumption, and matching users with content; in-network caching; and content-based service differentiation, instead of communication channels between hosts. ICN pushes many design principles from the Web to the network architecture by centering on what is relevant to the user and not where the content is located in the network. So, ICN manages contents and names to ensure their uniqueness in the network (i.e., because data are routed based on their names). The ICN communication model allows built-in native features aiming at optimizing and simplifying future content delivery architecture. The service providers should prepare their infrastructure capabilities to support efficient multicast data delivery as well as provide seamless mobile connectivity so users can move and the network can continue delivering data packets without interruption.

Typically, the ICN deployment schemes can be classified into three categories: (i) ICN over IP, which encapsulate ICN protocol data in IP (or UDP/TCP) packets or take ICN protocol information using IP options; (ii) ICN over L2, which completely replaces the IP layer and directly uses the data link protocol (such as PPP, Ethernet, IEEE 802.x) to deliver data between neighbors; and (iii) ICN over virtualized network, which exploits network virtualization technologies, such as SDN, to implement ICNs. Although these schemes have advantages and disadvantages, most works on ICN implementations focus on how to implement a particular ICN architecture. However, different ICN architectures employ different transmission techniques and packet format, which is not easy for the coexistence and interoperability of different ICNs. SDN and NFV are amazing approaches for improving the integration of ICN in the 5G network without deploying new ICN capable hardware.

Levering SDN for ICN requires a unified framework to facilitate the implementation and interoperability among different ICN architectures [9]. Such a content-centric framework should provide users network access to remote named resources, rather than to remote hosts [10]. The integration of ICN in the 5G network includes storage and execution capabilities to evolve the network from a dumb pipe transport toward added value intelligent network. Introduction of intelligence in ICN architecture improves the flexibility and scalability of content naming as well as enhances the performance of the QoS in the network. Intelligent ICN can also make it feasible to integrate mobile radio-aware ICN on the 5G networks.

Although supporting ICN-enabled SDN allows transforming the current network model into a simplified, programmable, and generic one, ICN still faces a number of challenges to its realization including routing computation, path labeling to discover the network topology and locate data in the network, and route assignments to route requests for data objects. Moreover, since ICN information should be inserted in each packet, the fragmentation of packets limits processing cost of the network resources. Mechanisms for caching objects in the network along the path need more investigation to deliver them more rapidly to an increasing number of users. Distribution of storage capabilities across the path with more elaborated content-routing algorithms is an open issue, so researchers have to cope with a proliferation of new and complex application contents and services, many of which are unknown today.

5.6 Mobile and Wireless Networks

The design of the future 5G systems should efficiently support a multitude of diverse services and introduce new methods making the network application service aware. The future 5G network architecture should be highly flexible for supporting traditional use cases as well as easy integration of future ones. Additionally, the 5G network will be able to handle user mobility, while the terminals will make the final choice among different access networks transparently. Mobile terminals will also hold intelligent components to make a choice of the best technology to connect, with respect to the constraints, and dynamically change the current access technology while guaranteeing the end-to-end connectivity.

5.6.1 Mobility Management

A key trend relates to mobility, as broadband mobile is expected to grow in the next decade. The future will encompass 1000 times more connected mobile device in the horizon 2020, all with different QoS requirements, which will interconnect to all kinds of heterogeneous and customized Internet-based services and applications. Accordingly, these developments demand rethinking about the network design, which leads to ask about the advantages of SDN in the most common wireless networking scenarios. It is also important to understand what the key challenges that exist in this realm are and how they can be addressed.

Presently, there is not much discussion regarding the mobility support in SDN. Software defined wireless networking (SDWN) would be an SDN technology for wireless/broadband networks that provides radio resource management (RRM), mobility management, and routing [11]. SDWN infrastructure can support composition to combine results of multiple modules into a single set of packet handling rules. For example, novel mobility management protocols should be provided to maintain session continuity from the application's perspective and network connectivity through dynamic channel configuration. Furthermore, mobility modules should be able to provide rapid client reassociation, load balancing, and policy management (i.e., charging, QoS, authentication, authorization, etc.)

Another important key challenge in wireless/broadband networks concerns multihoming. Multihoming means the attachment of end host to multiple networks at the same time, so users could freely move between wireless infrastructures while also supporting the provider. This approach would emerge by applying SDN capabilities to relay between the home network and edge networks. The future wireless/broadband system can be envisioned as a world in which mobile devices can move seamlessly between the wireless infrastructures in a trusted and secure manner. For example, as shown in Figure 5.4, in home networks, a virtualized residential gateway can improve service delivery between the core home network and the network-enabled devices. The target architecture emerges by applying SDN and NFV between the home gateway and the access network, moving most of the gateway functionality to a virtualized execution environment.

5.6.2 Ubiquitous Connectivity

As part of the future wireless networks, end users will need to communicate with each other and with the surrounding objects and machines, for example, sensors embedded in objects. Figure 5.5 shows how the cellular network will be completed by interaction with network topologies, including M2M, which will be completed with user/device-to-user/device communication at different levels of cooperation and coordination between different nodes.

Including all kinds of interactions of these ubiquitous systems will increasingly expand the network infrastructure, which will include new data services and applications, for example, smartphones and tablet with powerful multimedia capabilities, or even connected things surrounding the environment, such as building, roads, or car-to-car communication. Accordingly, essential design criteria to fulfill the requirements of future 5G systems are fairness between users over the covered area of ubiquitous systems, reduced latency, increased reliability, energy efficiency, and enhanced QoS and QoE requirements that originated from heterogeneous applications and services.

SDN paradigm can be deployed as higher layers of the protocol stack, as well as for wireless networks, such as the low-rate wireless personal area networks (LR-WPANs). Extending SDN to support LR-WPAN was considered impractical because these networks are highly constrained, that is, they require numerous low-cost nodes communicating over multiple hops to cover a large geographical area and duty cycles to provide low-energy consumption to operate for long lifetimes on modest batteries. Such an approach requires cross-layer optimization, data aggregation, and low software footprint due the limited amount of memory storage and CPU processing speeds.

Wireless SDN (WSDN) remains a key challenge for the future SDN-enabled networks. The controllers should provide an appropriate module to define the rules for LR-WPAN environment. WSDN controllers should provide flexibility to support node mobility, topology

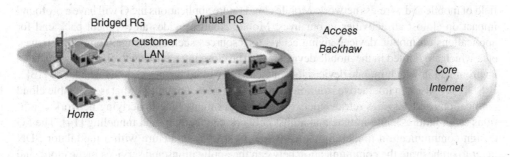

Figure 5.4 Cloud box virtualized residential gateway.

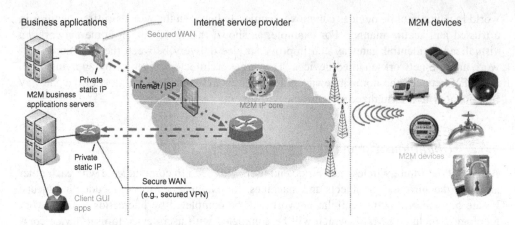

Figure 5.5   Integration of ubiquitous systems with 5G networks.

discovery, self-configuration, and self-organization. They also have to deal with link unreliability and robustness to the failure of generic nodes and the control node. Furthermore, although energy efficiency has been the target of diverse research works in the past, it remains an open issue that wireless IoT will face. The IoT ecosystem has become extremely complex and highly demanding in terms of robustness, performance, scalability, flexibility, and agility. The IoT will require new air interfaces, protocols, and models optimized for short and sporadic traffic pattern. SDN should significantly reduce the cost of powering the entire network and running the hardware as well as the software. Possible solutions are, for example, shutting off IoT components when they are idle, adapting link rates to be as minimal as possible, and introducing new energy-aware routing protocols [12]. In the latter case, the SDN controller collects utilization statistics for links to get visibility of flows in the network and forward flows according to these protocols. An important consideration that should be taken into account when designing energy-aware protocol is the need of the network to recover after failure while supporting automatic topology discovery at the same time.

5.6.3   Mobile Clouds

Mobile cloud computing is one of the technologies that are converging into a rapidly growing field of mobile and wireless network. Mobile cloud future applications in 5G will have a profound impact on almost all activities in our lives. Mobile cloud provides an excellent back end for applications on mobile devices giving access to resources such as storage, computing power, etc., which are limited in the mobile device itself. The close interaction with cloud may create an environment in which mobile devices look attached locally to the cloud with low latency [13].

SDN promises an interactive solution to implement new capabilities, that is, to enable cloud applications and services retrieve network topology, monitor the underlying network conditions (e.g., failures), and initiate and adjust network connectivity and tunneling [14]. The 5G design communication model aims to provide a global architecture with a modulator SDN layer to orchestrate the communication between the applications and services in the cloud and user's mobile terminal. Given the dynamic needs and supply of the network resource, with rich resources available in the cloud, mobile users can benefit from resource virtualization. The virtualization can abstract these dynamic mobile resources to accommodate the different requirements as elements move around within the mobile cloud.

Despite that SDN has some advantages such as resource sharing and session management, it incurs several limitations. In particular, because mobile users trigger repeatedly the embedded controller for marshaling and unmarshaling flow rules in OpenFlow messages, the overhead increases more significantly because of the limited computing capabilities and resources of mobile devices (i.e., extra memory consumption and extra latency). For example, mobile interactive applications (e.g., mobile gaming, virtual visits) require reliable connectivity to the cloud as well as low latency and impose higher bandwidth requirements from wireless access networks to cloud service. Furthermore, mobile users use cloud-connected devices through public and private MPG, which induces multidimensional limitations, including dynamic mobility management across HetNets, power saving, resource availability, and operating conditions, and further limits the movement of content across multiple devices and the cloud. Addressing these limitations simultaneously may increase device complexity, degrades the network performance, and causes connectivity dispersions. The key challenge for mobile clouds is how to transform physical access networks to multiple virtual and isolated networks, while maintaining and managing seamless connectivity.

5.7 Cooperative Cellular Networks

Another important paradigm that has recently gained a lot of attention as one of the most promising technologies in the next generation of wireless/cellular networking is multihop relay communication. Presently, cellular systems have a single direct link between the base station and the terminal. However, multihop networks require maintaining multilink between multiple transmitter and receiver to form multipath communication, the so-called multihop cooperative network. Compared to existing technology, which includes mechanisms for retransmission and multiple acknowledgments, multihop cooperative network can overcome these limitations by providing high-density access network. However, multihop cooperative network incurs several limitations and often suffers throughput penalties since it operates in half-duplex mode and therefore introduces insufficiency of the spectrum usage.

To increase the capacity of 5G systems, SDN can provide solutions to overcome the limitations of multihop wireless networks [15]. Indeed, SDN can provide advanced caching techniques to store data at the edge network to reach the required high capacity of 5G systems. One way to increase per-user capacity is to make cells small and bring the base station closer to the mobile client. In cellular communications, an architecture based on SDN techniques may give operators greater freedom to balance operational parameters, such as network resilience, service performance, and QoE. OpenFlow may work across different technologies (i.e., WiMAX, LTE, Wi-Fi) to provide rapid response to the subscriber mobility and avoid disruptions in the service. The decoupling between the radio network controller and the forwarding plane will enhance the performance of the base station.

Additionally, supporting many subscribers, frequent mobility, fine-grained measurement and control, and real-time adaptation introduces flexibility, scalability, and security challenges for future 5G systems architecture. SDN-enabled network devices should be able to provide scalability (i.e., increasing number of subscribers), frequent changes in user location (i.e., redirecting traffic to proxies), QoS (i.e., handling traffic with specific priority), and real-time adaptation to network conditions (i.e., load balancing). Cellular SDN networks should maintain subscriber information base (SIB) to translate subscriber attributes into switch rules to set up and reconfigure services flexibly.

However, the dynamic reconfiguration of a service needs a mechanism to handle notifications sent from middle boxes to the controller. Therefore, a deep packet inspection (DPI) engine would be required to enable finer-grained classification based on the application (i.e., such as Web, peer to peer, video, and VoIP traffic). DPI also would help to support intrusion detection and prevent systems that analyze packet contents to identify malicious traffic. Likewise, cellular controller protocols would enable the control of remote virtualized resources to simplify resources and mobility management [16]. The SDN controller would enable slicing the network into multiple tenants while enabling dynamic routing and traffic engineering, thereby easing hand-off management, minimizing delays, and potentially reducing packet loss. Such a cellular SDN controller [17] (as depicted in Fig. 5.6 [16]) would implement RRM APIs as northbound interfaces to simplify the QoS management (i.e., admission control, resource reservation, and interference management) and the resource provisioning. The controller may be enhanced by other techniques like header compression/decompression to reduce the overhead for applications with small packet payloads (e.g., VoIP packets). Compressing these packets before transmission on low-bandwidth links substantially lowers the overhead.

Cellular networks traditionally have been hierarchical with centralized control and data structures, which require high performance, custom hardware to process, and route traffic. The distributed control SDN model would be a key challenge for the evolution of SDN-enabled cellular network to provide high-performance, cost-effective, and distributed mobility management in cellular architecture. The increasing rollout of 5G technology could lead to an

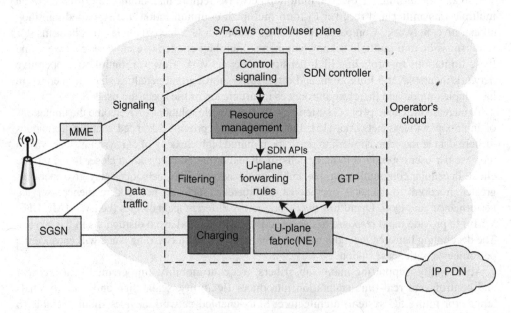

Figure 5.6 Fully virtualized SDN-enabled cellular network.

upsurge in SDN adoption. It seems to be possible to off-load a base station with the rising number of mobile clients requesting the network resources and provide load balancing strategy [18], for example, by providing multiple parallel transmissions. Furthermore, as for cloud partitioning and network slicing, it would be possible to divide the wireless traffic into several slices matching different traffic criteria, which may allow traffic isolation with respect to their patterns (i.e., VoIP, Data, Video, etc.). Such an approach may help to create virtual base stations and orchestrate the available resources among different mobile devices, thereby saving power and memory usage [19].

5.8 Unification of the Control Plane

Looking back to the development of the existing wireless communication technologies, it is easy to find that they were designed to provide new services in isolation. The future 5G cannot be defined in a single type of service or isolated services. Rather, it will provide a convergent network infrastructure that integrates multiple systems together. Weaving different access technologies together in a fluid fashion and creating smart gateways in transparent manner will be the goal that gives life to 5G. Leveraging SDN technologies for designing new control mechanisms and protocols for relocating functions and protocol entities will fulfill the new requirements of scale, latency, harmonization of protocol stacks between fixed and mobile (data and control planes), distributed mobility, energy efficiency, and unified access/ aggregation network for infrastructure simplification.

5.8.1 Bringing Fixed–Mobile Networking Together

The convergence of the fixed and mobile networks forms the backdrop for upgrades to the future networks. Both network infrastructures will represent the major part of investments for the network operators. The ultimate purpose is to offer better services over fixed and mobile networks with the best possible user's QoE while at the same time rationalizing and sharing fixed and mobile network infrastructures.

Although there are some initiatives to offer some degree of convergence alongside the emergence of IP-based services and IP Multimedia Subsystem (IMS), the convergence of fixed and mobile networks is a highly complex issue. Convergence is a trendy word because fixed and mobile networks were developed independently from each other and based on diferent technologies and protocols. Convergence is also synonymous to energy efficiency, because it is expected that the development of the future network will be based on ecosystem of close cooperation between fixed and mobile infrastructures. In addition to the ever higher-capacity trend, convergence of fixed and mobile networks is a highly complex issue, because it assumes certain trade-offs, so as to fully leverage off the benefits of moving different network functionalities and/or device equipment closer to each other and to different parts of the network, and should correspond to the behavior of end users who wish to remain agnostic about which technical infrastructure (3GPP, Wi-Fi, DSL, fiber) they may be using.

Fixed–mobile convergence can be segmented into two concrete approaches: structural convergence and functional convergence. Structural convergence concerns sharing fixed and mobile network equipments and infrastructures as much as possible. Functional convergence, that is, convergence of fixed and mobile network functions, relates to better distribution of various functions by distinguishing those that would be centralized from those that should be more distributed. This functional convergence is enabled by NFV and SDN. The decentralization of the network functions can be provided from the mobile core network down toward the access network, such as content delivery network (CDN) through the virtualization of home gateway to cloud functions, which are mostly dictated by the traffic optimization (e.g., latency, bandwidth, etc.).

Fixed–mobile convergence is also expected to evolve different stakeholder roles: the classical network providers will continue to play their central role, but there will be other stakeholders such over-the-top (OTT) providers, which should be allowed to vertically integrate with content providers and even the end users themselves, that will have a major involvement in the future evolving fixed–mobile ecosystem. In the context of SDN and NFV, the debate would be about exploring the network equipment that will be hosting the applications of other networks and the functions that can be migrated to the cloud.

5.8.2   Creating a Concerted Convergence of Packet–Optical Networks

Access technologies of future network systems would comprise various broadband transmission media such as optical fibers, millimeter-wave links, and so on. The expected impact in 5G wireless communication is to contribute to the emergence of new generations of optical transport networks, to cope with the expected significant traffic growth, and to meet the flexibility requirements. The convergence of packet and optical networks in the future 5G system makes it possible to reconfigure the optical network to support high-capacity data rate with a guaranteed end-to-end latency for on-demand applications such as network as a service (NaaS) [20].

Accordingly, the future generation of photonic communication would require programmable optical hardware to increase flexibility in the control plane and management plane of optical networks and enable the advent of software defined optical networking. The increased programmability of SDN creates an opportunity to address the challenges of unifying packet–optical circuit switching networks in a single converged infrastructure [21]. Unified software control of the physical layer is a key requirement for next-generation 5G wireless networks. The SDN-enabled optical cross-connects would be used to demonstrate the efficiency benefits of hybrid packet–optical circuit switching architectures for dynamic management of large flows, as well as the scalability and flexibility of high-capacity service provision, in data center applications.

Similarly, future SDN-enabled 5G systems should provide a convergence framework to make it more efficient to use the network resources, for example, by unifying the control and the management of these HetNets. The packet–optical networks may be unified by implementing two abstractions: (i) a common API abstraction (flow abstraction) at the control plane and (ii) a common map abstraction based on a data abstraction of a network-wide common map manipulated by a network API. As depicted in Figure 5.7, the common flow abstraction fits well with both networks and provides a common paradigm for control, by providing an abstraction of layers L2/L3/L4 packet headers as well as by L0/L1 circuit flows. The flow abstraction blurs the distinction between both technologies and processes them as flow of different granularity. The common map has full visibility into both packet and circuit network devices to interconnect network applications across both packets and circuits. Full visibility allows applications to joint and optimize network functions and services across multiple layers. The network functions would be implemented as simple and extensible centralized northbound interface to hide the details of state distribution from the applications.

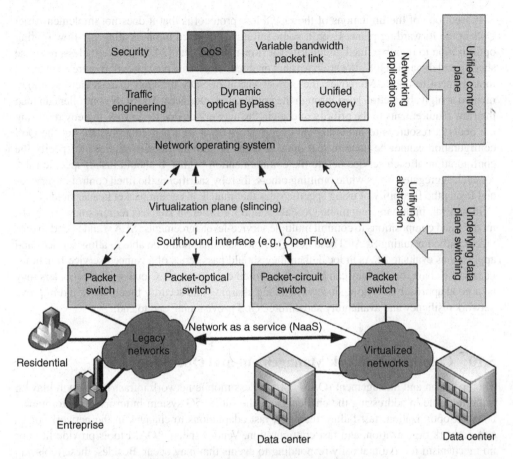

Figure 5.7 Unifying packet–optical data plane.

5.9 Supporting Automatic QoS Provisioning

The advanced 5G network infrastructure for the future Internet will include multiple HetNets that need sharing resources on all levels to meet the fast changing traffic patterns from different services and applications. Network operators should be able to predict the various traffic patterns as functions of the services provided by their networks. Service providers are evaluating implementations of storage and data traffic over a single network to meet the flexibility (e.g., the ability to accommodate short duration extra bandwidth requirements) and the efficient coexistence of multiple services [22, 23]. They have to cope with the large demands of QoS incoming from different wired and wireless devices, each with particular requirements. The QoS provisioning in the advanced SDN-enabled 5G networks is more complex and poses a real problem that need to be addressed. In particular, QoS automation should be supported at every wired and wireless technology that may share the same network slice. Although SDN allows creating different network slices in the same network infrastructure to provide a strict QoS, performance, and isolation required across applications without interfering with traffics in other slices, SDN does not provide the ways for automating QoS provisioning per application/per service.

Indeed, one of the limitations of the OpenFlow protocol is that it does not implement strict QoS in the forwarding plane. Even if some initiatives targeted implementing per-flow routing optimization to improve fined granularity for flow management [24, 25], nevertheless, resource sharing and dynamic QoS allocation was not enabled. Thus, data packets will require an external tool/protocol to do so. Moreover, the current vision of SDN the QoS management is implemented at high abstraction level through the northbound interfaces. The SDN controller can map the flow requirements to the priority queues in the network device it controls, thereby reserving the network resources to individual and aggregated flows in a particular switch, but the QoS configuration cannot be done in real time. A network administrator is required to specify the configuration of each service before the communication begins. It should install specific rules for each aggregated flows while omitting others, thereby sacrificing the fined control of services and losing the flexibility of using specific rules that match on certain packet header fields.

In general, improving automatic QoS allocation for different HetNets requires new methods, models, and compositions to commit multiple service-level agreements (SLAs) end to end to provide a unified resulting SLA. These new QoS mechanisms should be able to allow services and applications evaluate SLAs in local, then they should pass series of advanced service functions, i.e. service chain, before they can be used in unified environments. Cloud service providers may be a good approach to follow, since virtualizing charging and security functions would improve network resiliency and availability and enforce QoS provisioning end to end.

5.10   Cognitive Network Management and Operation

The operation and management (OAM) of wireless mobile network infrastructure will play an important role in addressing the challenges of the future 5G system in terms of performance, constant optimization, fast-failure recovery, fast adaptations to changes in the network loads, self-network organization, and fast configuration. Vendor-specific OAM tools provide little or no mechanism for automatically responding to events that may occur. Besides, these tools can provide their power in small and medium networks; in contrast, due to the excessive cost to deploy them in large-scale networks, they will be underutilized for. the future networks. Additionally, the existing OAM tools need to be individually configured and supervised by a human operator, which limits their flexibility. Since the network topologies are becoming more and more complex, manual configuration and deployment are getting less and less attention and are becoming impracticable. Also, the migration to high-speed networks (i.e., from 1 Gbps to 10 Gbps to 40 Gbps) creates further scalability challenges for the future OAM tools. In particular, diagnosing the network performance and bottlenecks without visibility into the traffic characteristics introduces new complexity with regard to the consistency of the network.

Accordingly, future 5G networks should be based on common network management and operation for mobile and wireless as well as for fixed network for economic network deployment and operation. Toward the automation of network OAM tasks, network operators must grapple vendor-specific configurations to implement complex high-level interfaces to manage and monitor network policies. Advanced intelligence should be developed for realizing the future OAM tools. The intelligence of OAM requires the development of new functional and system architecture, also taking the integration of both wireless and fixed networks into account. As SDN will be the bedrock for the future wireless/broadband networks, the OAM will be a key challenge for SDN-enabled networks.

Indeed, SDN introduces new possibilities for network management and monitoring capabilities that can improve performance, reduce the bottlenecks of the network, and enable debugging and troubleshooting of the control traffic. To avail of these possibilities, the future OAM tools should provide open and customizable interfaces to support event-driven model for SDN. SDN OAM tools should provide methodologies for the acquisition, analysis, and improvement of knowledge representing the semantics and operational goals and strategies, network properties, and automated reasoning for the alignment of different network functionality at runtime. To this end, high-level, declarative management languages will be required to ensure the consistency of the network states and detect failures in real time.

The expected impact of SDN OAM tools is to be able to scale to large-scale networks to deal with multiple controllers (i.e., in a distributed SDN model). They have to provide closed control loop functions dedicated to self-configuration, self-optimization, and self-healing. The control loop diagnostic and decision-making processes need to be adapted automatically, for example, by predicting the future actions based on the results of previous ones. This proactive capability will leverage the flexibility and programmability of the open SDN and improve their effectiveness and efficiency, thanks to cognitive processes that will enable creating more elastic network management either for the entire network or specific slices. The cognitive network management and operation approaches will develop a new management paradigm and investigate, develop, and verify processes, functions, algorithms, and solutions that enable future 5G networks to be self-managed. The cognitive OAM will include cognitive function orchestration and coordination and system verification for provisioning, optimization, and troubleshooting.

5.11 Role of Satellites in the 5G Networks

The 5G system had seen an increased demands on the backhaul with an increasing numbers of HetNets and small cells. Satellites will play a major role in the extension of 5G cellular network to new area such as ships on the sea and remote land area that are not covered by cellular networks. Also, high-throughput satellite communication (SatCom) systems will be able to complement terrestrial provision in an area where it is difficult to do so with other terrestrial cell such as LTE. Indeed, integrating satellites in the future 5G networks will be seen as an essential part of the terrestrial infrastructure to provide strategic solution for critical and life-saving services. Satellites would be able to collect and distribute data from clusters of sensors in the IoT and make them available to the terrestrial networks. Coupling SatCom systems with terrestrial cellular networks to integrate new use cases with satellites will provide a powerful new fusion enabling the innovation of services.

As SatCom systems can provide an overlay network, the integration of NFV/SDN would enable the inclusion of network node functions on board satellite to save on physical sites on the ground and open up new chances to improve network resiliency, security, and availability. Both SDN and NFV are complementary solutions in SatCom. SDN brings flexibility, automation and customization to the network. NFV brings agility in the delivery of services and reduces time-to-market development of new services. They also allow the dynamic reconfiguration of the network to give users the perception of infinite capacity of their applications. Satellites would provide a wide coverage area of wireless networks to extend the dense of terrestrial cells. They can provide larger cells in heterogeneous arrangement to supply critical and emergency services to take on and keep alive the network in cases of disasters. They also can be able to relieve terrestrial cells of signaling and management functions in a software defined network configuration. Satellites will be integrated to the terrestrial system to improve QoS as well as QoE to end users.

The future SatCom systems will be able to provide intelligent traffic routing among the delivery systems, caching high-capacity video to off-load the traffic from the terrestrial networks and thereby enable saving on valuable terrestrial spectrum. In particular, one of the key drivers of 5G network architecture is the lack of spectrum that would be used for the future wireless infrastructure, so frequency sharing between mobile and satellite systems can deliver major increases in the spectrum provided both sectors. Leveraging SatCom systems with techniques like SDN, NFV, and cognitive and software defined radio can be built into future systems to allow such frequency sharing.

The extension of SDN to satellites would provide an attractive perspective for the SatCom community. By exploiting SDN/NFV, satellite equipment will not be vendor specific; instead, they will be open, programmable, and reconfigurable platforms. SDN and NFV are expected to offer new cost-effective services, since SatCom operators will be offering the ability to monetize on their network while offering these future/expected services. For example, the emergence of C-RAN would enable virtualization of SatCom resources (i.e., ground equipments, aerospace access infrastructure); even more, the application of NFV and C-RAN to SatCom paves the way toward the full virtualization of satellite head ends, gateways/hubs, and even satellite terminals, thus entirely transforming SatCom infrastructure, enabling novel services, and optimizing resource usage.

Network virtualization is considered as the key enabler for the efficient integration of the satellite and terrestrial domains. Via the unified management of the virtualized satellite and terrestrial infrastructures, fully integrated end-to-end network slices can be provided, integrating heterogeneous segments in a seamless and federated way. Additionally, the integration of satellite within the 5G future network will extend the coverage of SatCom systems to support new services such as public transport service, vehicle to vehicle; surveillance with UAVs; and high-definition video monitoring, localization, and positioning. Moreover, nongeostationary satellites are actually investigated to achieve optimal networking and latency. Intelligent gateways can be designed to improve network resource use by providing hybridization of satellites and asymmetric digital subscriber line (ADSL) networks. Also, the virtualization can be used to provide the black box (flight data recorder) in the cloud for passenger's aircraft.

The role of satellites in the future 5G networks reveals many challenges to support flexible, programmable, and secure infrastructure. As satellites will be integrated in 5G broadband networks, they should enable extending the coverage of cellular backhaul while at the same time providing enhanced user-centric QoE, cost-effective user terminals, and energy efficiency. SatCom systems should continue to honor guaranteed service delivery to end users by providing higher throughput and low latency for interactive and immersive services independently from the user location. Additionally, the integration of satellites in 5G networks will introduce new challenges regarding the spectrum sharing. Since mobile terminal will use both terrestrial connectivity and satellite connection, mobile receivers should support both kinds of connectivity. Thus, multipolarized schemes are key challenge for satellite and context-aware multiuser detection. Techniques like SDN, NFV, and SDR (i.e., mobile terminals will have modulation and new error-control schemes that will be downloaded from the Internet on the run) are seen as a more challenging aspect of 5G networks, so they should be able to provide intelligent orchestration as well as smart antenna beam forming to enable and facilitate frequency sharing between terrestrial and satellite systems.

5.12   Conclusion

Evolution, convergence, and innovation are considered to be the technology routes toward 5G to meet a wide range of services and application requirements of the information society in 2020 and beyond. To that end, a network must be designed with the future in mind, so that hardware could be abstracted and dynamically utilized through virtualization technologies, which is why holistic SDN and NFV strategies are paramount.

The 5G network will be a combination of multisystems and multitechnologies that need to share the frequency spectrum as well as the physical infrastructure. Nevertheless, wireless and mobile networks will pose challenging issues regarding their integration in the future 5G wireless/mobile broadband world. Leveraging SDN and NFV for supporting and improving LTE networks remains an open issue that should address the way the network functions and the way components will be moved into a secured and virtualized cloud. SatCom system poses also challenging issues on how satellites will be integrated to the terrestrial backhaul wireless network, in such a way to provide heterogeneous segments in a seamless and federated way. Security is an open issue in SDN-enabled 5G networks as well. The programmability of SDN presents a complex set of problems facing the increasing vulnerabilities, which will change the dynamics around securing the wireless infrastructure.

References

[1] J. Hu and W. Lu, "Open wireless architecture—the core to 4G mobile communications," in Communication Technology Proceedings, ICCT 2003, 2003. 1337–1342.

[2] European Commission, "Horizon 2020: The new EU framework programme for research and innovation," 2012.

[3] Y. Saito, Y. Kishiyama, A. Benjebbour, T. Nakamura, A. Li, and K. Higuchi, "Non-orthogonal multiple access (NOMA) for cellular future radio access," in Vehicular Technology Conference (VTC Spring), 2013 IEEE 77th, 2013; p. 1–5.

[4] H. Kim and N. Feamster, "Improving network management with software defined networking," IEEE Commun Mag, vol. 51, no. 2, pp. 114–119, 2013.

[5] A. Ghosh, N. Mangalvedhe, R. Ratasuk, B. Mondal, M. Cudak, E. Visotsky, T. Thomas, J. Andrews, P. Xia, H. Jo, H. Dhillon, and T. Novlan, "Heterogeneous cellular networks: From theory to practice," IEEE Commun Mag, vol. 50, no. 6, pp. 54–64, 2012.

[6] Open Networking Foundation: "OpenFlow Specification 1.5.0". Available at https://www.opennetworking.org/images/stories/downloads/sdn-resources/onf-specifications/openflow/openflow-switch-v1.5.0.noipr.pdf (accessed December 19, 2014).

[7] G. Monteleone and P. Paglierani, "Session border controller virtualization towards 'service-defined' networks based on NFV and SDN," in IEEE SDN4FNS, 2013.

[8] W. John, K. Pentikousis, G. Agapiou, E. Jacob, M. Kind, A. Manzalini, F. Risso, D. Staessens, R. Steinert, and C. Meirosu, "Research directions in network service chaining," in Future Networks and Services (SDN4FNS), 2013 IEEE SDN for, 2013.

[9] W. Liu, J. Ren, and J. Wang, "A Unified Framework for Software-Defined Information-Centric Network". 2013. draft-icn-implementation-sdn-00. Accessed February 17, 2015 http://tools.ietf.org/html/draft-icn-implementation-sdn-00

[10] A. Detti, N. Blefari Melazzi, S. Salsano, and M. Pomposini, "CONET: A content centric inter-networking architecture," in Proceedings of the ACM SIGCOMM Workshop on Information-Centric Networking, 2011.

[11] D. Liu and H. Deng, "Mobility support in software defined networking," 2013.

[12] M. Jarschel and R. Pries, "An OpenFlow-based energy-efficient data center approach," in Proceedings of the ACM SIGCOMM 2012 Conference on Applications, Technologies, Architectures, and Protocols for Computer Communication, pp. 87–88. ACM, 2012.

[13] K.-H. Kim, S.-J. Lee, and P. Congdon, "On cloud-centric network architecture for multi-dimensional mobility," in Proceedings of the 1st MCC Workshop in Mobile Cloud Computing (MCC'12), 2012.

[14] P. Pan and T. Nadeau, "Software-defined network (SDN) problem statement and use cases for data center applications," in IEFT Internet Draft, 2011.

[15] X. Jin, L. E. Li, L. Vanbever, and J. Rexford, "SoftCell: Scalable and flexible cellular core network architecture," in CoNEXT, 2013.

[16] A. Basta, W. Kellerer, M. Hoffmann, K. Hoffmann, and E.-D. Schmidt, "A virtual SDN-enabled LTE EPC architecture: A case study for S-/P-gateways functions," in Future Networks and Services (SDN4FNS), 2013 IEEE SDN for, 2013.

[17] L. Erran Li, Z. M. Mao, and J. Rexford, "CellSDN: Software-defined cellular networks". In Techinical Report, Princeton University. 2012.

[18] A. Gudipati, D. Perry, L. E. Li, and S. Katti, "SoftRAN: Software defined radio access network," in ACM SIGCOMM Workshop on Hot Topics in Software Defined Networking, 2013.

[19] J. Kempf, B. Johansson, S. Pettersson, H. Luning, and T. Nilsson, "Moving the mobile evolved packet core to the cloud," in Wireless and Mobile Computing, Networking and Communications (WiMob), 2012 IEEE 8th International Conference on, 2012.

[20] T. Benson, A. Akella, A. Shaikh, and S. Sahu, "CloudNaaS: A cloud networking platform for enterprise applications," In Proceedings of the 2nd ACM Symposium on Cloud Computing (SOCC '11). ACM, New York, NY, USA, Article 8, 13 pages.

[21] J. Zhang, H. Yang, Y. Zhao, Y. Ji, H. Li, Y. Lin, G. Li, J. Han, Y. Lee, and T. Ma, "Experimental demonstration of elastic optical networks based on enhanced software defined networking (eSDN) for data center application," Opt Express, vol. 21, no. 22, 2013.

[22] S. Civanlar, M. Parlakisik, A. M. Tekalp, B. Gorkemli, B. Kaytaz, and E. Onem, "A QoS-enabled OpenFlow environment for scalable video streaming," in IEEE Globecom Workshops, 2010.

[23] H. E. Egilmez, B. Gorkemli, A. M. Tekalp, and S. Civanlar, "Scalable video streaming over OpenFlow networks: An optimization framework for QoS routing," in ICIP, 2011.

[24] H. E. Egilmez, S. T. Dane, B. Gorkemli, and A. M. Tekalp, "OpenQoS: OpenFlow controller design and test network for multimedia delivery with quality of service," NEM Summit, 2012.

[25] S. Civanlar, A. M. Tekalp, and H. E. Egilmez, "A distributed QoS routing architecture for scalable video streaming over multi-domain OpenFlow networks," in Proceedings of IEEE International Conference on Image Processing (ICIP 2012), 2012.

Part II

SDMN Architectures and Network Implementation

6

LTE Architecture Integration with SDN

Jose Costa-Requena, Raimo Kantola, Jesús Llorente Santos, Vicent Ferrer Guasch, Maël Kimmerlin, Antti Mikola, and Jukka Manner
Department of Communications and Networking, Aalto University, Espoo, Finland

6.1 Overview

This chapter proposes solutions to integrate software defined networking (SDN) technology with wide area mobile networks. The integration of SDN into mobile networks to become the software defined mobile network (SDMN) poses several architecture alternatives. To limit the discussion of the alternatives to a reasonable scope, in this chapter, we will discuss the issues and alternatives taking that the SDN uses the OpenFlow protocol.

We first need to define the proper location of the SDMN controller. It can be integrated with the Mobility Management Entity (MME) making the controller aware of mobility events, or it can be located in the Serving/Packet Gateway (S/P-GW) to control the transport network. The integration of SDN control with LTE network elements should follow a incremental process the idea being smooth deployment of SDN into a live mobile network. At best, the integration of SDN with LTE paves the way for 5G networks. For now, the objective is to keep using the current IP-based networks and add SDMN-based flexibility to the LTE network architecture.

Scalability, security, and resilience are key factors to be taken care of in order for SDMN to become the next infrastructure for 5G mobile networks. Finally, SDMN should bring some benefits both to mobile operators and end user. Section 6.2.1 presents the current LTE architecture. Section 6.2.2 discusses multiple options for the placement of the SDN controller. Section 6.2.3 introduces the vision for integrating SDN in the mobile network where mobile-specific functionalities are implemented as SDN applications. Section 6.3 studies the issues of ensuring scalability. Section 6.4 introduces the proposed security mechanisms, and Section 6.5 describes the benefits from the operator and the end-user points of view based on a simple

Software Defined Mobile Networks (SDMN): Beyond LTE Network Architecture, First Edition.
Edited by Madhusanka Liyanage, Andrei Gurtov, and Mika Ylianttila.
© 2015 John Wiley & Sons, Ltd. Published 2015 by John Wiley & Sons, Ltd.

scenario with built-in smart network services such as dynamic caching. Finally, Section 6.6 presents the research problems and conclusions.

6.2 Restructuring Mobile Networks to SDN

This section will describe the starting point of the restructuring, namely, the LTE network, then discuss the design alternatives, and finally propose a way of applying SDN concepts to LTE networks and beyond.

6.2.1 LTE Network: A Starting Point

Mobile networks consist of physical and logical entities. The physical layer is made of network routers (L3), switches (L2), and physical links (L1) with different technologies and topologies as shown in Figure 6.1. The logical layer consists of network elements (e.g., eNodeB, MME, S/P-GW, HSS, etc.) that perform the attachment of user devices, mobility, and transport of data from mobile devices across the mobile network in such a way that mobility is hidden from the core Internet. The physical layer (L2 and L3) provides the connectivity and transport functionality to the logical layers that implement the mobile-specific control functions. The access network consists mainly of the eNodeBs that provide the radio access to the user equipment (UE). The backhaul consists of all the network switches for aggregating the traffic from the access network and provides the connectivity toward the core network. Finally, all the connection services, mobility services, and billing functionality are implemented by the network elements (i.e., MME, S/P-GW, PCRF, HSS) located in the core network.

Mobility management in LTE mobile networks is the critical functionality, and any new technology that handles mobility events has to deliver a reliable and low-latency handover. Mobility in LTE networks is implemented through different methods depending on whether it is within same radio technology, that is, intra-E-UTRAN, or to different radio access technologies (RAT). In this work, we focus in the intra-E-UTRAN where the handover can be performed following two procedures. One procedure consists of X2-based handovers where there is a connection between the source and the target eNBs through which the handover

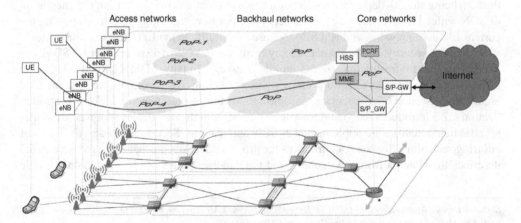

Figure 6.1 Physical and logical layers in mobile network. *refers to the router with connectivity to public Internet

operation is negotiated and then reported to the associated MME. The second procedure is S1-based handover; there is no direct connection between the source and the target eNBs, and therefore the handover operation is negotiated via their respective MMEs. In our analysis, we consider in more detail the scenario where the new target eNB is in the same tracking area ID (TAI) associated to the same Mobility Management Entity (MME). Other scenario consists of the case where the new TAI belongs to the same MME. In the former scenario, mobility management uses the S1 MME interface between the eNB and the MME.

A fundamental problem in the IP from the mobility point of view is that the IP address identifying the node fixes its location to a certain anchoring point. This occurs because the IP address also has the role of the routing locator. The common solution in mobile networks consists of using tunneling. UE IP packets are tunneled over the GPRS Tunneling Protocol (GTP). The GTP tunnels are established between the eNB and the S/P-GW. A GTP tunnel uniquely identifies traffic flows that receive a common QoS treatment between a UE and a P-GW. A traffic flow template (TFT) is used for mapping traffic to an underlying bearer. The GTP tunnel endpoint identifier (TEID) unambiguously identifies the tunnel endpoint of a user data packet, separates (identifies) the users, and also separates the bearers of a certain user as depicted in Figure 6.2a.

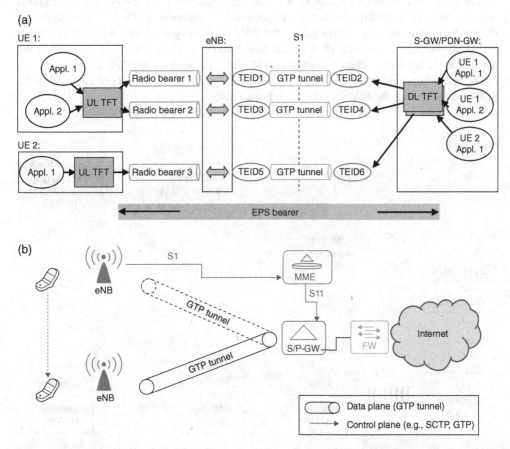

Figure 6.2 (a) Tunneling of user traffic over mobile networks and (b) handover process controlled from MME through S1 and communicated to S/P-GW to recreate a GTP tunnel.

When an UE moves to a new eNB, the GTP tunnel has to be recreated between the new eNB and the S/P-GW, while the inner data flow keeps using the original UE IP address.

The handover process is initiated and managed through the S1 interface as shown in Figure 6.2b. MME is aware of the mobility process and communicates with the S/P-GW to recreate the GTP tunnel between the new eNB and the S/P-GW as shown in Figure 6.2b.

6.2.2 Options for Location of the SDMN Controller

Integrating SDN into mobile networks to become SDMN can be done in several ways: (i) controller can be integrated with the MME in order to be aware of mobility events, or (ii) controller can be integrated with the S/P-GW to control the transport network.

Figure 6.3 shows the current LTE architecture, which allows multiple options for integrating the SDN controller.

Figure 6.4 describes one option of integrating SDN into the LTE architecture. This option consists of decoupling the S/P-GW into the control and the data planes. The control part of the S/P-GW (i.e., S/P-GWc) provides IP address allocation for the UE and is responsible for applying the TFT to the user data flows. The data plane of the S/P-GW (i.e., S/P-GWu) provides the GTP tunneling termination endpoint and the anchoring of the GTP tunnels during the handover process. The control part of the S/P-GW is integrated with the SDN controller and sends the TFT to the S/P-GWu, which then enforces it as data filtering. The rest of the network elements are not changed, and the MME interacts with the S/P-GWc.

The second option for integration SDN in the LTE architecture consists of embedding the SDN controller with the MME as shown in Figure 6.4b.

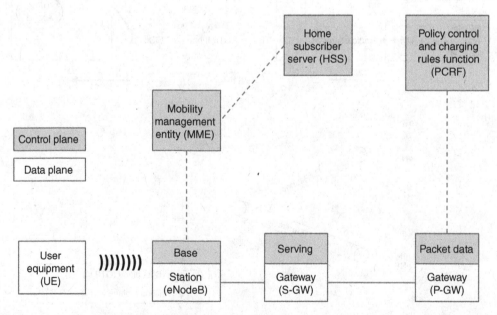

Figure 6.3 LTE mobile architecture.

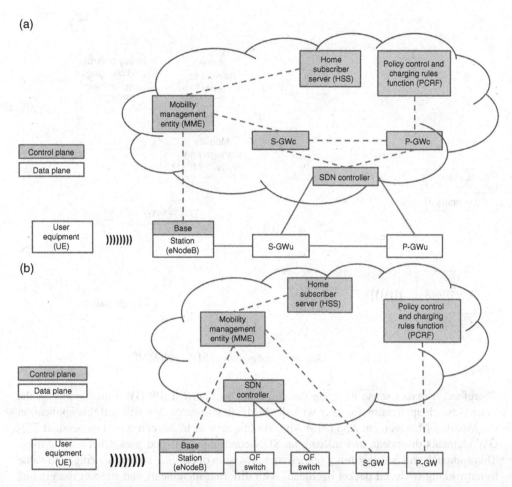

Figure 6.4 (a) Integration of SDN with S/P-GW and (b) integration of SDN with MME.

This option allows the SDN controller to learn about mobility events directly from the MME allowing to apply new rules in the switching nodes to reestablish routing paths in an optimal manner. In Figure 6.4b, the DP based on OF between eNB and S/P-GW need to understand GTP.

Integrating SDN controller functionality with MME provides a smooth integration in the long term as well as a disruptive solution in mobile networks. The issue for the SDN integration is how to support mobile-specific protocols with OpenFlow. Since OpenFlow does not support GTP and modifying the OpenFlow for such support would lead to loosing economies of scale in the switching elements as well as make the integration of caching and network monitoring functions into the network more costly and cumbersome, it is reasonable to study the options of replacing the GTP-u with standard data communication protocols such as variants of Ethernet and MPLS. The resulting mobility solution, instead of tunneling over IP, would be based on SDN-controlled switched paths.

In SDMN, the control plane is moved out of the basic networking elements into centralized servers—these servers resemble classical anchor points used in many mobility protocols.

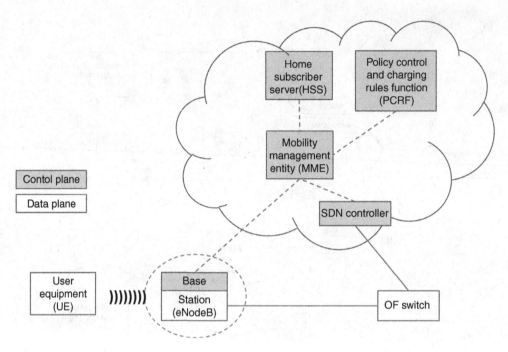

Figure 6.5 Disruptive integration of SDN with MME.

Therefore, it makes sense to group the controller and current S/P-GW functionality in the same network application together with the MME functionality. We will call this application the Mobility Management App (MM App). In this approach, the currently independent S/P-GW elements disappear, and instead, an SDN-controlled switched packet network is used. This approach will add flexibility (e.g., easier provision of caching and monitoring) and value to networking (reduced packet overhead) with different increments and support the gradual introduction of high network throughputs, optimal flow management, and traffic engineering possibilities. Figure 6.5 shows the integration of mobility management with SDN controller. The challenge for the SDN control is to meet the necessary delay requirement in order to achieve seamless terminal mobility without too much signaling overhead. Another challenge is to meet all the functional requirements with the constraints set by the OpenFlow protocol.

Mobility is a critical aspect of mobile networks, which requires specific functionality in the network elements. Having tight linkage between the MME and the SDN controllers gives the best chance that the time-constrained functions of mobility, such as seamless handovers, are handled efficiently from the SDN controller.

6.2.3   Vision of SDN in LTE Networks

Starting from the late 1990s, the 3GPP has been taking steps toward a clear separation of data and control planes and the respective elements in the architecture. We propose to take this concept to the next level following the SDN paradigm. Figure 6.6 presents the 5G network control as a group of SDN applications. They are the Base Station App, Backhaul App, MM

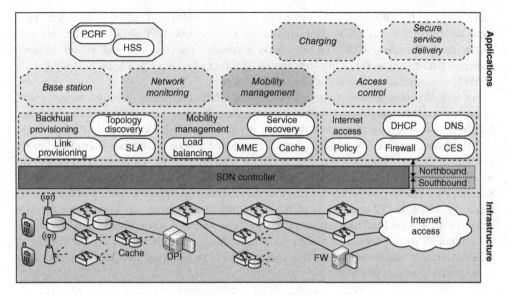

Figure 6.6   5G as a group of SDN Apps.

App, Monitoring App, Access App, and Secure Service Delivery App. The network applications are orchestrated via the Controller Northbound API. Multiple SDN applications operate without conflicts.

The Base Station App runs the control software that is now vertically integrated with the eNB. The physical base stations under its control consist of an antenna, a band-pass filter, and an Ethernet card for backhaul connectivity.

6.2.3.1   MM App

The MM App implements mobility as a service (MaaS) and is a reactive function. MaaS appears on the upstream interface of the OF switch (mOFS) that offers its service to the Access App. When a mobile device moves from the area of one eNB to an area of another, the rule in the mOFS for the device may need to be modified and a new rule may need to be created in the new eOFS where eOFS is the first aggregation point for the connected eNBs. If the new eNB is under the same eOFS as the previous one, then it is enough to modify an existing rule in the eOFS. We also need to take care of balancing the load across the alternative paths between an eNB and a particular mOFS. The MM App chooses the path for a device. The load balancing decision is made based on input from the Network Monitoring App. In any case, it is desirable that the point of attachment of a mobile to the Internet is fixed while it stays under the coverage of the current mobile network. To make this possible, every eNB in a mobile network has several preprovisioned paths to every mOFS.

The MM App incorporates the MME. In addition, it needs to manage the quality of service for each user, to balance the load among the alternative paths across the aggregation network, and to route the user to a cache, when possible. Among the currently popular SDN concepts, OpenFlow is the most prominent. It does not support GTP tunneling. Therefore, it makes sense to study alternative ways of carrying the user plane traffic from eNBs to the Internet and from the Internet to the eNBs. One alternative is to replace GTP data plane tunnels with carrier-grade Ethernet (CGE) encapsulation methods. The result would be an Ethernet switched path-based mobility implementation. Another alternative would be to study the use of MPLS encapsulation between eNBs and the Internet.

For mobility management, for many procedures, eNBs need to be directly connected with some tens of their neighbors over the X2 interface. For this reason, in an Ethernet-based solution, a suitable way to connect the eNBs up in the network hierarchy is 802.1ad. In the 802.1ad frame, one VLAN tag identifies a user under the eNB, while the other identifies the destination such as "Internet" or an X2 neighbor. The Internet VLAN is switched to 802.1ah by eOFS. The X2 VLANs are either locally switched in eOFS to destination eNBs or using 802.1ah the switching to the needed neighboring eNB can take place higher up in the network hierarchy. Since the X2 interface is used for mobility management, we propose that the MM App will request the Backhaul App to provision these switched paths. Finally, it is convenient that, although packet forwarding is based on switching, each eNB has either IP routing or Ethernet routing functionality for the X2 interface. For this purpose, the MM App may need to assign IP addresses to eNBs.

6.2.3.2 Access App

In one physical mobile network, there may be many Access Apps. In that case, an Access App is owned and operated by a particular mobile virtual network operator (MVNO). Putting mobility aside, it is the Access App that is responsible for providing the data services to mobile users. Key properties of the Access App include providing Internet access, firewalling unwanted traffic, and providing access to premium content.

In 5G, we propose moving from simple network firewalls that apply the same rules to every customer, to cooperative firewalls with user specific admission rules managed by the extended 3GPP policy management architecture. This is justified by the need to manage the reachability of the device per application and per user without cumbersome NAT traversal. It is also justified by the need to block all packets with a spoofed source address and all DDoS packets from reaching the mobile network, consuming any air interface capacity and disturbing the power saving sleep mode of the mobile device. In the proposed solution, it suffices for all mobile devices to have just a private address. Therefore, scaling to any number of users and devices in 5G does not depend on the success of IPv6.

A task of the Access App is to assign an IP address for a mobile device. This can be a private address. Thus, the Access App provides the point of attachment to the Internet and to the service delivery networks to each mobile device by controlling an OF switch (iOFS) that connects the mOFS to the Internet. The point of attachment should be as stable as possible while the mobile moves or even roams to foreign networks. For an incoming flow, the Access App will handle firewalling and request downstream load balancing from the MM App. We believe that all flow admission should be managed by a policy that is part of the subscription information of the user. Policies could be managed by an extended 3GPP policy and charging management architecture. Policies can be dynamic, that is, treat different remote hosts differently based on reputation produced by a trust management system. Moreover, a cooperative firewall can make queries to, for example, the sender's firewall or certification authorities before making the final admission decision. This would dissolve the boundary between closed and open networks, managing all flow admission by the policy. A mobile device under the cooperative firewall is reachable using the host fully qualified domain name (FQDN), a suitable identity and the routing locator of the iOFS controlled by the Access App.

Traffic through the service delivery network is also tunneled. A binding state managed by the firewall ties the service delivery tunnel and the mobile backhaul tunnel together at the iOFS. The Realm Gateway (RG), a component of the Access App, can admit traffic directly from the legacy Internet without cumbersome NAT traversal.

6.2.3.3 Secure Service Delivery App

The final SDN App on the path to the communication partner is the Secure Service Delivery App. By the service delivery network, we mean the network that connects two mobile networks or a mobile network with a fixed customer network or with a data center that has the desired applications or the desired content. We suggest that by applying SDN concepts to service delivery, we can seek benefits such as securing the process of service delivery and maximally benefiting from the economies of scale of cheap switches and generic hardware for control processing. The minimum goals of the service delivery network are to prevent source address spoofing and DDoS and admit only legitimate traffic (e.g., only I can turn on my own sauna connected to Internet of Things).

6.3 Mobile Backhaul Scaling

The driving factors for the proposed design are as follows: (i) The Base Station App is rather delay sensitive due to both radio and application aspects. Therefore, we believe that the control software of a physical base station must reside rather close to it although it can be separated to a distinct node. (ii) The goal of mobility management is to provide seamless handovers.

Most data applications can tolerate connection loss during a handover for several hundreds of milliseconds. Interactive voice can tolerate a loss of connection for several tens of milliseconds. Applications like gaming benefit from delay reductions to the area of 10 ms. Therefore, the control functions for mobility management can reside a few hundred kilometers from the base station but not on a different continent (the propagation delay over a round-trip of 200 km in fiber is 1 ms) and only rarely in a different country.

(iii) The delay requirements for the Access App are mostly determined by the perceived service responsiveness of the network. By this, we mean factors such as voice session setup time or network contribution to system response time. For the Access App, another important driving requirement, in addition to delay, is scalability. The Nokia white paper [1] predicts that by 2020 a mobile user will consume 1 Gbyte of data per day. From this, we can calculate that scaling the access network to tens and hundreds of millions of mobile users is a significant challenge with the technology that we expect to be available by 2020: for example, 100 million customers will use about 1000 100GE interfaces when there is no caching.

To ease the challenge, caching has to be used maximally in mobile access. Also, we propose that most popular content of content providers and from content distribution networks should be collocated in the same sites (i.e., data centers) as the Access App. Therefore, another way to look at the access network controlled by the Access App is that it, in fact, is a set of telco data centers and the network that connects them. In practice, a very significant part of the tens of terabits of traffic consumed by the mobile users will be served from these data centers. Moreover, by connecting large cache servers to the CGE nodes serving hundreds of thousands to a million customers each and smaller caches to the eNBs, we can save on the number of mOFS and iOFS switches and on the required performance of these switches making the design presented in Figure 6.8b even more feasible.

Let us model the scaling of a 5G mobile backhaul network for 100 million customers with the assumption of 1 gigabyte of dedicated traffic capacity per user per day. We further assume that each eNB serves 100 to 1000 customers and that the backplanes of the backhaul data plane nodes are below 4 TiB/s. Figure 6.7 shows an overprovisioned network design without caching. It is an example that we use to study the requirements. The figure shows link and backplane speeds and average link loads when 24 h traffic is transmitted in 6 h.

To let the SDN applications manage the network, we place an OpenFlow switch (eOFS) as the first aggregation element to which eNBs are connected. The eNB will terminate the protocol stack over the air interface and send all traffic from a user to an Ethernet VLAN using 802.1ad encapsulation. The eOFS will tag the packets from the user toward the Internet: a suitable encapsulation is 802.1ah. The second OpenFlow switch (mOFS) is required before the entry point to the Internet. The mOFS will tag and route the packets from the Internet to the right eNB and the right mobile device. For traffic aggregation from many eOFSs to a few mOFS, we will use CGE switches since they are simpler than OF switches.

We can isolate each mobile into its own subnet using 802.1ah. The traffic from eNBs to the point where MaaS is offered to the Access App and back can be switched through the described network. For the purpose of load balancing, each link to the Internet must be reachable from each eNB over several paths (e.g., eight). This is because it is beneficial to keep the point of attachment of a mobile to the Internet stable while it stays in the area of the mobile network.

Given the assumptions, it is easy to calculate that the required tag length for marking the path between the two OFS is in the order of 20–24. If the eNBs are small on average, a single 802.1ah path from the eNB to mOFS would lead to longer tags (around 29 bits). Normally, by deploying eOFS to which eNBs are connected, the path tag can, for example, be the 24-bit I-SID in 802.1ah. An alternative to 802.1ah is MPLS-TP. One of the tasks of the Mobile Backhaul App is to set up and manage the service routing in the CGE switches between eOFS and mOFS and take care of fault recovery in this network. In the 802.1ah encapsulation, the I-SID value marks the path between an eOFS and mOFS. The B-VLAN tag can be used to

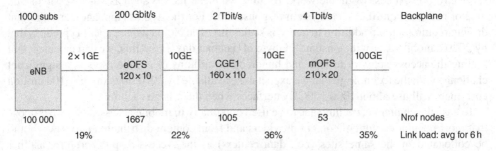

Figure 6.7 Network scaling example.

separate traffic of different virtual operators if necessary and finally the C-VLAN tag as network identifier within the operator, to choose the right mOFS and GW, since the address provisioning already allocates different networks (subnetting/29) per UE. In case of MPLS encapsulation, several MPLS labels would be needed.

When integrating SDN in mobile networks, scalability is another major issue to be considered. Figure 6.9 represents the aggregation of data paths interconnecting different tracking areas with the corresponding GW element for that mobile operator. Figure 6.8a shows a possible mobile network topology where OF switches are used to aggregate traffic in different parts of the mobile backhaul.

In LTE, when the UE attaches to the mobile network through the eNodeB, the MME will request default context from the S/P-GW that will assign an IP address to the UE and will establish a tunnel between the eNodeB and the OFS-GWx. Figure 6.9 shows the usage of 802.1ad and 802.1ah in the mobile network to allow carrying the data between the eNodeB and the GW but still using existing Ethernet switches in the network. During the attachment, both an uplink and downlink between the eNodeB and the GW are established.

The aim of the uplink is to communicate the originating eNB with the specific point of attachment to the Internet selected by the MME for a given UE. We aim at offering MaaS transparently to the MVO, so that any standard routing equipment can be used for connecting to public networks.

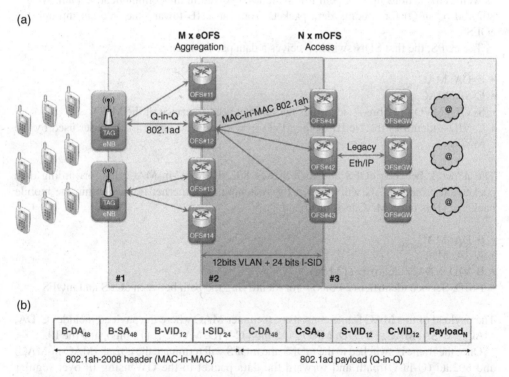

Figure 6.8 (a) Aggregation of tracking areas in mobile access network and (b) Ethernet packet encapsulation for 802.1ah and 802.1ad.

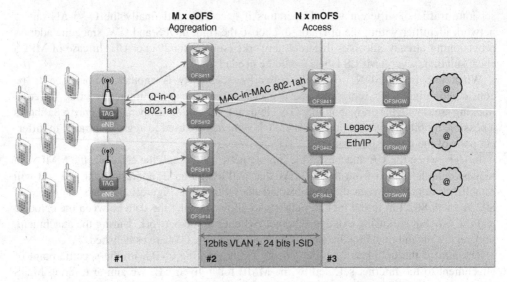

Figure 6.9 Direct L2 path eNB to OFS-GW w 802.1ah (MAC-in-MAC).

Now, we describe a possible usage of the different fields in the encapsulation of Figure 6.8 as well as the actions in the OpenFlow switches involved in the communication path. We use 802.1ad (Q-in-Q) for sending data packets from the eNB toward the Internet through the eOFS.

The eOFS, the first SDN switch, receives a data packet with:

- C-DA: MAC_{GW}
- C-SA: MAC_{UE}
- S-VID: MVO_{ID} (12 bits)—identifies the Internet access service of the MVO
- C-VID—identifies the NetID of the MVO (private address space of this user used by the MVO)

The network between eOFS and mOFS uses 802.ah (MAC-in-MAC) for forwarding data packets. The outer MAC will be used for switching the Ethernet frames within the mobile backhaul. This outer MAC uses upstream:

- B-DA: MAC_{mOFS}
- B-SA: MAC_{eOFS}
- B-VID: VLAN identifier (12 bits)
- I-SID: Service identifier (24 bits)—for identifying the path between eOFS and mOFS

The payload of this MAC frame consists of the inner MAC where the payload includes C-DA, MAC_{GW} C-SA, MAC_{UE} S-VID, MVO_{ID} (12 bits) C-VID, MVO_{NetID} (12 bits), and the IP_{UE}.

Once the frame is switched to the mOFS, the mOFS will terminate the 802.ah (MAC-in-MAC) and 802.ad (Q-in-Q) path and forward the data packet to the GW using IP over regular Ethernet. This enables using standard IP routers for connecting to public networks. The SDN-Ctrl and mOFS are responsible for maintaining an updated status of the location of the UEs.

In the downstream direction, the Internet GW can send data to the specific eNB where the UE is located. We have the reverse process in the downstream: we create in mOFS the 802.ah (MAC-in-MAC) and 802.ad (Q-in-Q) frames out of the received packets from the GW that use IP over Ethernet. We use 802.ah (MAC-in-MAC) for forwarding data from mOFS toward eOFS. The combination of B-DA+VLAN tag (B-VID)+service identifier (I-SID) determines the L2 paths. The structure of the MAC would be:

- B-DA: MAC_{eOFS}
- B-SA: MAC_{mOFS}
- B-VID: VLAN identifier (12 bits)
- I-SID: Service identifier (24 bits)

The payload of the outer MAC will include the inner MAC consisting of:

- C-DA: MAC_{UE}
- C-SA: MAC_{GW}
- S-VID: MVO_{ID} (12 bits)
- C-VID: MVO_{NetID} (12 bits) and the IP packet of the UE

We terminate the 802.ah (MAC-in-MAC) path in the eOFS exposing the 802.ad (Q-in-Q) for further analysis. The SDN-Ctrl and eOFS are responsible for maintaining an updated status of the location of the UEs for the downstream traffic. The matching state in eOFS is based on the C-DA, S-VID, and C-VID that determine the current eNB where the UE is located and the packet will be forwarded to the correct eNB.

For scalability of the proposed usage of different identifiers, considering that each mobile device will consume 8 IP addresses, with private addresses of the range 10.x.y.z, we can identify 2 million×212=233 devices. We can allocate several S-VID values for one MVO because it is unlikely that in one physical network there would be thousands of MVOs. The C-VID identifies the network within a given MVO. The scalability with 2 million hosts per network is the product of S-VID×C-VID (networks)×2M (UEs). This example forwarding system is summarized in Figure 6.9.

6.4   Security and Distributed FW

The current Internet model makes it possible for a host to send data packets to any destination address, whether these packets are wanted or not. A receiver is unable to establish a set of requirements a remote sender must comply with prior to sending any data packets. As a result, unwanted traffic can only be discarded by the receiver upon arrival. In addition, source address spoofing benefits DDoS attackers because it makes it difficult to attribute evidence of antisocial behavior to the originating hosts or networks. Spammers, hackers, fraudsters, and other malicious users rely on botnets for their activity that destroys network value to the majority of the users.

From Prisoner's Dilemma tournaments, we know that cooperative strategies can become dominant in societies where actors can efficiently share their opinions about the trustworthiness of other actors and where the interactions are unending. A precondition is that obstinate violators of rules, who simply will refuse to cooperate, can be curbed in some way. Based on this result, we argue that host-based stand-alone solutions cannot tackle the wide variety of threats that exist today. Traditionally, firewalls have been deployed to protect both hosts and networks, by executing a set of rules ordered in a predefined fashion, based mostly on local information and data gathered from deep packet inspection of data traffic at different protocol layers. The result is ultimately limited to accepting or dropping a given connection.

In modern and future Internet, most hosts use wireless connections and personal battery-powered devices. The devices may also sense or manipulate objects in the real world creating many new threats compared to the traditional Internet. It would be desirable to block all packets with spoofed source addresses, all DDoS packets, as well as all unwanted packets from reaching the air interface and even more desirable from reaching the mobile device where a host-based firewall would have to wake up the device to process any unwanted packet and as a result exhaust the battery while it would be doing a perfect job for the security of the device. It would also be desirable to admit to the mobile device only flows that come from authorized parties. At the same time, for the developer of legitimate applications, it would be desirable to offer a network that automatically manages reachability without the need for application level NAT traversal mechanisms.

To meet the above goals, we propose to extend the functionality of traditional stand-alone state-full firewalls to policy-based cooperative firewalls. The policies are then defined in the firewall nodes, usually located at the network edges where we now have a network address translator or a generic gateway (S/P-GW). By embedding the firewall functionality in these nodes, it is possible to acquire a wider and more coherent view of the network that otherwise would not be possible with a stand-alone host-based solution. For example, the receiver's policy may require the following conditions to be met before flow admission:

- The address of the sender's edge node is not spoofed.
- Present a certificate of the sender's edge node.
- Present a stable and verifiable identity of the sending host.
- The sender's network is not blacklisted.
- The sender itself is not blacklisted.

In addition to the above, the firewall can also propose the remote edge to quench a source (your host x is DDoSing me, stop it) or report a reflection attack to an uninfected host that is exploited by an attacker for hiding its own identity.

Under such conditions, if the receiver suffers any kind of attack, it is always possible to attribute blame on the sender's network or the particular host that was initiating the connection. It is also possible to block all traffic from hosts that are known to distribute malicious code or act on behalf of an attacker. Making all the checks of identity apply to all communications is, however, too costly and not feasible. Therefore, we argue that all communication should be managed by policy. By efficiently collecting and aggregating evidence and distributing the results to cooperative firewalls, their blocking policy can become dynamic and react in an accurate and fine-grained manner to new malicious actors.

Under this new paradigm, the firewall adopts then the functionality of communication trust broker for the hosts that it protects. The access to these hosts is only granted upon successful policy negotiation. The mission of the policies is to define a number of prerequisites that need to be fulfilled prior to accepting a new flow between the communicating parties. As a result, unwanted traffic can be effectively blocked closer to the source, and repeated unsuccessful policy negotiations can be attributed to specific users, thus discouraging attackers with malicious intents.

6.4.1   Customer Edge Switching

To demonstrate the feasibility of the ideas presented in Section 6.4, we have created the technique of customer edge switching (CES). It offers a cooperative firewall with dynamic policy management as a virtualizable security software entity. Network awareness motivates the necessity of dynamic policies as they enable different levels of security and a fine-grained response to attacks. The required knowledge can be acquired based on local information as a result of policy negotiations or deep packet inspection, as well as from the collaboration of other connected devices or a global reporting system for security and trust.

It is possible to leverage virtualization for spawning dedicated firewall instances in the cloud. The number of these instances is determined by the amount of traffic and mobility procedures. Virtualization gives a flexible network provisioning by dynamically adjusting the amount of resources available. The adoption of SDN solutions facilitates the deployment of new services in a virtualized framework. New security modules such as CES can be colocated with the S/P-GW elements. (Note that one of the core functions of P-GW is address assignment). If a firewall must give promises about host addresses to remote parties, it is best that the firewall itself assigns the addresses. Therefore, it is logical that a cooperative firewall should be integrated with the P-GW.

As SDN brings the possibility of dynamic flow modifications, therefore the level of security can also be adjusted in a fine-grained manner. For example, an initially trusted flow may benefit from minimal intrusive security policy, but on the event of security threats, this flow may be resubmitted to more extensive DPI analysis on different firewall instances and to a honey pot. This further analysis allows to collect further evidence or ultimately trigger a BGP update to create a sinkhole in the network and disconnect the attacking network for a given time.

Virtualization directly benefits the operators in the sense that it can improve their efficiency by dynamically allocating more resources during busy hours and reducing them during idle times. Energy consumption can also be decreased by shutting links down during the idle times.

6.4.2   RG

When the networks of both communicating hosts have adopted CES, we can ensure a clearly better level of security compared to the state of the art. When the sender is not behind a CES node but either has a globally unique IP address or is behind a NAT or NAPT, we must provide interworking for the servers that are behind a CES node (for the case of servers with globally unique IP addresses, a CES not trivially acts as a NAT). For this interworking case, we developed the RG that is able to dynamically admit flows to the servers that it is serving from any legacy Internet hosts [2]. Upon a DNS query, the key algorithm (circular address pool) dynamically reserves an RG outbound address for a short period (we use 2 s in the demonstrators) of time. Upon the arrival of the next new flow, this reservation is removed and the address is released for the next DNS query. Additional information about the expected flow (such as its

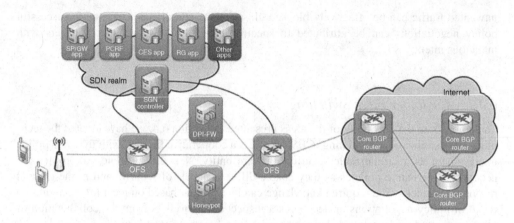

Figure 6.10   Integration of CES in SDN.

port) can be configured in the DNS leaf node that resides in the RG itself making the reserved state available just for the expected new flow and thus making hijacking of this state difficult. The RG can also translate between IPv4 and IPv6. To secure the RG, we developed a number of heuristics that, for example, protect the RG from DDoS attacks that use spoofed DNS queries to powerful DNS servers that serve queries from any hosts and may thus end up being used as reflectors for the purpose of DDoSing hosts behind an RG or any other hosts as shown in Figure 6.10.

6.5   SDN and LTE Integration Benefits

The integration of SDN in LTE networks provides benefits in terms of CAPEX and OPEX since control functionality is implemented by cloud services and will thus benefit from commodity computing facilities. Another benefit emerges from the fact that the transport network is simplified by removing the GTP tunnel for the data plane. In the future, this allows using commodity off-the-shelf OpenFlow switches, which will provide the required data forwarding features controlled from the cloud.

From the different options of integrating SDN with LTE, the proposal of putting the MME and the SDN controller in one SDN App brings several benefits. The controller needs to have the necessary information about the location of the UE and the associated mobile operator as well as the necessary attachment and handover events. Therefore, the controller should be integrated with the MME and S/P-GW to receive those events and establish the required MAC-in-MAC and Q-in-Q mappings. Moreover, this integration results in the next disruption where data plane is managed from a single MME/controller element. The evolution toward this architecture can take place progressively where the MME will keep its current interfaces for receiving the signaling through the S1-MME interface. The MME maintains the current standard process and establishes GTP tunnels between legacy eNodeB and S/P-GW. Simultaneously, the MME can include the new SDN functionality and establish communications between the new model eNodeBs and IP routers directly at layer 2 without GTP tunneling. In this scenario, the same MME, when receiving the signaling from an SDN-based eNodeB through the S1-MME interface, will establish the connection with the

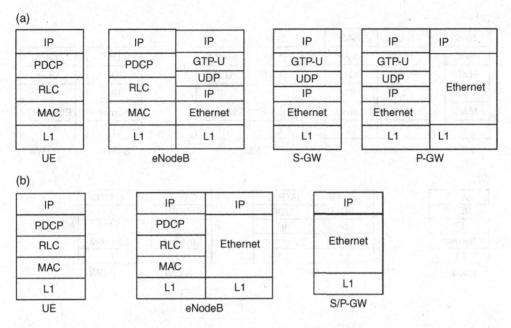

Figure 6.11   (a) User plane networking stack in LTE and (b) user data plane networking stack in LTE with SDN control.

termination SDN switch over L2 using TUN interfaces. The networking stack currently used for the user plane is depicted in Figure 6.11a. The radio layers are terminated in the eNodeB from where GTP is used up to the S-GW and the P-GW that provides the bridge to the public Internet.

The usage of 802.1ad in the backhaul and integration of MME with the SDN controller allows the removal of GTP. This will result in the simplification of the stack in the eNodeB that terminates the radio layers and includes an Ethernet switch toward the rest of the network in the backhaul as shown in Figure 6.11b. Moreover, the S/P-GW is simplified after removing the GTP-u and consists of a simple Ethernet switch and IP router toward the public Internet. In this architecture, the mobility is performed by the SDN controller.

This architecture leads to an optimized transport network as well as a scalable control plane that converges into single network application: MME with embedded SDN controller functions. This MME would run either in dedicated HW or as a cloud service to allow launching multiple instances as needed to overcome scalability limitations of having the functionality in a physical network element. The MME on the other hand will continue supporting their current networking stack as depicted in Figure 6.12. This approach allows smooth transition. The MME would be able to manage current network elements, that is, eNodeB and S/P-GW, but the integration with the SDN allows managing the new eNodeB and S/P-GW where GTP has been removed.

In addition to the integration of the MME with the SDN and the simplification of the backhaul network by removing GTP from the network elements (i.e., eNodeB and S/P-GW), the network has to be flattened. The network elements are normally located in the core network.

(a)

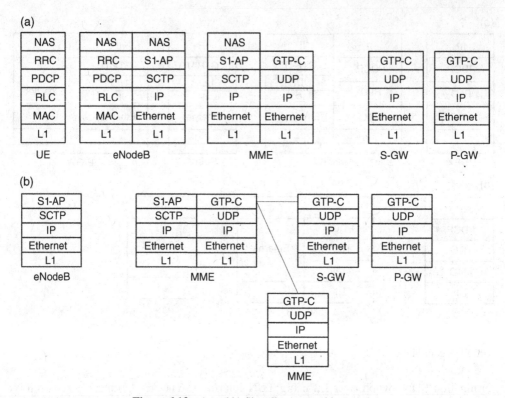

(b)

Figure 6.12 (a and b) Signaling networking stack.

Instead, the network needs to be flattened so the network elements are located closer to the eNodeB in the backhaul. This allows deployment of stand-alone access networks with their own network elements. The coordination of multiple access networks would be done using a centralized database, and the handover between the MMEs located in each access network would be carried out through the S10 interface. The signaling networking stack remains the same to interact with legacy LTE network elements such as S/P-GW and other MMEs as shown in Figure 6.12b.

6.6 SDN and LTE Integration Benefits for End Users

The integration of SDN in LTE networks brings certain benefits to mobile operators in the sense that EPC functionality can be virtualized and moved to commodity servers, which means reductions of CAPEX and OPEX. However, the end users can also benefit from adopting SDN as described in the following sections.

• Content Caching

The always increasing role of content delivery networks (CDN) in the Internet traffic shows a clear shift to content consumption. CDNs leverage the power-law nature of content popularity distribution where many users request popular contents within a short period of time. Therefore, storing a copy of popular contents in caches placed at the proximity of end users reduces server load, decreases network congestion, and lowers delay. In the context of 5G mobile networks, placing caches directly at the edge would be of great benefit for the network. Still, placing caches exclusively at the base stations is not the ultimate solution as the number of users that would use the cache would be too limited to really benefit from demand patterns. Instead, we advocate the usage of multistage caches with a rather small general-purpose least recently used (LRU) cache collocated in every base station to absorb retransmission events with high temporal locality demands (e.g., live streaming) and large caches spread over the different points of presence. The spreading of the caches allows to aggregate traffic of a large portion of users, therefore reducing the traffic in the core and consequently also the ISP connection charges of the MO. By looking at the reasonable backplane speeds in our 100 M 5G network, we deduce that it is economical to connect large caching servers behind the CGE switches each serving close to 1 million users. Cache hits would then reduce the number of required high cost mOFS and iOFS switches as well as reduce Internet connection charges of the mobile operator. Since a CGE switch is capable of decoding the 802.1ah, caching nodes can be easily connected to these switches directly.

The usage of SDN with the proposed integration facilitates the dynamic relocation of the cache based on the number of users. We created a pilot to demonstrate the effect on the network when moving the cache. We use HTTP live streaming for the video file and performed tests with the architecture presented in Figure 6.13.

We performed various tests with increasing number of users (up to 16 per eNB) to check the variability of the caching effectiveness based on number of users in different locations.

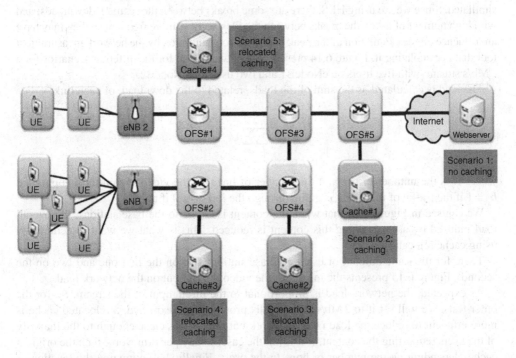

Figure 6.13   Architecture of the testbed with SDN integrated with LTE and caching.

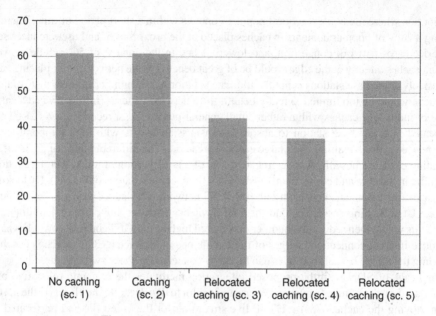

Figure 6.14 Load for different types of caching.

We simulated a user streaming a video by an HTTP download of the segments with a bandwidth limitation to 2 Mb/s and multiplying the bandwidth limitation by the number of users simulated. Since we are using HLS, there are some breaks between the segment downloads; and with high number of users, the breaks between the files become more frequent so they may have an influence on the calculation of the peak bandwidth consumption by the network measurement tool that is calculating it. Figure 6.14 presents the network load for the different scenarios for a 2 Mb/s stream, with five users on eNodeB1 and two users on eNodeB2.

The load is calculated as the sum of the loads, related to the download, of each link:

$$L = \sum_{j=1}^{N}\sum_{i=1}^{n} a_i b_{ij}$$

where N is the number of users, n the number of links, a_x the video stream throughput, and $b_{xy} = 1$ if the traffic of the user x is going through the link y or 0 if not.

We can see in Figure 6.14 that when the content is closer to the base stations, the overall load induced by users fetching this content is reduced. This is what we want to achieve by using cache relocation.

Then, for the same amount of users per base station (five on the first one and two on the second), Figure 6.15 presents the impact of the video throughput on the network load.

As expected, the network load is proportional to the throughput of the stream. So for the other tests, we will set it to 2 Mb/s. The results prove that, as expected, a relocated cache is more efficient to reduce the load of the network but only if it is close enough to the majority of the users requesting the content. If it is not the case, it may perform worse than the original cache, depending on the number of hops to the users. Finally, by optimizing the location of caching, consequent bandwidth can be saved.

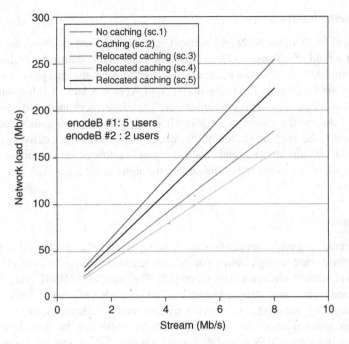

Figure 6.15   Network load depending on the video throughput.

6.7   Related Work and Research Questions

Some concepts for SDMNs were already discussed in previous publications [3–10]. One of the pioneering SDN proposals for wireless mobile networks was OpenRoads [6], an open architecture that can be deployed on campus-like environments to enable handovers between heterogeneous wireless networks. The SoftCell [3] and CellSDN [4] target at cellular core networks to improve the scalability and flexibility on both the data and the control plane by applying enhancement techniques including multidimensional aggregation of forwarding rules and caching packet classifiers and policy tags at local agents. The FluidNet [7] proposes a scalable and lightweight framework for cloud-based radio access networks, which improves both performance and resource usage by applying a set of algorithms to dynamically reconfigure the fronthaul. For the radio access dimension, OpenRadio [8] introduces a novel design for a wireless data plane with modular and declarative programming interfaces that offers the flexibility to implement protocol optimization on off-the-shelf wireless chips. The SoftRAN [9] focuses on the radio access network and proposes a software defined centralized control plane to abstract access resources as a virtual base station. The 3GPP also proposes the self-organizing networks (SON) [10] to enable the network self-configuration and self-optimization. Our work on 5G SDN takes the vision forward and advocates the necessity of integrating SDN to the upcoming 5G networks.

Some components of our SDMN vision were already covered without the mobile network context. Shirali-Shahreza et al. published a conceptually very similar OF-based approach for sampling that was motivated by security aspects and even demanded changes in the OF protocol [11].

6.7.1   Research Problems

While the lower levels of the SDN architectures have gained most attention, the SDN applications are still a field of research [12]. SDMN requires scalable controller architecture with a good northbound API to serve as a transparency layer between the data plane and the network applications. One weakness of existing northbound APIs is a lack of information about the state of the network devices at the controller side. Therefore, mechanisms to gain packet-level information, same as the existence of a northbound API, are undoubtedly parts of future networks. Finally, the implementation of the concept of secure service delivery and east–west interface in SDN would be beneficial. The latter would allow control communications cloud to cloud from socket to socket rather than going through the switches each time.

6.7.2   Impact

Provided the remaining problems can be solved, the separation of control and data planes has the potential to provide cost savings from capacity sharing and provide economies of scale from the virtualization of network elements in the cloud [12]. The usage of SDN will bring down the costs of acquiring and maintaining standard switches. The separation of control from data plane will lead to the usage of general-purpose switches without mobile dedicated solutions.

SDN brings new business models and opportunities with new business roles. One of the major business impacts of SDN in mobile networks is that current network equipment vendors are likely to change their role from "equipment vendor" to a software vendor. The vendor markets will be organized into horizontal layers. SDNM will also bring new possibilities: the logical evolution is that the mobile network operator (MNO) will drive the SDMN adoption as optimization of its current infrastructure. The adoption of SDN will lead the MNO to deploy or lease its own cloud to run its control plane functions, independently of network device vendors. The MNO will benefit from the potential cost reduction when using general-purpose and standardized hardware in both the user plane elements when using OpenFlow and Ethernet switches and in the control plane cloud platform.

Mobile operators need to closely cooperate with new entrants such as cloud providers (e.g., Amazon, Google to share premises, etc.) in order to provide a better customer experience.

For operating the networks, three principal business roles with distinct competences can be identified: (i) mobility management including frequency licenses and use, towers, base station sites, and understanding mobility patterns; (ii) providing connectivity between sites; and (iii) dealing with the end customers, providing them the services and the user experience they want. These roles map rather nicely into the breakdown of network functions to SDN applications in Figure 6.6. Once SDN is deployed, it becomes feasible to reshuffle the roles of present-day incumbent, mobile, mobile virtual operators, and content providers in such a way that efficient competition is ensured on the market.

6.8   Conclusions

We propose to use SDN in 5G mobile networks as the solution for needed scaling to the increased traffic demand and to the number of users and applications with acceptable cost and the necessary level of control. In this chapter, we conclude that for modeling the 5G as a software defined network, a group of SDN applications (e.g., Base Station, Backhaul, Mobility Management, Monitoring, Caching, Access, and Service Delivery App) is required. We also describe a set of use cases, namely, caching and firewalling, that provide evidence of feasibility of SDN in 5G networks. Scalability is analyzed in detail to ensure SDN can be deployed in mobile networks.

References

[1] NSN White Paper, "Technology Vision 2020, Technology Vision for the Gigabit Experience," 2013. Available at: http://networks.nokia.com/file/26156/technology-vision-2020-white-paper (accessed May 21, 2015).

[2] R. Kantola, "Customer Edge Switching." Available at: www.re2ee.org (accessed on February 18, 2015).

[3] X. Jin, L. E. Li, L. Vanbever, J. Rexford, "SoftCell: Taking Control of Cellular Core Networks," 2013. Available at: http://arxiv.org/abs/1305.3568 (accessed on February 18, 2015).

[4] L. E. Li, Z. M. Mao, J. Rexford, "Toward Software-Defined Cellular Networks," in Proceedings of the EWSDN, 2012. Software Defined Networking (EWSDN), 2012 European Workshop on Darmstadt, October 25–26, 2012, IEEE ISBN 978-1-4673-4554-5.

[5] A. Basta, W. Kellerer, M. Hoffmann, K. Hoffmann, E.-D. Schmidt, "A Virtual SDN-Enabled LTE EPC Architecture: A Case Study for S-/P-Gateways Functions," in Proceedings of the SDN4FNS, Trento, Italy, 2013.

[6] K-K Yap, R. Sherwood, M. Kobayashi, T-Y. Huang, M. Chan, N. Handigol, N. McKeown, G. Parulkar, "Blueprint for Introducing Innovation into Wireless Mobile Networks," in Proceedings of the ACM VISA, 2010.

[7] K. Sundaresan, M. Y. Arslan, S. Singh, S. Rangarajan, S. V. Krishnamurthy, "FluidNet: A Flexible Cloud-Based Radio Access Network for Small Cells," in Proceedings of ACM MobiCom, 2013.

[8] M. Bansal, J. Mehlman, S. Katti, P. Levis, "OpenRadio: A Programmable Wireless Dataplane," in Proceedings of the ACM HotSDN, 2012.

[9] A. Gudipati, D. Perry, L. E. Li, S. Katti, "SoftRAN: Software Defined Radio Access Network," in Proceedings of ACM HotSDN, 2013.

[10] 3GPP, Self-Organizing Networks (SON) Policy Network Resource Model (NRM) Integration Reference Point (IRP), 2013.

[11] Y. G. Sajad Shirali-Shahreza, Efficient Implementation of Security Applications in OpenFlow Controller with FleXam," in Proceedings of the IEEE 21st Annual Symposium on High-Performance Interconnects, 2013.

[12] Gartner Report, "Hype Cycle for Networking and Communications," 2013. Available at: https://www.gartner.com/doc/2560815/hype-cycle-networking-communications- (accessed May 21, 2015).

7

EPC in the Cloud

James Kempf and Kumar Balachandran
Ericsson Research, San Jose, CA, USA

7.1 Introduction

The architecture of the Internet is being fundamentally restructured by cloud computing and software defined networking (SDN) [1–3]. Cloud computing and SDN conceptually insert computation into the center of the network and separate control and execution in a way that decentralizes execution.[1] Control can then be flexibly exercised, often from a central entity running in a cloud data center. SDN and cloud find applicability in next-generation operator networks through network function virtualization (NFV) [4–6]. SDN for operator networks and NFV are currently under much discussion in the literature and in fora specifically dedicated to standardizing their interfaces [7, 8].

SDN and NFV have relevance for the mobile network operator too, specifically with respect to the Evolved Packet Core (EPC) [9]. Additionally, SDN and NFV can realize true convergence of various different kinds of core networks, transport backbones, and services; businesses can now flexibly deploy common services across different operator domains such as broadband Ethernet, mobile networks, and broadcast media networks. Such convergence will ideally enable services to transcend the complications of working across different security, identity, and mobility mechanisms, while the lower-level implementations of such functionality are themselves altered over time to a common framework. SDN and NFV have the potential to substantially change how operators deploy and manage mobile and fixed networks and, in addition, how operators develop and offer services to their users.

---

[1] This is meant to be normative rather than stricture.

Software Defined Mobile Networks (SDMN): Beyond LTE Network Architecture, First Edition.
Edited by Madhusanka Liyanage, Andrei Gurtov, and Mika Ylianttila.
© 2015 John Wiley & Sons, Ltd. Published 2015 by John Wiley & Sons, Ltd.

7.1.1 Origins and Evolution of SDN

SDN started with the introduction of OpenFlow 1.0 in 2009 [10]. OpenFlow initially featured a simple flow switch model where flows are a time sequence of packets identified by having a packet header that matches a pattern. Only unicast packets having Ethernet headers tagged with simple 802.1Q VLAN [11], IPv4, and TCP/UDP were supported. The latest version of OpenFlow 1.4, published in 2013, supports broadcast/multicast packets with headers having carrier Ethernet, Multiprotocol Label Switching (MPLS), IPv6, and a variety of transport protocols [12]. An extension mechanism supports experimental or vendor-specific extensions.

The OpenFlow architecture has not changed much over the course of its development. The control plane is centralized in a controller where flow forwarding on the user plane is programmed, unlike a standard IP-routed network. The controller sets next-hop forwarding on the switches using the OpenFlow protocol over a secure channel. The user plane could be implemented as a softswitch rather than in hardware. In data center applications, softswitches are the predominant way in which OpenFlow is deployed, as in, for example, the open virtual switch (OVS) softswitch [13]. Softswitches are not as common in networking applications outside the data center.

In an OpenFlow switch, packets enter through hardware switch ports into a pipeline consisting of a series of forwarding tables. Each table has conceptually six columns:

- A rules column, containing a pattern that matches fields in the packet header, the input port, and the metadata from previous tables. The header field patterns support exactly matching the pattern or a matching to a wildcard expression for which a part of the field can be anything.
- A priority column, which specifies the priority order of this pattern with respect to other patterns that match a header.
- A counters column, which indicates the OpenFlow counters that should be updated if the packet matches the pattern. The OpenFlow protocol supports messages for querying these counters to obtain statistics on flows running through the switch.
- An action column, which specifies the actions that must be executed should the packet header match the pattern. Examples of actions include rewriting the packet header in some fashion, forwarding the packet to an output port, or sending the packet to the next table in the pipeline for further processing. The packet can also be dropped, which is the default action if no pattern matches, or forwarded to the controller.[2]

The headers of incoming packets are matched against the match column, and if any patterns match, the actions associated with the top priority rule are collected. If the actions include sending the packet to an output port, then the actions are executed; otherwise, the packet is sent to the next table and the actions are executed when the packet exits the packet processing pipeline. An action can collect part of the header for metadata and pass that along through the pipeline to use in matching during the next phase. The switch design also includes group tables to support programming multicast and broadcast flows and meter tables to support quality of service (QoS) on flows.

---

[2] OpenFlow 1.0 by default sent the packet to the controller if no pattern matched. This led to some confusion initially about whether OpenFlow could handle large volume traffic flows, but the "first packet to the controller" meme was never really a fundamental part of the design and has since been deprecated. The ability to forward packets to the controller has been kept as an option however since it is extremely useful for routing control plane packets to the controller that are incoming from networks outside the OpenFlow domain.

While SDN started with OpenFlow, the concept has since broadened beyond a single protocol and switch design to include a variety of different systems and protocols. For example, the Contrail data center network virtualization system [14] uses XMPP[3] [15], a protocol developed for instant messaging, to program the user plane elements. In addition, centralization of the control plane has led to a merging of the control and management planes into a single controller at the architectural level [16], with control plane decisions having a characteristic time constant for decision making that is shorter than the management plane.

All SDN systems have the following two key principles in common:

- Separation of Control Plane from User Plane: The control plane is handled by a centralized controller, which then programs routes into the user plane with a protocol connecting the controller to the packet switching and routing elements. The controller performs forwarding and routing calculations for the network. The interface between the controller and the user plane elements is arbitrarily called the *southbound interface*, merely because it usually appears on the bottom of diagrams showing an SDN controller. Newer controllers, such as OpenDaylight [16], feature support for installing and deploying multiple control and management protocols on the southbound side.
- Abstraction of the User Plane Control into a Collection of Application Programming Interfaces (APIs) and Objects Exposed to Programmers Constructing Applications that Control Forwarding and Routing: An API can be in Java, Python, or a Representational State Transfer (REST) format [17] for remote procedure calls (RPCs). These APIs allow a programmer to control and manage the user plane while limiting the possibility of misconfigurations and errors by encapsulating correct behavior in abstractions. Some SDN protocols, for example, OpenFlow, also feature extensive support via an API for fine-grained measurement beyond that available in legacy network management protocols such as SNMP. The interface between the programmer and the SDN controller is called the *northbound interface* for obvious reasons.

These principles distinguish an SDN network from a traditional distributed IP routing control plane. In a traditional IP routing network, forwarding and routing calculations are done by the individual forwarding elements, and network management is controlled by vendor-specific command line interface scripts that don't hide the network complexity. In contrast, the abstractions presented by the SDN controller to the programmer are typically less complex and more consistent. This simplifies the job of constructing correct and understandable network management and control programs. Network management thus becomes a matter of program development against a standardized API.

The application to OpenFlow/SDN to mobile networks, and, in particular the EPC, is the subject of Section 7.3.

7.1.2 NFV and Its Application

NFV emerged out of a rethinking of how to deploy communication system applications involved in running an operator network, specifically on how to apply the advances in enterprise and Web services deployment technology over the past ten years to communication

---

[3] Extensible Messaging and Presence Protocol.

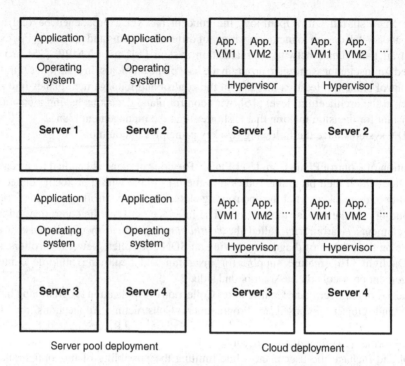

Figure 7.1   Comparison of the server pool versus cloud deployment patterns.

system applications. Many core network subsystems, for example, the Internet Multimedia Subsystem (IMS) [18] and the EPC, are implemented in software. The applications in these subsystems are deployed directly on servers, with only the operating system between the communication system application and the server (see the left-hand side of Figure 7.1). Server pools dedicated to particular functions can be configured to handle extra load, but this kind of overprovisioning has been found to be costly and wasteful of power in enterprise and Web applications. Server pools also scale poorly in enterprises, requiring an additional piece of hardware to be installed if more application capacity is necessary.[4] In addition, since the clock rate of processors reached an upper limit of around 3 GHz in the early 2000s, chip manufacturers have been increasing the number of processor cores available on a chip as a way to continue densifying the amount of processing power. Most operating systems are written to manage hardware threads, but they typically don't handle separate processing cores efficiently.

Enterprise and Web applications have begun to deploy on top of a virtualized platform in large data centers. The operating system and application are packaged as a software image called a *virtual machine* (VM).[5] The VM is elevated from being deployed directly on dedicated hardware to being deployed on a *hypervisor*, a software virtualization layer that manages the interaction with the hardware. The hypervisor has been written to efficiently handle processors

---

[4] In telecommunication networks, it is difficult to say that this difference is very big, but a network operator can create services that naturally balance the load over time.

[5] They may also be deployed in special operating system processes in VMs called containers that have extra protection to ensure isolation between applications.

with multiple hardware cores. The result is a clean separation of concerns between the physical hardware and the VM. This kind of deployment paradigm is generically called *cloud computing* and is outlined on the right side of Figure 7.1. Virtualization has been applied to networking and storage resources in addition to computation. Here as well, a software virtualization layer is interposed between applications and the actual physical resources, allowing sharing of the physical resources between multiple users.

In enterprise cloud deployments, increased hardware utilization efficiency is achieved by *oversubscription*, scheduling more VMs to a server than cores available to execute them. This ensures that the server is kept busy more than 80–90% of the time, rather than less than 30% as is typical of nonvirtualized server deployments. In public cloud (utility computing) deployments, such as Amazon Web Services (AWS) [19], oversubscription is less common because it would put the public cloud operator at risk of violating the service-level agreement toward customers. Public data centers also support *multitenancy*, where multiple individuals and organizations share computational resources with isolation enforced between the different tenants. Isolation ensures that each tenant sees a slice of the compute, storage, and networking resource that they have contracted for and that their slice incurs no interference from other tenants. Cloud operating systems such as OpenStack [20] manage the deployment of multitenant virtualized compute/ storage/network infrastructure at a high level, enforcing isolation between the tenants.

Deployment of the application is managed by an *orchestration* system that monitors the load on the running application VMs. If the orchestration system detects excessive loading by an application, it can *scale out* the application by starting up new application VMs to handle the additional load.[6] Idle VMs can similarly be deactivated. A server that is not running any VMs can operate in low power mode or even be powered down, saving operating cost and reducing carbon footprint.

Additionally, the orchestration system can arrange for a pool of application VMs to handle traffic forwarded through a front-end *load balancer*. The load balancer sprays user traffic packets to all the active VMs, reducing the load on any one. Load balancing is facilitated by constructing applications so that they are *stateless*, with all the user state held in a back-end database whose data consistency is protected by transactions. The only state held in the application is state that can easily be reconstructed by a short user interaction, such as shopping cart contents. These kinds of applications are often called *three-tier applications*. The client interaction software, typically in a browser, is the first tier, the stateless application containing the business logic is the second tier, and the database forms the third tier. The load balancer is seen as part of the routing infrastructure.

The NFV manifesto [4, 5] advocates the cloud computing paradigm for communication system applications. The following technical and business benefits were identified:

- A reduction in capital expenditure for specialized hardware. Subsequent study showed the benefits to be minimal since many communication system applications were already running on standardized IT components.

---

[6] In contrast, traditional server-based scale-up applications require the enterprise or Web operator to purchase and install a larger server when more capacity is required, a costly and time consuming proposition. Telecommunications applications deployed using the traditional server pool architecture use a scale-up paradigm to handle load and for redundancy.

- A reduction in operating expenditure and carbon footprint by more efficiently utilizing hardware to reduce power consumption and by enabling use of the underlying hardware and software platform for multiple applications.
- Faster innovation in service development and deployment and the ability to target services more narrowly at particular geographic areas and customer demographics without incurring multi-year-long development cycles.
- Ability for multiple tenants to utilize the software and hardware platform, so a variety of applications could be deployed on the same infrastructure.
- Opening up the infrastructure procurement process in operators to new entrants, both commercial and nonprofit (such as academics), since the barrier of entry becomes lower due to the use of software rather than hardware.

In Sections 7.2 and 7.3, we discuss how an NFV deployment of the EPC might evolve.

7.1.3   SDN and Cross-Domain Service Development

Most network operators generate the bulk of their revenue from services purchased by enterprise and individual customers, for example, enterprise VPNs or wireless plans. Routing and forwarding are simply a means toward providing the connectivity needed to enable those services. To achieve the level of innovation sought by the NFV manifesto authors, service development and deployment need to be simplified and integrated much more tightly across the operator's network than is currently the case. Functions not specifically involved in the immediate delivery of packets such as identity management, policy management, and charging and billing need to become as available as services like routing and forwarding are for transport SDN. Access to functionality from different domains (cloud, fixed WAN, and mobile core) needs to be simplified. While routing and forwarding have received almost all of the attention in the SDN community, the topic of how to achieve a simplified platform for service development and deployment has been mostly ignored.

This aspect of managing an operator's network is the domain of OSS/BSS systems.[7] Today, such systems typically require extensive human intervention. A customer service representative takes a call for a service and starts an order flow that may require a technician to drive to the customer site and install a piece of equipment or change a switch setting. Once the technician has accomplished the task, the customer service representative must notify the customer that their service is ready. Troubleshooting may be required to resolve issues. Provisioning a service such as an enterprise VPN[8] can often take weeks or months. Accessing the business logic involved in charging and billing systems is often complex and differs depending on the particular domain of the network (WAN,[9] mobile core, and cloud), further complicating service development.

Ericsson's service provider SDN (SP-SDN) [21] is an approach to rapid and flexible cross-domain service creation that complements SDN and NFV. The SP-SDN features Web service APIs crafted with abstractions representing objects and operations involved in service creation,

---

[7] OSS, operations support system; BSS, business support system.
[8] Virtual private network.
[9] Wide area network.

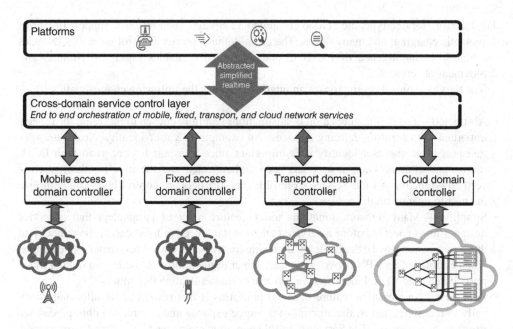

Figure 7.2   SP-SDN architectural concept.

deployment, and management, just like the SDN controller provides for routing and forwarding. These APIs expose network functionality at the service layer rather than the transport layer and in many cases can be based on functionality provided by SDN or NFV if it is available. But given the existence of a large installed base and legacy software controlling it, SP-SDN also offers the potential to simplify network service creation based on existing legacy equipment and software.

Figure 7.2 illustrates the SP-SDN architectural concept. The different operational domains in the operator network are located at the bottom: data center, wide area networking (IP and transport), and mobile and fixed access. Each of these has a set of transport control functions involving networking that are reflected up through their respective transport controller as a collection of APIs for configuring and managing transport. Ideally, the transport control and management functions will be virtualized, and the APIs will reflect a collection of useful abstractions, but the Service Control layer can work with legacy networks as well. In addition, the data center includes APIs for controlling VM execution and placement, for example, using a cloud operating system such as OpenStack.

The cross-domain Service Control layer spans across the transport domains. Each transport domain is likely to have its own transport controller, reflecting the specific technical and administrative aspects of controlling that domain. For example, the transport controller for the mobile domain will reflect the need for mobility. Similarly, the data center controller will need to coordinate allocation and deployment of networking and computing resources. Some domains might share a controller if the demands of control are similar enough. In addition, the Service Control layer encompasses other control aspects of network services that do not involve transport: access control, for example, the radio scheduler, the analytics related to location and routes, etc.; policy control, for example, QoS settings based on the time of the day, location, service type, etc.; cloud compute and storage, such business support functions as dynamic charging; and many others. The cross-domain Service Control layer thus includes the high-level intelligence of the EPC and additionally enables rapid provisioning and deployment of services.

The Service Control layer exposes an interface that has the following characteristics:

- Abstracted—The API features a set of abstractions carefully chosen to represent the objects important for operators defining services. An example is a user identity. At the network transport layer, the user identity isn't important since transport is concerned with flows, routes, and circuits. Once a user has been authenticated and the user's authorization is verified—functions of the Service Control layer—the network control layer can authorize the user to operate on the network.
- Simplified—Many network functions today feature a lot of parameters that a service designer must specify before a service can be instantiated. In most cases, the majority of these parameters may be duplicates or derivable from a service-level parameter. For example, when establishing a VPN between the data center domain through the IP and optical transport domain, the VLAN identifier for a single customer is often the same.
- Real time—An attractive feature of cloud platforms is that resources are allocated elastically in real time; that is, the amount of resource expands and contracts within predefined limits to meet demand. The Service Control layer allows services to be defined, provisioned, and deployed through a customer portal in a few minutes, rather than taking hours, days, or even months of time.

The APIs that the Service Control layer exposes to clients are not the traditional protocol APIs from previous generation network architectures nor are they the command line interface APIs exposed by network equipment, but rather Web APIs, for example, REST APIs. In addition, Secure Sockets Layer (SSL) security and Web-based authentication can easily be added to an interface if security is necessary. Many tools exist for conveniently programming Web APIs in commonly used languages such as Java and Python.

The actual content of the Service Control layer APIs—the abstractions, objects, and operations that are exposed—will evolve from specific use cases. As commonalities between use cases become better understood, abstractions will emerge in the same manner as in software engineering, where, for example, function calls developed out of a need to abstract common operations into parameterized chunks of code. Abstraction has considerably simplified reasoning about and developing and deploying software.

SP-SDN is an application of the service-oriented architecture (SOA) [22] concept from the enterprise and Web worlds to communication services. In a system designed according to SOA, discrete software components provide functionality to other components as services. These components are distinguished by having well-defined interfaces so that the services can be deployed on any platform. The interfaces do not allow programmers access to the internal implementation, allowing the implementation to be changed for optimization purposes. The interfaces are typically implemented as RPCs using the REST or SOAP[10] [23] HTTP format. Architecting a system according to SOA principles allows system

---

[10] Simple object access protocol.

components to be combined in arbitrary ways, exactly the kind of flexibility that is needed for the speedy innovation that NFV advocates.

Section 7.4 describes application of the SOA principle to the EPC.

7.2   EPC in the Cloud Version 1.0

The initial version of a virtualized EPC will follow the same pattern as with enterprise cloud software deployments: the existing applications will be lifted up, packaged in VMs, and deployed on a virtualized platform with their functionally exposed through HTTP APIs. Both mobile control plane applications such as the Mobility Management Entity (MME), the Home Subscriber Server (HSS), and the Policy and Charging Rules Function (PCRF) as well as the Serving/Packet Gateways (S/P-GWs), which have both a control plane and a user plane routing function, will deploy applications in VMs. Figure 7.3a contains a high-level schematic of how such a deployment pattern looks.

In this type of deployment pattern, both the control and user plane flows run through a data center.

Given the existing state of cloud operating system and orchestration software, represented by OpenStack, there are a variety of technical problems involved in deploying the EPC on a cloud platform [24]. Two in particular stand out:

* The P-GW manages session state in the Packet Data Protocol (PDP) context [25]. Unlike enterprise 3-tier applications, the PDP context can't realistically be managed through a simple database since it must be constantly consulted when managing packet data flows. It also can't be quickly reconstructed from a user session if the P-GW fails. Thus, the auto-scaling and reliability pattern for managing 3-tier applications won't carry over to managing MME, S-GW, and other EPC entities like the PCRF and the HSS[11] where substantial system state must be quickly available.
* The virtual networking capabilities of cloud operating systems are rudimentary and not capable of supporting the demands of challenging EPC applications. The enhanced bearer capability of the EPC requires the network to handle QoS properly. Cloud operating system network virtualization usually only handles best effort traffic. Connectivity options into cloud data centers are also relatively limited, confined primarily to best effort Internet service or occasionally to enterprise VPN service, and setup has long lead times. In the worst case, the EPC might need to manage multiple BGP autonomous systems (ASes) in the cloud if the operator supports mobile virtual network operators (MVNOs).

These problems are already receiving attention in the OpenStack consortium as NFV deployments begin to roll out. The existing telecommunications style mechanisms for handling EPC redundancy and scalability such as OpenSAF [26] can probably address the first point in the short run, but more sophisticated distributed system techniques for managing state will be required to support the more flexible redundancy and scalability potential of cloud deployments. The second point is already being addressed by some

---

[11] Home Subscriber Server.

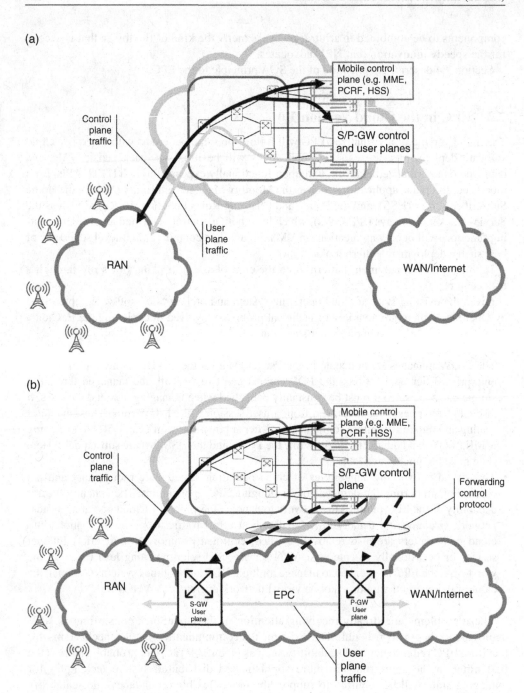

Figure 7.3 Different EPC cloud deployment patterns. (a) Control and user plane in cloud. (b) Control plane only in cloud.

ongoing work in OpenStack on QoS for virtual networks. However, the more complex challenges of managing a multi-AS virtual cloud have yet to be addressed.

An issue that stands out with respect to the virtualized EPC is the performance of software forwarding on standard server hardware.[12] While the control plane entities logically don't require high-performance forwarding, the user plane entities such as the S/P-GWs do. The canonical software forwarding entity in most open source cloud deployments is the OVS [27]. OVS provides OpenFlow support as well as support for the ability to configure IP tunnels, one of several ways that the cloud operating system virtualizes the physical network, using the OVSDB protocol [28], a database management protocol specifically designed for configuring OVS. Performance of OVS even when optimized is quite variable with packet size. OVS can almost achieve 10G line rate performance for 1024-byte packets; however, the performance for 64-byte packets struggles to reach 1G [29]. Optimizations can improve forwarding above OVS performance [30]. Recent work indicates that an optimized version of Intel's Data Plane Development Kit (DPDK) [31], a collection of specialized libraries designed specifically for accelerating user plane applications and recent work, can easily achieve line speed on a 10G network interface card (NIC) [32]. However, mere achievement of high line speeds is not sufficient. Switching issues must still be addressed by OVS [33].

7.3   EPC in the Cloud Version 2.0?

The next step would go beyond simply moving the existing EPC network functions to a cloud by moving the EPC onto an SDN substrate, where control and user plane are completely separated as in Figure 7.3b. In this case, the control plane flows run into the control plane entities in the cloud, but the user plane flows run through dedicated switching hardware controlled by a protocol from the control plane entities. Certain use cases become easier to support if the control plane is SDN based. And as discussed in the previous section, the performance of forwarding on standardized Intel hardware may ultimately become enough of an issue to again recommend hardware specialized for forwarding. User plane devices located remotely may be easier to deploy than a data center. One possible implementation strategy could involve extending OpenFlow to handle routing of GTP[13] tunneled flows [34], which we will use as an example for purposes of discussion. Whether or not the EPC in the Cloud Version 2.0 is deployed depends on how quickly the performance of software-based switching becomes unacceptable and how important the use cases become that are difficult to implement using the centralized scheme. Here, we discuss a single use case, UE multihoming. Further use cases can be found in Ref. [34].

7.3.1   UE Multihoming

Support for multihomed UEs can add complexity, especially in mobile networks [35]. The problem is that the IP address acts as both a routing locator and an endpoint identifier.

---

[12] Indeed, the first routers deployed in the 1980s and early 1990s were implemented in software on minicomputers such as the VAX, and later on Sun Microsystems workstations. It was only later as the amount of traffic increased beyond what could be handled by software at line speed that routers were built with dedicated hardware.
[13] GPRS Tunneling Protocol.

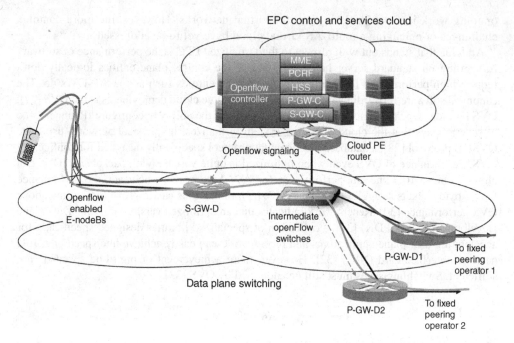

Figure 7.4 Upstream multihoming with OpenFlow. Kempf et al. (2012) [34]. Reproduced with permission from IEEE. © IEEE.

Unless the UE has two addresses, it is difficult in a standard distributed control plane IP-routed network to have a UE be multihomed. Aside from the issue of IPv4 address scarcity, the EPC uses the IP address as an endpoint identifier in GTP. With OpenFlow, however, multihoming is considerably simpler. OpenFlow treats IP addresses as pure endpoint identifiers, discarding the routing topology internally. Forwarding is done according to the programmed OpenFlow rules and not based on the IP network longest prefix matching. As a consequence, the EPC can advertise different sets of subnet prefixes externally and have different kinds of traffic directed to different gateways. Alternative schemes for supporting upstream multihoming sometimes don't provide adequate provision for operator accounting and charging. The mobile network PCRF needs to be involved in all decision making about where to place flows. OpenFlow even provides a collection of statistics that can be utilized for accounting and charging purposes.

As Figure 7.4 illustrates, the upper flow goes through upstream provider 1, while the lower flow goes through upstream provider 2. The OpenFlow controller and PCRF set up GTP tunnels to the P-GWs user plane with different IP addresses for the mobile UE externally, and these are rewritten at the P-GW to point to the same IP address on the mobile UE. The mapping is handled based on the application. The same technique could be used for handling multiple wireless interfaces. Techniques that do not use SDN for supporting multiple wireless interfaces in the network are quite complex [36, 37], while techniques requiring changes on the end nodes run up against the diversity of end nodes in today's mobile networks.

7.3.2   The EPC on SDN: OpenFlow Example

Using OpenFlow as an example control/user plane separation protocol, we can see how the EPC could be redesigned on an SDN platform. The following subsections describe the modifications needed in the OpenFlow switch architecture to support GTP TEID routing and the changes in the EPC architecture that an OpenFlow substrate would enable.

7.3.2.1   Switch Architectural Modifications

OpenFlow 1.4 supports an n-tuple of header match fields in the flow table, with one additional field matching metadata for communication of data collected by matches between tables. Unlike earlier versions of OpenFlow, the header field tuple size is not fixed because multiple MPLS labels can be pushed onto the header. The user plane protocols that are supported are Ethernet including carrier Ethernet (802.1aq) [38], MPLS, IPv4 and IPv6, and the IP L4 protocols (TCP, UDP, SCTP, DCCP, ARP, and ICMP). GTP TEID routing extends the tuple with two additional fields: the 2-byte GTP header flag field and the 4-byte GTP TEID field. The header flag fields are used to distinguish between GTP-U packets that are subject to fast path OpenFlow GTP TEID routing and other types of GTP packets including some GTP-U packets that need to be handled by the slow path on the switch.

In addition to the flow table extension, GTP TEID routing requires the addition of virtual ports to support encapsulation and decapsulation [39]. A *virtual port* is an abstraction that handles complex header manipulation specific to particular protocols. Virtual ports are particularly useful for tunneling protocols because they hide the complexities of the tunnel header manipulations from the forwarding pipeline implementation. On input, a virtual port accepts packets from a physical port or another virtual port; processes them to add, remove, or modify a tunnel header; and then passes the packets along to the next virtual port or inserts them into the flow table classifier pipeline. On output, virtual ports become the target for forwarding rules exactly as for physical ports, and they similarly add, remove, or modify a tunnel header, and then pass the packet along to another virtual port or to a physical port for output.

Virtual ports for processing GTP-U tunnel packets are needed on the Serving Gateway, the PDN Gateway, and on the wired network interfaces of the eNodeB.[14] GTP virtual ports are configured from the OpenFlow controller using a configuration protocol. The details of the configuration protocol are switch dependent; for an example of a standardized configuration protocol, see Ref. [40]. The configuration protocol must support messages that perform the following functions:

- Allow the controller to query for and return an indication whether the switch supports GTP fast path virtual ports and what virtual port numbers are used for fast path and slow path GTP-U processing.
- Allow the controller to instantiate a GTP-U fast path virtual port within a switch data path for use in the OpenFlow table *Set-Output-Port* action, and bind a GTP-U virtual port to a physical port.

---

[14] Scheduling constraints for the wireless interface could also be incorporated into the scheme in some fashion.

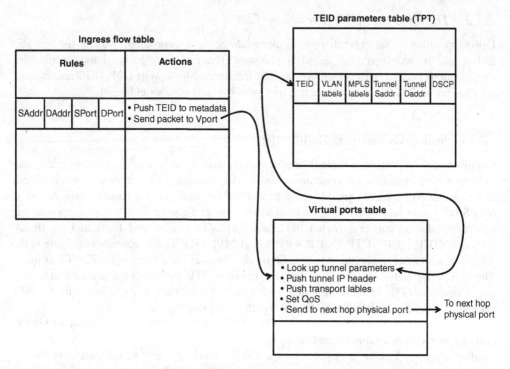

Figure 7.5   Tunnel ingress GTP OpenFlow gateway architecture.

The controller instantiates an encapsulation virtual port for each physical port on the eNodeB wired interface and on the interfaces of the S/P-GWs that may forward packets received from outside the EPC (i.e., from the UE or the Internet) to inside, and a decapsulation port that may forward packets from inside the EPC to outside.

Figure 7.5 contains a diagram of the switch architecture for the tunnel ingress side of the GTP OpenFlow gateway. An OpenFlow 1.4 GTP encapsulation gateway maintains a hash table mapping GTP TEIDs into the tunnel header fields for their bearers, called the TEID parameters table (TPT). The hash table stores the TEID, VLAN tags for the tunnel (if any) and MPLS labels for the tunnel (if any), the tunnel source and destination IP addresses, and any DSCP markings for QoS. The TEID hash keys are calculated using a suitable hash algorithm with low collision frequency.

The table maintains one such row for each GTP TEID/bearer originating on the gateway. The TEID field contains the GTP TEID for the tunnel. The VLAN tags and MPLS labels, if used by the EPC transport network, are ordered within the corresponding label fields and define transport network tunnels into which the packet can be routed. The labels also include the VLAN priority bits and MPLS traffic class bits. The tunnel origin source IP address contains the address on the encapsulating gateway to which any control traffic involving the tunnel should be directed (e.g., error indications). The tunnel end destination IP address field contains the IP address of the gateway to which the tunneled packet should be routed, where the packet will be decapsulated and removed from the GTP tunnel. The QoS DSCP field contains the DiffServ code point, if any, for the bearer. This field may be empty if the bearer is a default bearer with best effort QoS but will contain nonzero values if the bearer QoS is more than best effort. An OpenFlow GTP gateway also supports three slow path software ports for GTP traffic that is not handled by GTP fast path routing. Slow path forwarding is handled by the switch control plane software.

Tunnels are managed by the OpenFlow controller in the following way. In response to a GTP-C control packet requesting that a tunnel be set up, the OpenFlow controller programs a gateway switch on the tunnel ingress side to install rules and actions in the flow table and TPT entries for routing packets into GTP tunnels via a fast path GTP encapsulation virtual port. The rules match the packet filter for the input side of GTP tunnel's bearer. Typically, this will be a 4-tuple of IP source address, IP destination address, UDP/TCP/SCTP source port, and UDP/TCP/SCTP destination port. The IP source address and destination address are typically the addresses for user plane traffic, that is, a UE or Internet service with which a UE is transacting, and similarly with the port numbers. An action is installed in the flow table to forward the packet to a virtual port bound to the next-hop physical port and to write the tunnel's GTP TEID directly into the metadata.

When a packet header matches the packet filter fields in a GTP TEID routing, the GTP TEID is written into the lower 32 bits of the metadata and the packet is directed to the virtual port. The virtual port calculates the hash of the TEID and looks up the tunnel header information in the TPT. The virtual port then constructs a GTP tunnel header and encapsulates the packet. Any VLAN tags or MPLS labels are pushed onto the packet to ensure proper transport routing, and any DSCP bits or VLAN priority bits are set in the IP or MAC tunnel headers to ensure proper QoS. The encapsulated packet is then forwarded out the bound physical port.

On the egress side of the GTP tunnel, the OpenFlow controller installs rules and actions for routing GTP encapsulated packets out of GTP tunnels. The rules match the GTP header flags and the GTP tunnel endpoint IP address for the packet as follows:

- The IP destination address is the IP address of GTP tunnel termination that is on the switch.
- The IP protocol type is UDP (17).
- The UDP destination port is the GTP-U destination port (2152).
- The GTP header fields match a GTP-U packet without any extension headers.

If the rules match, the action is to forward to the virtual port. The virtual port simply removes the GTP tunnel header and any transport labels and forwards the user plane payload out the bound physical port.

Figure 7.6 contains a diagram of the switch architecture for the tunnel egress GTP OpenFlow gateway. Both gateway and nongateway switches in the EPC can also utilize GTP TEID routing to route packets in individual GTP tunnels to a specific destination. A flow table rule matches on the TEID for the tunnel in question, and the action forwards the tunneled packet out the next-hop port.

7.3.2.2   Architectural Modifications to the EPC on an OpenFlow SDN Substrate

GTP TEID routing allows the GTP control plane part of the S/P-GWs to be, in effect, separated from the gateway and situated in the OpenFlow controller as an OpenFlow application. Note that in almost all cases, the user plane packets do not need to be redirected to the controller. Flow routes are computed with policy applied by the PCRF, which itself may be

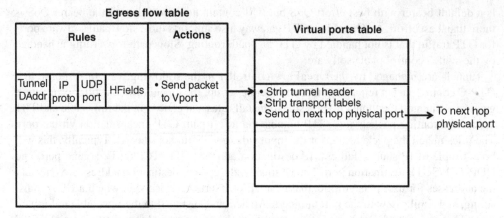

Figure 7.6 Tunnel egress GTP OpenFlow gateway architecture.

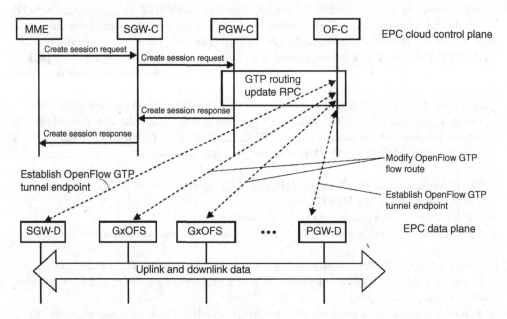

Figure 7.7 GTP *Create_Session_*Request on OpenFlow. Kempf et al. (2012) [34]. Reproduced with permission from IEEE. © IEEE.

structured on top of an SDN platform [41]. We present an example of how EPC control operations can be modified by this capability. In the example, we assume that communication between the gateway GTP control plane entities and the OpenFlow controller, both running in the cloud, happen via RPC.

The example is the creation of a new bearer and associated GTP tunnels. Bearers and GTP tunnels are set up using the GTP-C *Create_Session_Request* message. This procedure is used in a variety of message sequences, for example, in the E-UTRAN *Initial_Attach* procedure described in Section 5.3.2.1 of Ref. [9]. In Figure 7.7, the OpenFlow message flows for the *Create_Session_Request* message are shown. In the figure and following discussion, OF-C is the OpenFlow controller, SGW-C is the Serving Gateway control plane entity, PGW-C is the PDN Gateway control plane entity, SGW-D is the Serving Gateway GTP enhanced OpenFlow switch, PGW-D is the PDN Gateway GTP enhanced OpenFlow switch, and GxOFS is a nongateway GTP enhanced OpenFlow switch.

The MME sends a *Create_Session_Request* to the SGW-C, and the SGW-C sends the request to the PGW-C. The PGW-C calls into the OF-C through a *GTP_Routing_Update* RPC, requesting the OF-C to establish a new GTP tunnel endpoint at the SGW-D and PGW-D and to install routes for the new GTP bearer/tunnel on intermediate switches, if necessary.

The OF-C issues a sequence of OpenFlow messages to the appropriate GTP enhanced OpenFlow switches on the user plane. The sequence begins with an *OFP_BARRIER_REQUEST* to ensure that there are no pending messages that might influence processing of the following messages. Then, an *OFPT_FLOW_MOD* message is issued, with a GTP extension to the match field. The message specifies actions and instructions to establish a flow route for the GTP tunnel that encapsulates and decapsulates the packets through the appropriate virtual port. In addition, immediately following the *OFPT_FLOW_MOD* message, the OF-C issues an OpenFlow vendor extension message to the gateways containing the TPT entries for the encapsulation virtual port. The two OpenFlow messages are followed by an *OFPT_BARRIER_REQUEST* message to force the gateways to process the flow route and TEID hash table update before proceeding.

Prior to returning from the *GTP_Routing_Update* RPC, the OF-C also issues GTP flow routing updates to any GTP extended OpenFlow switches (GxOFSs) that need to be involved in customized GTP flow routing. The messages in these updates consist of an *OFP_BARRIER_REQUEST* followed by an *OFPT_FLOW_MOD* message containing the GTP match extension for the new GTP flow the actions and instructions described in Section 2.2.1 for customized GTP flow routing. A final *OFP_BARRIER_REQUEST* forces the switch to process the change before responding. The flow routes on any GxOFSs are installed after installing the GTP tunnel endpoint route on the SGW-D and prior to installing the GTP tunnel endpoint route on the PGW-D, as illustrated in the figure. The OF-C does not respond to the PGW-C RPC until all flow routing updates have been accomplished.

Once the RPCs have returned, the PGW-C and SGW-C return *Create_Session_Response* messages. When the MME receives such a response, it can signal the eNodeB with the *Initial_Context_Setup_Request/Attach_Accept* message indicating that the terminal is free to start using the bearer/tunnel.

Similar results can be achieved for other GTP-C operations.

7.4 Incorporating Mobile Services into Cross-Domain Orchestration with SP-SDN

In order to support integrated, cross-domain service development, the EPC services need to be encapsulated by APIs that represent usable abstractions to allow service programmers access. Here, we discuss an example of how an API on the PCRF can enable flexible creation and control of cross-domain services. The specific example is enabling on-demand, policy-driven QoS for mobile cloud services [21].

3G and 4G mobile networks provide the ability to control mobile service QoS through the PCRF. The PCRF communicates with the Subscriber Policy Repository (SPR), which contains the subscription records for users, to authenticate and authorize services, and the S/P-GWs to implement routing policy, such as marking packets with DiffServ code points for enhanced

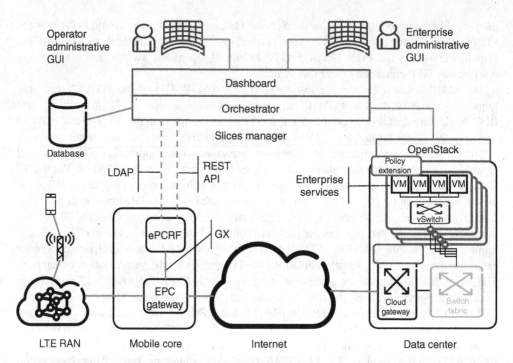

Figure 7.8   On-demand policy-driven QoS architecture for end-to-end network slices.

QoS [42]. It is possible to implement the PCRF according to SOC design principles in an enhanced function that we denote as the enhanced PCRF (ePCRF). The ePCRF would be structured as a service with a REST API to provide control access for QoS policy to a cross-domain controller. One desirable capability is the ability to relate policy control to charging. Such a capability would realize a flexible approach to policy and charging for mobile networks, and it can be combined with other services. Initial investigations in this direction have been demonstrated [43].

In the cloud domain, an OpenStack cloud gateway allows the cross-domain controller to configure notification of mobile service access when mobile devices connect to cloud services. Figure 7.8 illustrates the on-demand policy-driven QoS architecture. An orchestration layer implements the cross-domain Service Control function and provides user interface (UI) access to the network operator and enterprise customers with enterprise accounts in the cloud and the mobile network. The cloud data center supplies network, compute, and storage resources for enterprise services, including mobile cloud services such as streaming video, which may need enhanced QoS in the mobile network. The cloud data center is connected to the wide area network through a cloud gateway, essentially a software router in a VM, which includes the mobile service access notification enhancement.

In what follows, we provide some ideas on the possible enhancements to the PCRF; these directions are however by no means agreed on. The ePCRF REST API would expose abstractions for the following objects in the mobile network:

• The capabilities available for a particular subscriber: Depending on their subscription plan, a subscriber and service may have different rights to QoS.

- A mobile session identifier to determine which session is being controlled: Sessions are associated with subscribers and their rights to enhanced QoS treatment are determined by the subscriber profile.
- A packet flow within a session that may be entitled to enhanced QoS treatment: Packet flows are identified by a maximum requested bandwidth, a priority, a resource reservation, and the GTP bearer identification 5-tuple—the source IP address, the source port, the destination IP address, the destination port, and the protocol.

The REST API can further support operations to get the capabilities associated with a subscriber, to create and delete a session, and to obtain session status. It also supports a full set of operations on flows: add and remove a flow, update a flow, and get the status of a flow. As an example of mobile service orchestration, consider a use case where an enterprise has an account with a mobile operator and an OpenStack cloud service provider (the cloud and mobile network could be under the control of the same organization or different organizations, e.g., a public cloud). The mobile operator sets up an account for the enterprise through the mobile operator GUI, and the Orchestrator utilizes the OpenStack Keystone service to obtain tokens from OpenStack for configuring cloud resources. These are stored in the Orchestration database. The enterprise administrator adds mobile services entitled to enhanced QoS through the enterprise administrator's GUI. The Orchestrator obtains the global IP address of these services from the OpenStack Nova service. The service IP address along with the name and other information are stored in the Orchestration database. The enterprise administrator also adds subscribers entitled to obtain enhanced QoS for particular services. The subscribers' International Mobile Subscriber Identity (IMSI) and policy profile are obtained from the ePCRF and recorded. When a subscriber starts a device, the ePCRF reports the IP address of the device, identified by the IMSI, to the Orchestrator. When a device entitled to enhanced QoS service accesses a mobile cloud service, the mobile cloud service access notifier in the OpenStack cloud gateway triggers and reports the source and destination IP addresses to the Orchestrator. The Orchestrator issues an *updateFlow* message to the ePCRF to update the QoS privileges on the flow, and the flow is moved to an enhanced 3GPP bearer.

7.5 Summary and Conclusions

The next 10 years are likely to see more changes in the mobile packet core than in past ten years. The LTE packet core is in many ways an evolution of the original GPRS service that was provided with GSM and was introduced in 2000 [44]. The pressures of the expected media traffic volume are likely to require higher volume forwarding performance. The operator requirements for simplified cross-domain service construction to support rapid innovation on the same scale as Internet service providers are likely to drive a restructuring of the EPC software into services. Yet some aspects of the EPC are unlikely to change much if at all. Although using SDN may remove the need for having a physical mobility anchor where the IP address has both locator and identifier function, GTP provides a virtualization solution for mobile networks similar to VXLAN[15] [45] and GRE[16] [46] in fixed networks and therefore is likely to remain if in somewhat modified form. Mobile operators will continue to want the

---

[15] Virtual extensible LAN.
[16] Generic routing encapsulation.

ability to flexibly provide different QoS to different traffic types and to charge users for the service in a flexible manner, so some version of the PCRF and HSS is likely to remain. And of course the unparalleled ability of the EPC to efficiently support multiple radio types is required. These aspects of the EPC have been very successful and are worth preserving. In this chapter, we've discussed a few options about how the EPC might evolve, but the real contours of its shape are as yet unclear.

References

[1] J. Kempf, P. Nikander, and H. Green, "Innovation and the Next Generation Internet," Infocom IEEE Workshops on Computer Communications Workshops, March 2010.

[2] N. McKeown, "Software Defined Networking," Keynote talk, INFOCOM, April 2009. Available at: http://www.cs.rutgers.edu/~badri/552dir/papers/intro/nick09.pdf (accessed on February 18, 2015).

[3] P. Hui and T. Koponen, "Report on the 2012 Dagstuhl Seminar on Software Defined Networking," September 2012. Available at: http://vesta.informatik.rwth-aachen.de/opus/volltexte/2013/3789/pdf/dagrep_v002_i009_p095_s12363.pdf (accessed on February 18, 2015).

[4] M. Chiosi, D. Clarke, P. Willis, A. Reid, J. Feger, M. Bugenhagen, W. Khan, M. Fargano, C. Cui, H. Deng, J. Benitez, U. Michel, H. Damker, K. Ogaki, T. Matsuzaki, M. Fukui, K. Shimano, D. Delisle, Q. Loudier, C. Kolias, I. Guardini, E. Demaria, R. Minerva, A. Manzalini, D. López, F. Salguero, F. Ruhl, P. Sen, "Network Functions Virtualization," Position paper presented at SDN and OpenFlow World Congress, 2012, 16 pp. Available at: http://portal.etsi.org/NFV/NFV_White_Paper.pdf (accessed on February 18, 2015).

[5] M. Chiosi, S. Wright, D. Clarke, P. Willis, C. Donley, L. Johnson, M. Bugenhagen, J. Feger, W. Khan, C. Cui, H. Deng, C. Chen, L. Baohua, S. Zhenqiang, X. Zhou, C. Jia, J. Benitez, U. Michel, K. Martiny, T. Nakamura, A. Khan, J. Marques, K. Ogaki, T. Matsuzaki, K. Ok, E. Paik, K. Shimano, K. Obana, B. Chatras, C. Kolias, J. Carapinha, DK Lee, K. Kim, S. Matsushima, F. Feisullin, M. Brunner, E. Demaria, A. Pinnola, D. López, F. Salguero, P. Waldemar, P. Grønsund, G. Millstein, F. Ruhl, P. Sen, A. Malis, S. Sabater, A. Neal, "Network Functions Virtualization 2," Position paper presented at SDN and OpenFlow World Congress, 2013, 16pp. Available at: http://portal.etsi.org/nfv/nfv_white_paper2.pdf (accessed on February 18, 2015).

[6] ETSI, "NFV," 2013. Available at: http://portal.etsi.org/portal/server.pt/community/NFV/367 (accessed on February 18, 2015).

[7] Open Network Foundation, "Software Defined Networking Definition," September 2013. Available at: https://www.opennetworking.org/sdn-resources/sdn-definition (accessed on February 18, 2015).

[8] ETSI, "Network Functions Virtualization," September 2013. Available at: http://www.etsi.org/technologies-clusters/technologies/689-network-functions-virtualisation (accessed on February 18, 2015).

[9] "LTE: General Packet Radio Service (GPRS) Enhancements for Evolved Universal Terrestrial Radio Access Network (E-UTRAN) Access," Release 10, 3GPP, Version 10.5.0, TS 123.401, 2011.

[10] "OpenFlow Switch Specification: Version 1.0 (Wire Protocol 0x01)," Open Network Foundation, December 2009. Available at: https://www.opennetworking.org/images/stories/downloads/sdn-resources/onf-specifications/openflow/openflow-spec-v1.0.0.pdf (accessed on February 18, 2015).

[11] IEEE Std. 802.1Q-2011, "Media Access Control (MAC) Bridges and Virtual Bridged Local Area Networks," Institute of Electrical and Electronics Engineers, 2011.

[12] "OpenFlow Switch Specification: Version 1.4 (Wire Protocol 0x05)," Open Network Foundation, October 2013. Available at: https://www.opennetworking.org/images/stories/downloads/sdn-resources/onf-specifications/openflow/openflow-spec-v1.4.0.pdf (accessed on February 18, 2015).

[13] Open Virtual Switch. Available at: http://www.openvswitch.org (accessed on January 20, 2015).

[14] A. Singla and B. Rijsman, "Contrail Architecture," Juniper Networks, 2013. Available at: http://www.juniper.net/us/en/local/pdf/whitepapers/2000535-en.pdf (accessed on February 18, 2015).

[15] P. Saint-Andre, "Extensible Messaging and Presence Protocol (XMPP): Core," RFC 6120, Internet Engineering Task Force, March 2011.

[16] OpenDaylight Consortium, "OpenDaylight," 2013. Available at: http://www.opendaylight.org/ (accessed on February 18, 2015).

[17] Wikipedia, "Representational State Transfer," 2013. Available at: http://en.wikipedia.org/wiki/Representational_state_transfer (accessed on February 18, 2015).

[18] "IP Multimedia Subsystem (IMS); Stage 2," 3GPP, TS 23.228, Release 9, 2010.

[19] Amazon Web Services (AWS), "Cloud Computing Services," 2014. Available at: http://aws.amazon.com/ (accessed on February 18, 2015).

[20] OpenStack Foundation, "OpenStack Open Source Cloud Computing," 2013. Available at: http://www.open stack.org/ (accessed on February 18, 2015).

[21] J. Kempf, M. Körling, S. Baucke, S. Touati, V. McClelland, I. Más, and O. Bäckman, "Fostering Rapid, Cross-domain Service Innovation in Operator Networks through Service Provider SDN," Proceedings of the ICC, June 2014.

[22] Wikipedia, "Service Oriented Architecture," 2014. Available at: http://en.wikipedia.org/wiki/Service-oriented_ architecture (accessed on February 18, 2015).

[23] Wikipedia, "SOAP," 2013. Available at: http://en.wikipedia.org/wiki/SOAP (accessed on February 18, 2015).

[24] G. Karagiannis, A. Jamakovicy, A. Edmondsz, C. Paradax, T. Metsch, D. Pichonk, M. Corici, S. Ruffinoyy, A. Gomesy, P. S. Crostazz, T. M. Bohnertz, "Mobile Cloud Networking: Virtualisation of Cellular Networks". Available at: http://www.iam.unibe.ch/~jamakovic/MCN_ICT2014_IEEE.pdf (accessed on January 20, 2015).

[25] "Digital cellular telecommunications system (Phase 2+); Universal Mobile Telecommunications System (UMTS); General Packet Radio Service (GPRS); GPRS Tunnelling Protocol (GTP) across the Gn and Gp interface", 3GPP, TS 129 060 version 10.1.0, 2011.

[26] OpenSAF: The Open Service Availability Framework. Available at: http://www.opensaf.org/ (accessed January 20, 2015).

[27] Open vSwitch. Available at: http://www.openvswitch.org/ (accessed January 20, 2015).

[28] B. Pfaff and B. Davie, "The Open vSwitch Database Management Protocol," RFC 7047, Internet Engineering Task Force, December 2013.

[29] M. Honda, F. Huici, G. Lettieri, L. Rizzo, and S. Niccolini, "Accelerating Software Switches with Netmap," Proceedings of the European Workshop on SDN, 2013. Available at: http://www.ewsdn.eu/previous/presenta tions/Presentations_2013/mswitch-ewsdn.pdf (accessed on February 18, 2015).

[30] L. Rizzo, "Netmap: A Novel Framework for Fast Packet I/O," Proceedings of the USENIX ATC Conference, 2012.

[31] Intel Corporation, "Intel® Data Plane Development Kit (Intel® DPDK) Overview Packet Processing on Intel® Architecture," December 2012. Available at: http://www.intel.com/content/dam/www/public/us/en/documents/ presentation/dpdk-packet-processing-ia-overview-presentation.pdf(accessed on February 18, 2015).

[32] 6WIND, "6WIND Continues 195 Gbps Accelerated Virtual Switch Demo at SDN and NFV Summit in Paris." Available at: http://www.6wind.com/blog/6wind-continues-195-gbps-accelerated-virtual-switch-demo-at-nfv-and-sdn-summit-in-paris-march-18-21/(accessed on February 18, 2015).

[33] G. Pongrácz, L. Molnár, Z. L. Kis, and Z. Turányi, "Cheap Silicon: A Myth or Reality? Picking the Right Data Plane Hardware for Software Defined Networking," Proceedings of HotSDN, 2013.

[34] J. Kempf, B. Johansson, S. Pettersson, H. Lüning, and T. Nilsson, "Moving the Mobile Evolved Packet Core to the Cloud," Proceedings of the IEEE Wireless and Mobility Conference, November 2012.

[35] A. Mihailovic, G. Leijonhufvud, and T. Suihko, "Providing Multi-Homing Support in IP Access Networks," The 13th International Symposium on Personal, Indoor, and Mobile Radio Communications, 2002.

[36] S. Ahson and M. Ilyas, Fixed Mobile Convergence Handbook. Boca Raton: CRC Press, 2011.

[37] S. Radosavac, J. Kempf, and U. Kozat, "Security Challenges for the Current Internet Architecture: Can Network Virtualization Help?," NetEcon '08: Workshop on the Economics of Network Systems and Computation, 2008.

[38] Metro Ethernet Forum, "Metro Ethernet Network Architecture Framework—Part 1: Generic Framework," March 2004. Available at: http://www.metroethernetforum.org/Assets/Technical_Specifications/PDF/MEF4. pdf (accessed on February 18, 2015).

[39] J. Kempf, S. Whyte, J. Ellithorpe, P. Kazemian, M. Haitjema, N. Beheshti, S. Stuart, and H. Green, "OpenFlow MPLS and the Open Source Label Switched Router," Proceedings of the International Teletraffic Conference, IEEE, San Francisco, September 2011.

[40] "OF-CONFIG 1.2: OpenFlow Management and Configuration Protocol," Open Network Foundation, ONF TS-016, 2014.

[41] M. Amani, T. Mahmoodi, M. Tatipamula, and H. Aghvami, "Programmable Policies for Data Offloading in LTE Network," Proceedings of ICC, June 2014.

[42] "Universal Mobile Telecommunications System (UMTS): Policy and Charging Control over Rx Reference Point," 3GPP, TS 129 214 version 7.4.0, 2008.

[43] F. Castro, I. M. Forster, A. Mar, A. S. Merino, J. J. Pastor, and G. S. Robinson, "SAPC: Ericsson's Convergent Policy Controller," Ericsson Review, (January), 2010.

[44] 3GPP, "GPRS and EDGE." Available at: http://www.3gpp.org/technologies/keywords-acronyms/102-gprs-edge (accessed on January 20, 2015).

[45] M. Mahalingam, D. Dutt, K. Duda, P. Agarwal, L. Kreeger, T. Sridhar, M. Bursell, and C. Wright, "VXLAN: A Framework for Overlaying Virtualized Layer 2 Networks Over Layer 3 Networks," Internet Draft, work in progress. Available at: https://datatracker.ietf.org/doc/draft-mahalingam-dutt-dcops-vxlan/?include_text=1 (accessed January 20, 2015).

[46] S. Hanks, T. Li, D. Farinacci, and P. Traina, "Generic Routing Encapsulation (GRE)," RFC 1701, Internet Engineering Task Force, October 1994. Available at: http://www.rfc-editor.org/rfc/rfc1701.txt (accessed January 20, 2015).

8

The Controller Placement Problem in Software Defined Mobile Networks (SDMN)

Hakan Selvi,[1] Selcan Güner,[1] Gürkan Gür,[2] and Fatih Alagöz[1]

[1] SATLAB, Department of Computer Engineering, Bogazici University, Istanbul, Turkey
[2] Provus—A MasterCard Company, Istanbul, Turkey

8.1 Introduction

Traditional networks consist of a large variety of network nodes such as switches, routers, hubs, different network appliances, complicated protocols and interfaces, which are defined in detail through standardization. However, these systems provide limited ways to develop and adopt new network features and capabilities once deployed. Therefore, this semistatic architecture poses a challenge against adaptation to meet the requirements of today's network operators and end users. To facilitate required network evolution, the idea of programmable networks and software defined networking (SDN) has been proposed [1]. This approach is devised to simplify network management and enable innovation through network programmability. In the SDN architecture, the control and data planes are decoupled and the network intelligence is logically centralized in software-based controllers. An SDN controller provides a programmatic interface to the network, where applications can be written to perform management tasks and offer new functionalities. The control is centralized and applications are written as if the network is a unified system. While this simplifies policy enforcement and management tasks, the binding must be closely maintained between the control and the network forwarding elements [1]. For instance, an OpenFlow controller sets up OpenFlow devices in the network, maintains topology information, and monitors the network status. The controller performs all the control and management functions. The information of host locations and external paths are also managed by the controller.

Software Defined Mobile Networks (SDMN): Beyond LTE Network Architecture, First Edition.
Edited by Madhusanka Liyanage, Andrei Gurtov, and Mika Ylianttila.
© 2015 John Wiley & Sons, Ltd. Published 2015 by John Wiley & Sons, Ltd.

Itsends configuration messages to all switches to set the entire path. The port for the flow to be forwarded to or other actions like dropping packets is defined by the OpenFlow controller [2].

SDN paradigm provides not only facilitation of network evolution via centralized control and simplified algorithms and programmability by enabling deployment of third-party applications but also elimination of middleboxes and rapid depreciation of network devices [3]. Since the underlying network infrastructure is isolated from the applications being executed via an open interface on the network devices, they are transformed into uncomplicated packet forwarding devices [1]. Therefore, the controller-related aspects of the software defined network are paramount for addressing the emerging intricacies of SDN-based systems.

Likewise the wired networks, current mobile networks[1] suffer from complex control plane protocols, difficulties in deployment of new technologies, vendor-specific configuration interfaces, and inflexible and expensive equipment [4]. Although the demand of smart wireless devices has experienced a quantum leap and the mobile data explosion is challenging the mobile networks, the mobile network infrastructure is not adapting to these conditions in a sufficient and flexible manner [5]. In that regard, the concept of SDMN is expected to be instrumental and alter the network architecture of the current LTE (3GPP) networks and, accordingly, of emerging mobile systems drastically [6]. Although SDMN paradigm will facilitate new degrees of freedom for traffic, resource, and mobility management, it will also bring forth profound issues such as security, system complexity, and scalability. The controller placement-related challenges also emerge as critical factors on the feasibility of SDMN.

In this chapter, we discuss important aspects of the Controller Placement Problem (CPP) in SDMN. First, we briefly introduce the SDN controller concept and describe the problem. Second, we discuss the characteristics of SMDN contrasted with wired networks and optimization parameters/metrics of CPP in SDMN. Then we present available solution methodologies and analyze relevant algorithms in terms of performance metrics. Finally, we conclude with some research directions and open problems for CPP in SDMN context.

8.2   SDN and Mobile Networks

The current 3GPP LTE standard defines cellular 4G networks and is updated regularly as releases with a perspective of 5G networks. The mobile network is separated into two strata with current LTE architecture: a packet-only data plane and a management plane to manage mobility, policies, and charging rules. Data plane consists of base stations (eNodeB), Serving Gateways (S-GW), and Packet Data Network Gateway (P-GW). Mobility Management Entity (MME), Policy Charging and Rules Function (PCRF), and Home Subscriber Server (HSS) constitute the management plane [7]. In LTE technology, the network mechanisms execute in a manner as shown in Figure 8.1. The S-GW serves as a local mobility anchor that enables seamless communication when the user moves from one base station to another. It tunnels traffic to the P-GW. It enforces quality of service (QoS) policies and monitors traffic to perform billing. The P-GW also connects to the Internet and other cellular data networks and acts as a firewall that blocks unwanted traffic. The policies at the P-GW can be very fine grained based on various parameters such as roaming status of the user, properties of the user equipment, usage caps in the service contract, and parental controls [4].

---

[1] In this work, we refer to infrastructure-based mobile networks when we use "mobile networks" term.

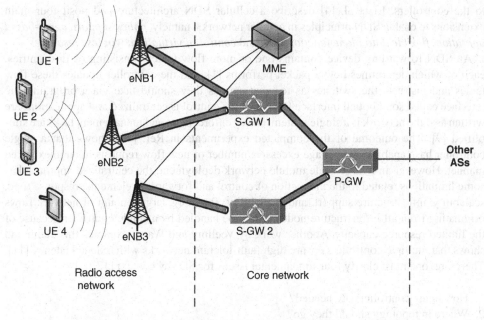

Figure 8.1   Existing LTE architecture (Adapted from Ref. 4).

Although this mobile communication architecture yields to easier management, it still has several limitations. Centralizing data plane functions such as monitoring and QoS functionality at P-GW node introduces scalability challenges due to pecuniary reasons. In that regard, applying SDN principle leads to a flattening of the data plane by simplifying its elements into pure forwarding elements and exporting the control plane intelligence to a remote controller node. This change makes possible to create cheaper equipments and reduces the scalability pressure on P-GW [4]. Additionally, SDN principles are expected to provide flexibility, openness, and programmability to the mobile networks. By means of this approach, mobile network operators can innovate inside their domain more easily with less dependence on UE vendors and service providers [6].

From the perspective of SDN and mobile networks, there are two possible paths for SDN integration to current mobile networks:

• Evolutionary: This is a more probable scenario since there is a huge installed base of mobile networks, and they are envisaged to evolve rather than being completely replaced to meet the requirements of 5G networks. Network virtualization and content-centric operation are expected to be more intrinsic for future networks. This trend is also rendering SDN principles more favorable. The SDN integration is going to be intertwined with these changes.
• Clean-slate approach: Clean-slate design and greenfield deployments provide more degrees of freedom since they are not subject to constraints posed by incumbent systems. However, they are more costly and difficult to implement. Although the design and specification of mobile networks according to SDN paradigm poses substantial issues such as security, scalability, and performance, it is more challenging in practical terms than those theoretical aspects.

The main enabler construct for realizing SMDN is a distributed layer of sensors and actuators embedded in mobile network tiers for enabling centralized control and intelligence migration to the controllers. Li et al. [4] describe a cellular SDN architecture and posit four main extensions to enable SDN principles in cellular networks, namely, *policy support*, *agent-based operation*, *flexible data plane functionality*, and *control of virtualized wireless resources*.

An SDN forwarding device contains one or more flow tables consisting of flow entries, each of which determines how a packet performs [1], and the controller updates these flow tables and instructs the switches as to what actions they should take via a programmatic interface called southbound interface [8]. Since the control is centralized and applications are written as if the network is a single system, policy enforcement and management tasks are simplified [9]. The outcome of the completed experiments in Ref. [10] shows that a single controller has capability to manage excessive number of new flow requests in an unexpected manner. However, in a large-scale mobile network deployment, the centralized approach has some limitations related to the interaction of control and forwarding elements, response time, scalability, infrastructure support, and availability. Typically, large amount of network flows originating from all infrastructure nodes cannot be handled by a single controller because of the limited resource capacity. Another work by Voellmy and Wang promotes this claim and shows that multiple controllers ensure high fault-tolerant networks with reduced latency [11]. Therefore, one must clarify four fundamental issues for SDMN context [12]:

1. How many controllers are needed?
2. Where in topology should they go?
3. How should they interact?
4. How will the mobile network evolution toward SDN be reflected in this problem? This effect can be on a multitude of aspects such as standardization, problem formulation, or solution methodologies.

The answers of these essential questions depend on the network topology among other user-imposed requirements. From the latency perspective, a single controller would be mostly adequate. On the other hand, fault tolerance and scalability concerns impel researchers to consider using multiple controllers in networks [12].

Moreover, the architecture of mobile networks and its provisioning is a challenging and complex task due to kaleidoscopic information about the network's topology [13]. Dynamically changing topology is an inherent characteristic of mobile networks. Thus, the deployment of multiple controllers requires being in harmony with related "on-the-fly" network units. Nevertheless, the controllers should perform synchronously to maintain a consistent view of the network [14]. If redundant controller(s) is integrated into the system, additional communication overhead will occur. Thus, the location of controllers should be optimal. Lastly, mobile network traffic can fluctuate over time; controller placement scheme should regard dynamic rearrangement of the number and the location of controllers [9].

8.3   Performance Objectives for SDMN Controller Placement

In this section, we will examine various controller requirements that affect the network state and algorithm efficiency. It is harder for fully distributed control planes to fail compared to centralized planes. However, for controller placement, there is a trade-off among performance objectives of the placement algorithm. Some algorithms could place controllers to maximize

Table 8.1   Performance objectives and their effects on networks

	Scalability	Reliability	Latency	Resilience
Fault tolerance		✓		✓
Service delay	✓		✓	
Utilization	✓	✓	✓	

fault tolerance, or some could minimize the propagation delay or distance to the nth closest controller [12]. The general performance objectives for control placement and their related effects on the network can be seen in Table 8.1. To minimize the delay of network-based services, scalability and latency are considered. For scalability, service delay may be traded, while latency minimization directly benefits service delay. For utilization of the network, the effects of the scalability, reliability, and latency are taken into account. Fault tolerance is directly affected by reliability and resilience objectives.

8.3.1   Scalability

For networks with more than one controller, a controller may become overloaded if the switches mapped to this controller have large number of flows. However, the remaining controllers may operate underutilized. It is instrumental to shift load across controllers over time depending on the temporal and spatial variation in traffic conditions. Static controller assignment can result in suboptimal performance since no switch will be mapped into a less loaded controller. The replacement of the controller can help to improve performance of overprovisioned controllers. Instead of using static mapping, elastic controller architecture can be used to map controller to balance the load as it reflects performance.

8.3.2   Reliability

According to IEEE, *reliability* is defined as "the probability that a system will perform its intended functions without failure, within design parameters, under specific operating conditions, and for a specific period of time" [15]. If the connection between controller and the forwarding planes is broken because of the network failures, some switches will be left without any controller and thus will be disabled in SDN-based networks. Network availability should be ensured to assure the reliability of SDN. Therefore, improving reliability is important to prevent disconnection between controller and the switch or between controllers. To reflect the reliability of the SDN controller and to find the most reliable controller placement for SDN, a metric can be defined as the expected percentage of the valid control paths when network failures happen [12]. A control path is defined as the route set between switches and their controllers and between controllers. Consistency of the network should also be ensured when multiple controllers are in the network.

Each control path uses existing connection between switches. If control path is represented as a logical link, SDN control network is responsible to enable a healthy communication between switches and their controllers, which is a requirement for control paths to be valid. The failure of control paths, which means the connection is broken between switch and its controller or among controllers, results in the case where control network will lose its functionality. If the number of the control paths is too large, forwarding service may fail, which causes severe problems. So to define a controller placement algorithm, reliability must be considered as a placement metric. To formulate reliability better, various statistical and empirical approached are possible [12].

The optimization target to define a reliable network is to minimize the expected percentage of control path loss. To maximize reliability of SDN, several placement algorithms are developed to automate controller placement decision that work in a reliability-aware manner [10]. These algorithms are described in Section 4.1.

8.3.3 Latency

Latency is simply "the delay between the time the data is sent from its origin and received at its destination" [16] and a critical QoS metric for communication networks. It is more important for multimedia communications since that kind of traffic is delay sensitive. For instance, one of the envisaged requirements of 5G networks is to have a delay smaller than 1 ms, which implies an order of reduction compared to 4G networks. For large-scale networks, single controller deployment is typically not sufficient to reach adequate performance due to various factors. However, when the network has several controllers, a new matter of contention emerges. Since several controllers maintain the control logic of the network, these controllers need to communicate with each other to maintain network consistency. The latency between controllers has to be considered, especially if controller communication traffic is frequent [17]. The latency in that setting comprises of processing, transmission, and propagation latencies.

Even though latency between controllers is considered during placement, the reaction of the remote controller and the time to pass for delivering the reaction to a switch bound the overall network performance. It can be called as *propagation latency*, which should be at reasonable levels for speed and stability. It is enough for propagation delay to become infeasible for real-time tasks or slow down unacceptably even for small delay. Adding some intelligence to switches can reduce these delays. However, this method adds complexity to the system, and it is against the idea that SDN uses a simple and dump switch model.

Controller placement algorithms are developed to minimize latencies or maximize some latency-based parameters, which are defined as:

• *Average-case latency*: If the network is simplified as a network graph, the connections between components represent edges. The weight of the edge represents propagation latency. The average propagation latency L_{avg} for a placement of controllers is average of minimum propagation latencies on each edge. $d(v, s)$ is the shortest path from node $v \in V$ to node $s \in V$:

$$L_{avg}(S') = \frac{1}{n} \sum_{v \in V} \min_{s \in S'} d(v, s)$$

- *Worst-case latency*: This value is defined as the maximum propagation delay from node to controller:

$$L_{wc}(S') = \max_{(v \in V)} \min_{(s \in S')} d(v, s)$$

- *Nodes within a latency bound:* Instead of minimizing the average or worst case, it might be better to place controllers in a way that it maximizes the number of nodes within a latency bound. This approach is called as *maximum cover*. For most topologies, adding controllers yields slightly less than proportional reduction [12].

8.3.4 Resilience

A good controller placement should minimize latencies between nodes and controllers or among controllers. However, minimizing latency is not always sufficient. According to Ref. [17], the placement of the controller should also meet some resilience constraints. These constraints are defined in this section.

8.3.4.1 Controller Failures

Using more than one controller not only decreases latencies but also increases tolerance of the network to failures in case the controllers stop working. In a related work [18], it is assumed that a node is not able to route anymore and becomes practically off if it loses its connection to the controller. However, Hock et al. [17] suppose that in case a controller is out of order, all the switches assigned to the failed controller can be reassigned to the second closest controller by using a backup assignment or signaling-based shortest path routing. Until the last controller survives, all nodes are functional in this way. Although resilience is considered, this solution will probably increase the latency of the reassigned nodes and their new controller. The new controller may be much further away compared to the previous one. This situation would result in higher latency. The described failure scenario is an example of the worst case, since the last surviving controller is located furthest from the center of the network such that some of the nodes need to pass through the whole network to reach the controller. However, a placement algorithm to increase resilience should also consider this worst-case scenario during failure-free routing.

8.3.4.2 Network Disruption

In a network, not only controller failures occur. Network components, links, and nodes may also suffer damage, which is more important to consider because the topology itself is changed. Because of link failures, paths between some nodes are severed. This situation causes nodes to be assigned to some other controllers even though latencies may increase. Additionally, some parts of the network may be in danger because of these link failures, and many nodes cannot be assigned to any controller. Although these nodes may be working and are able to perform forwarding operations, they cannot get control messages from any controller.

Link failures prevent rerouting of nodes even if they are physically connected, since path will no longer be available.

8.3.4.3   Load Imbalance

If the nodes are assigned to the nearest controller using latency as a metric or shortest path distance between the node and controller, there may be situations when some controllers are overloaded due to excessive traffic flow. There can be an imbalance in the number of nodes per controller in the network. Typically, the higher the number of nodes attached to a controller, the greater the load on that controller. The increase in number of node-to-controller requests in the network induces additional delay due to queuing at the controller system. Resilience for controller placement requires nodes of different controllers to be well-balanced.

8.3.4.4   Intercontroller Latency

Since single controller is not sufficient to ensure resilience in a network, if any two controllers are far away from each other, the messages from one to the other need to pass through the entire network, which increases intercontroller latency.

8.4   CPP

The controller placement strategies affect every aspect of SDN, from node-to-controller latencies to network availability and from operational costs to performance. In addition, the peculiarities of SMDN such as mobility compared to wired networks complicate the problem structure. Optimizing every variable in this problem is NP hard, so it is very important to find an efficient controller placement algorithm [19]. Heller et al. concluded that finding optimal solution is computationally feasible but within failure-free scenarios in [12]. They took into consideration just latency requirements like *average-case latency* and *worst-case latency*, and reasonably, they presented that in most topologies, one single controller is enough to fulfill the existing latency requirements. If we broaden our viewpoint to CPP with various goals, which include reliability, network resilience, fault tolerance, or load balancing, many more controllers communicating with each other are necessary to meet these resilience requirements [17].

In Figure 8.2, we depict various factors or parameters in CPP setting. The fundamental factors determining the solution space are the location and number of controllers and their communication requirements. Moreover, the SDN controllers may be of different characteristics entailing processing capability, supported communication primitives, and intelligence. This condition implies nonidentical controllers leading to heterogeneity. The network characteristics also affect the problem structure drastically. For our focus in this chapter, this phenomenon is paramount since mobile networks exhibit peculiar inherent characteristics due to mobility, wireless transmission, and dynamic network structure. For any algorithm or scheme for CPP, there are practical constraints such as complexity. Although these factors are assumed to be operative for relatively asynchronous execution of controller, CPP algorithm may need to be more active and executed frequently when network function virtualization becomes more common.

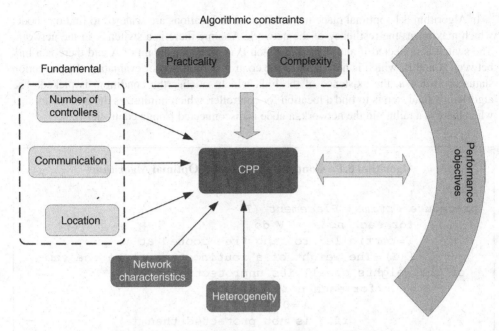

Figure 8.2   CPP parameters and performance objectives.

Since a single controller cannot handle large amount of flows within the network with desired performance capabilities due to capacity limitation, multiple controllers are employed for better network management. However, one must specify the number of controllers to use and the locations of them in the network architecture [19]. While trying to find these important questions' answers, the main objective should be not only to minimize the latencies between nodes and controller but also to maximize resilience by fulfilling certain constraints.

8.4.1   Placement of Controllers

The separation of forwarding and control planes allows forwarding plane to be simple with the controller plane entailing and managing the network intelligence. However, this separation can impair system performance, e.g. reduce reliability of the communication. Thus, in network design, placement of controller(s) should be considered in a way that it achieves both reliability and performance.

8.4.1.1   Single Controller Placement

In single controller placement (SCP), the aim is to place a single controller according to some predefined objective. A very common objective is to assure a high level of resilience 'i.e. to protect the controller from being disconnected from nodes. This is important due to the fact that the first requirement for successful operation of a controller is to keep it capable of communicating with its peers in the network.

In Algorithm 8.1 (optimal placement), all possible locations are scanned to find one node, which maximizes the resiliency of the network. Assume there is a switch A in the network. The switch is protected if and only if a switch B is not downstream of A and there is a link between A and B, which is not a part of the controller routing. For evaluating the protection status of switch u, the existence of the link, which satisfies the condition, is picked. The algorithm's final step is to find a location for controller, which minimizes the probability that when there is a failure in the network, a node is disconnected from a controller [20].

Algorithm 8.1   Controller Placement: Optimal Algorithm

```
procedure Optimal Placement (T)
        for each node v∈V do
        T=controller routing tree rooted at v
        Γ(T) =The weight of a routing tree to be the sum
  of the weights of all its unprotected nodes
              for each node u≠v do
                  W = 0
                if u is not protected then
                    W =number of downstream nodes of u in T
                end if
                Γ(T) = Γ(T) + W
            end for
            controller location=node v with minimum Γ(T)
        end for
```

When the network size is large so that searching among all locations is not practical, a heuristic method is required. Algorithm 8.2 is a heuristic method that selects the node with the largest number of directly connected nodes. $D'(v)$ denotes the number of the protected neighbors of the node. The algorithm continues until it finds a node with the maximum number of protected neighbors [20].

Algorithm 8.2   Controller Placement: Greedy Algorithm

```
procedure Greedy Placement ()
    Sort nodes in V such that D(v(1)) ≥ D(v(2)) ≥
D(v(n))
    controller location = v(1)
    for i = 1 to n do
        A=set of neighbors of node v(i)
        D'(v(i))=number of members of A that are
connected to other
```

```
                          members, either directly or through one hop
        other
                  than the controller.
                if D'(v(i)) > D'(controller location) then
                  controller location = v(i)
                end if
                if (D'(v(i)) == D(v(i)) then
                     break;
                end if
            end for
```

In previous two algorithms, no controller routing is considered. Any arbitrary routing tree can be chosen to maximize the protection of the network against component failures or optimizing performance. Since finding a routing that maximizes the protection of the network for any controller location is an NP-hard problem, algorithms can be used to find a suboptimal solution. Algorithm 8.3 is a resilience-improved routing scheme. The algorithm starts with a shortest path tree and modifies the tree to add to the resilience of the network. Iteration is continued until no further improvement is possible that increases the resiliency [20]. From all these three algorithms, Greedy Routing Tree algorithm performs better than the two other according to Ref. [20].

Algorithm 8.3   Greedy Routing Tree (GRT) Algorithm

```
procedure Routing Greedy (G, controller loc)
                       T = shortest-path tree
                       i = 1
                       repeat
                       for nodes v with d(v,
      controller) == i do
                               if v is the only node
      with d(v,controller)==i then
                                          next;
                           end if
                           for every node u ∈ V \
      {downstream nodes of v} and
                             (v, u)∈ E and (v, u)∉ T
      do
                                   if d(u,
      controller) ≤ d(v, controller) then
                           T'(u) = tree built by replacing (v,
```

```
                                              upstream node of v in
  T) by (v,u)
                                    if (T') < Γ(T) then
                                        replace T by T'
                                            end if
                                  end if
                            end for
                  end for
                  i = i + 1
                  until all nodes checked
```

8.4.1.2   Multiple Controller Placement

As the size of the network increases, using a single controller reduces the reliability and degrades the performance of the network. Therefore, multiple controllers are employed for better network availability and management. The CPP for this setting is denoted as multiple controller placement (MCP). As in SCP, in case of deploying multiple controllers, consistency of the network state must be achieved in the communication between controllers. The answer to where to place controllers depends on the metric choices and network topology itself. To place the multiple controllers efficiently, the CPP should have a near-optimal solution once the optimal solution is unattainable.

Graph-Theoretic MCP Problem Formulation

The CPP is inherently suitable for graph-theoretic modeling since it addresses a node selection problem in network graph. If we define the network as a graph (V, E), V is the set of nodes and $E \subseteq V \times V$ is the set of links. Let n be the number of nodes $n = |V|$. Network nodes and links are typically assumed to fail independently:

- p is the failure probability for each physical component $l \in V \cup E$.
- $path_{st}$ is the shortest path from s to t, given that s and t are any two nodes.
- $V_c \subseteq V$ is the set of candidate location where controllers can be placed.
- $M_c \subseteq V$ denotes the set of controllers to be placed in the network.
- M and $P(M)$ denote the number and a possible topological placement of these controllers, respectively.

To reduce propagation delay, each switch is connected to its nearest controller using the shortest path algorithm. If several shortest paths exist, the most reliable one is picked. A good placement should maximize the existing connectivity among the switches. All controllers can be connected to all switches forming a mesh. However, this will increase the complexity and the deployment cost. It will decrease the scalability of the network since the network size grows as switches spread across the geographic locations. To maximize network resilience and connectivity, to increase scalability, and to decrease probability of failure, controllers should be placed accordingly, which is an optimization problem [18].

   In the following part, we describe and discuss some MCP algorithms studied in literature.

Random Placement

Although this is usually not a practical algorithm, it is typically used as a baseline case for performance evaluation. In random placement algorithm, each candidate location may have a uniform probability of hosting a controller. In that case, RP algorithm randomly chooses k locations among all potential sites, where $k = 1$ for single controller. Another option is to utilize a biased probability distribution, which reflects a preference among potential controller locations. This scheme is instrumental to concentrate controller deployments to specific network segments.

Greedy Algorithms

Greedy algorithms adopt the locally optimal solution at each stage of the algorithm's run. Although a greedy algorithm does not necessarily produce an optimal solution, it may yield locally optimal solutions that approximate a global optimal one in a reasonable amount of time.

Algorithm 8.4   l-w-Greedy Controller Placement Algorithm [19]

procedure l-w-greedy Controller Placement
    Sort potential location V_c in descending order of node
  failure properties, the
    first $w|Vc|$ elements of which is denoted as array L_c
    if $k \leq 1$ **then**
        Choose among all sets M' from L_c with $|M'| = k$ the set
  M'' with maximum ∂
        return set M''
    end if
    Set M' to be the most reliable placement of size l
    while $|M'| \leq k$ **do**
        Among all set X of 1 element in M' and among all
  set Y of $l+1$ elements
        in $L_c - M' + X$, choose sets X, Y with maximum ∂
        $M' = M' + Y - X$
    end while
return set M'

The MCP problem naturally lends itself to greedy approaches since controllers can be placed one by one during the solution. Hu et al. [19] describe l-w-greedy algorithm (Algorithm 8.4) where controllers are placed iteratively in a greedy way. k controllers are needed to be replaced among $|V|$ potential locations. A list of potential locations is generated, which are then ranked increasingly according to failure probabilities of switches. One location at a time is selected from $w|V|(0 < w \leq 1)$. For first iteration $l = 0$, the algorithm computes the cost associated with each candidate location under the assumption that connections from all switches converge at that location. Location with the highest value is picked. In the second iteration, the algorithm searches for a second controller with the highest cost from candidate locations. The algorithm is iterated until all k controllers have been chosen and placed.

For $l > 0$, after l controllers have placed, the algorithm allows for l steps backtracking in each subsequent iteration. All possible combinations are checked of removing l of already placed controllers and replacing them with $l + 1$ new controllers [21].

Metaheuristics

A metaheuristic is a higher-level heuristic designed to find, generate, or determine a lower-level heuristic that may provide a suboptimal, albeit sufficiently good, solution to an optimization problem. They are typically utilized when the optimization problem is too complex for the computational resources or incomplete or imperfect information is available. Some examples are tabu search, evolutionary computation, genetic algorithms, and particle swarm optimization.

Hu et al. [21] investigate various CPP algorithms including simulated annealing (SA) metaheuristic. SA is a probabilistic method for global minimum of a cost function that may possess several local minima [22]. Although SA is a known technique for global optimization problems, the key of effective usage is optimizing the configuration of the algorithm. It is important to reduce the search space and converge to the vicinity of optimal placement rapidly.

For the MCP problem, SA can be devised as follows:

1. *Initial state*: Place k controllers at the k most reliable locations.
2. *Initial temperature*: To make any neighbor solution to be acceptable, the initial temperature T_0 should be a large value. P_0 is acceptance probability in the first k iterations. Δ_0 is the cost difference between the best and the worst solutions obtained in Y executions of the random placement. T_0 can be computed by $-|\Delta_0|/\ln P_0$.
3. *Neighborhood structure*: $P(M)$ denotes a possible placement of k controllers. x_c controller location of $P(M)$. x_k location of $(V - P(M))$. The best exchange x_k in $(V - P(M))$ is defined such that $\Delta_{ck} = \min_j \in (V - P(M)) \Delta_{cj}$ where Δ_{ij} is the reduction in the objective function that is obtained when $x_i \in (V - P(M))$. The cycle of the algorithm is completed when all x_cs in $P(M)$ are examined.
4. *Temperature function*: The temperature decreases exponentially. That is, $T_{new} = \alpha T_{old}$ [21].

Brute Force

With a brute-force approach, all possible combinations of k controllers in every potential location are calculated. Then, the combination with the best cost is picked. This approach is exhaustive and optimal result is obtained after an extremely long execution time even for small networks. Feasible solution can be found by the brute-force algorithm, but it is infeasible to run a brute force to completion, which can take weeks to complete for large topologies [9].

Experimental Results

Figure 8.3 shows the cumulative distribution (CDF) of the relative performance of the algorithms on the Internet2 OS3E (Open Science, Scholarship and Services Exchange) topology according to Ref. [21]. The result of the algorithm comparison is 2-1-greedy and 1-1-greedy, and SA performs the best. SA performs better than 2-1-greedy, which finds better placement than 1-1-greedy. The random placement has the worst performance as expected.

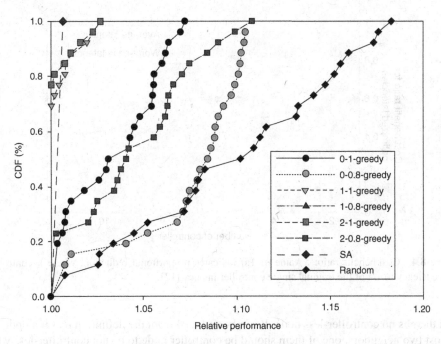

Figure 8.3   The CDF of relative performance of the placement algorithms on OS3E topology [21].

8.4.2   *Number of Required Controllers*

If the employed controllers are located efficiently but without a predetermined number, one should also find the answer of the following challenging question: how many controllers should we use in order to meet our objectives? Apparently, the answer is a variable according to trade-off considerations between related objectives/metrics. Heller et al. [12] studied to get a result from just latency point of view. Figure 8.4 shows that although the effect of controller numbers varies from average-case latency to worst-case latency, increasing controller numbers implies a proportionally reduction in both. Hu et al. approach the problem from a different viewpoint and focuses on reliability as the main concern [21]. The results of their experiments showed that the optimizations on different topologies provide very similar results. Using too few (even a single) controllers reduced reliability expectedly. However, the results also show that after a certain controller proportion in the network, additional controllers and expected path loss is inversely correlated because if large numbers of controllers are employed redundantly, too many controller paths between controllers cause low reliability.

According to Hock et al. [17], if one considers fulfilling more resilience constraints, which are addressed in Section 8.3.4, he must locate the controllers in a controller-failure and network-disruption tolerated way. Thus, there must be no controller-less node in the network, and "a node is considered controller-less if it is still working and part of a working subtopology (consisting of at least one more node), but cannot reach any controller. Nodes that are still working, but cut off without any working neighbors, are not considered to be controller-less." as defined in Ref. [17]. Therefore, number of controllers should be increased

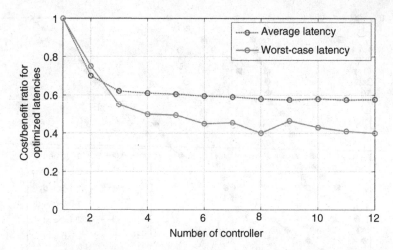

Figure 8.4 Cost/benefit ratio: a value of 1.0 indicates proportional reduction, where k controllers reduce latency to $1/k$ of the original single controller latency [12].

until there is no controller-less node. It can be deduced from the definition that if a node has at most two neighbors, one of them should be controller node to be not controller-less when both neighbors fail. (They limit defects to the case of two simultaneous failures in that work. Because if more than two arbitrary failures happen concurrently, the topology can be completely disrupted and no controller placement would help it anymore.) The experiment results show that on Internet2 OS3E topology, the number of controller-less nodes decreases with increasing number of controllers (k) and it is possible to eliminate all controller-less nodes in all one and two failure scenarios with a number of seven controllers.

So to calculate the number of required controllers, the network must be divided into virtual subtopologies consisting of at least two nodes that can be totally cut off from the remaining part of the entire network by at most two link/node failures, and one of the internal nodes must be controller node [17]. Then, we can find the number of necessary controllers in two phases:

1. Find all possible subtopologies of at least two nodes, which do not include any smaller subtopology in itself. Since all found subtopologies need a separate controller, the maximum count of necessary controllers is 8 in Figure 8.5.
2. Since it is aimed to use a minimum number of controllers covering all subtopologies, try to minimize the number, which is found in (1). There are three intersected subtopologies in the network as shown at the upper right-hand corner of Figure 8.5. Two controllers are sufficient to manage these three subnetworks. Thus, the minimum required controller number is 7.

There are 34 nodes in Figure 8.5, so there are $(34/7) = 5.4$ million possible placements with seven controllers, but there are two possible controller nodes for each subtopology, and three possibilities for intersected subtopologies reduce possible controller placements to $2^5 \times 3 = 96$. However, the best placement of these 96 possibilities in terms of maximum overall node-to-controller latencies is colored by red and the magnitude is 44.9% of diameter

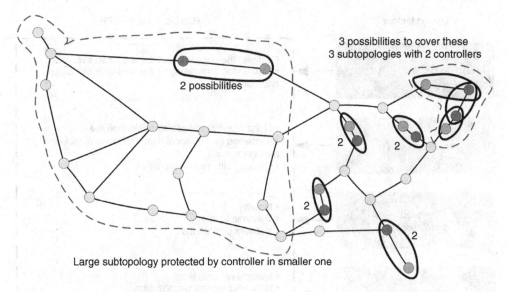

3 possibilities to cover these
3 subtopologies with 2 controllers

2 possibilities

Large subtopology protected by controller in smaller one

Figure 8.5 Subtopologies needing a controller to eliminate controller-less nodes [17]. There are $2^5 \times 3 = 96$ potential placements in total.

of the network, which is shown 22.5% in [12] regardless of any resilience constraint. That indicates that there is a trade-off between resilience constraints and latencies [21]. When optimizing for average latency in OS3E topology, the best placement for a single controller also provides the optimal reliability. However, if the network uses more controllers, optimizing for latencies reduces reliability by ~13.7%; in a similar vein, optimizing reliability increases the latencies, but it is possible to find an equilibrium point where it provides latency and reliability constraints [12].

8.4.3   CPP and Mobile Networks

The ramifications of mobile networks on CPP are depicted in Figure 8.6. For mobile networks, resilience is a critical issue considering the multitiered structure especially for heterogeneous wireless networks. Moreover, load variance is much larger, which affects the processing and response times of the controller. The load variance and characteristics of mobile networks are required to be integrated into these problem definitions. The practicality of the algorithm needs to be considered considering the information and protocol data exchange of mobile network specs.

The complexity factor has two aspects: *offline* corresponding to the complexity of the algorithm execution versus *online* corresponding to the complexity of the controller signaling and operation in the mobile network. Moreover, the complexity is higher due to the diversity of network nodes and mobile end devices. This is also reflected in the heterogeneity-related challenges caused by multiple network tiers and network function virtualization in the system.

Design factors   Ramifications on CPP

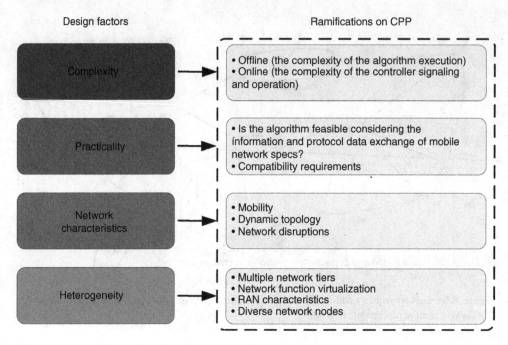

Figure 8.6   Ramifications on CPP due to mobile network domain.

The mobility attribute of mobile networks results in a dynamic topology and potential network disruptions in addition to load variance. This challenge needs to be addressed via adaptive and dynamic controller provisioning in the network. It complicates the problem structure with spatiotemporal changes in system parameters used in placement algorithms.

8.5   Conclusion

Current mobile networks suffer from complex control plane protocols, difficulties in deployment of new technologies, vendor-specific configuration interfaces, and expensive equipment. However, the wireless applications and services have become indispensable with the ever-increasing traffic volumes and bit rates. Therefore, mobile network infrastructure is supposed to adapt and evolve to address this stringent level of requirements in a sufficient and flexible manner. In that regard, the concept of SDMN is expected to be instrumental and emerge as an integral part of future mobile networks. Although SDN paradigm will facilitate new degrees of freedom in mobile networks, it will also bring forth profound issues related to mobile network characteristics. In that regard, the centralized controller and how to place it is a key issue for SDMN design and operation. Therefore, the controller placement-related challenges emerge as critical elements for the feasibility of SDMN. For practical SDMN, the controller placement algorithms have to be devised, which considers scalability, complexity, mobile network characteristics, and compatibility with generic SDN systems.

References

[1] Mendonca, M., Nunes, B. A. A., Nguyen, X., Obraczka, K., and Turletti, T. (2014) A Survey of Software-Defined Networking: Past, Present, and Future of Programmable Networks. IEEE Communications Surveys and Tutorials, vol. 99, pp. 1–18.

[2] Fernandez, M.P. (2013) Comparing OpenFlow Controller Paradigms Scalability: Reactive and Proactive. IEEE 27th International Conference on Advanced Information Networking and Applications (AINA), pp. 1009–1016.

[3] Limoncelli, T. A. (2012) Openflow: A Radical New Idea in Networking. Communications of the ACM, vol. 55 no. 8: 42–4.

[4] Li, E., Mao, Z. M., and Rexford, J. (2012) Towards Software Defined Cellular Networks. Software Defined Networking (EWSDN), European Workshop 2012, Darmstadt, Germany.

[5] Cisco Visual Networking Index (VNI): Global Mobile Data Traffic Forecast, 2013–2018 Report. http://www. cisco.com/c/en/us/solutions/collateral/service-provider/visual-networking-index-vni/white_paper_c11-520862. html. Accessed January 21, 2015.

[6] Pentikousis, K., Wang, Y., and Hu, W. (2013) MobileFlow: Toward Software-Defined Mobile Networks. IEEE Communications Magazine, vol. 51, no. 7, 44–53.

[7] Mahmoodi, T. and Seetharaman, S. (2014) On Using a SDN-Based Control Plane in 5G Mobile Networks. Wireless World Research Forum, meeting 32, Marrakech, Morocco.

[8] Ashton, M. and Associates (2013). Ten Things to Look for in an SDN Controller. https://www.necam.com/ Docs/?id=23865bd4-f10a-49f7-b6be-a17c61ad6fff. Accessed January 21, 2015.

[9] Bari, M. F., Roy, A. R., Chowdhury, S. R., Zhang, Q., Zhani, M. F., Ahmed, R., and Boutaba, R. (2013) Dynamic Controller Provisioning in Software Defined Networks, Network and Service Management (CNSM), 2013 9th International Conference on, 18–25, Zürich,Switzerland.

[10] Tootoonchian, A., Gorbunov, S., Ganjali, Y., Casado, M., and Sherwood, R. (2012) On Controller Performance in Software-Defined Networks. In USENIX Workshop on Hot Topics in Management of Internet, Cloud, and Enterprise Networks and Services (Hot–ICE), vol. 54.

[11] Voellmy, A. and Wang, J. (2012) Scalable Software Defined Network Controllers. Proceedings of the ACM SIGCOMM 2012 conference on Applications, technologies, architectures, and protocols for computer communication, SIGCOMM'12, pp. 289–290, New York, NY, USA.

[12] Heller, B., Sherwood, R., and McKeown, N. (2012) The Controller Placement Problem. ACM HotSDN 2012, pp. 7–12.

[13] Mülec, G., Vasiu, R., and Frigura-Iliasa, F. (2013) Distributed Flow Controller for Mobile Ad-Hoc Networks. 8th IEEE International Symposium on Applied Computational Intelligence and Informatics, pp. 143–146. Timisoara, Romania.

[14] Levin, D., Wundsam, A., Heller, B., Handigol, N., and Feldmann, A. (2012) Logically Centralized?: State Distribution Trade-Offs in Software Defined Networks. ACM HotSDN 2012, pp. 1–6.

[15] IEEE. (1999) IEEE standard for communication-based train control (CBTC) performance and functional requirements. IEEE Std 1474.1-1999, New York, USA.

[16] IEEE. (2005) IEEE standard communication delivery time performance requirements for electric power substation automation. IEEE Std 1646-2004, New York, USA.

[17] Hock, D., Hartmann, M., Gebert, S., Jarschel, M., Zinner, T., and Tran-Gia, Phuoc (2013) Pareto-Optimal Resilient Controller Placement in SDN-based Core Networks. Proceeding of the 25th Int. Teletraffic Congress (ITC), Shangai, China.

[18] Zhang, Y., Beheshti, N., and Tatipamula, M. (2011) On Resilience of Split-Architecture Networks. IEEE GLOBECOM 2011, pp. 1–6.

[19] Hu, Y., Wendong, W., Gong, X., Que, X., and Siduan, C. (2012) On the Placement of Controllers in Software-Defined Networks, The Journal of China Universities of Posts and Telecommunications, vol. 19, no. 2, pp. 92–97.

[20] Behesti, N. and Zhang, Y. (2012) Fast Failover for Control Traffic in Software-Defined Networks. Next Generation Networking and Internet Symposium. IEEE GLOBECOM 2012, Anaheim, CA, USA, pp. 2665–2670.

[21] Hu, Y., Wendong, W., Gong, X., Que, X., and Shiduan, C. (2013) Reliability-aware Controller Placement for Software-Defined Networks. IFIP/IEEE International Symposium on Integrated Network Management (IM2013), Ghent, Belgium.

[22] Bertimas, D. and Tsitsiklis, J. (1993) Simulated Annealing. Statistical Science, vol. 8, no.1, pp. 10–15.

9

Technology Evolution in Mobile Networks
Case of Open IaaS Cloud Platforms

Antti Tolonen and Sakari Luukkainen

Department of Computer Science and Engineering, Aalto University, Espoo, Finland

9.1 Introduction

The current upgrade of mobile networks to LTE technology is a significant technological change that might facilitate an opportunity to introduce novel complementary technologies to the networks. Mobile network operators must invest in new network solutions to remain competitive in terms of mobile data transfer speeds. However, they struggle with the increasing cost of dedicated network hardware and declining revenues. Therefore, there is a clear need for new technologies that, on the one hand, provide increased network performance and value and, on the other hand, lower the overall cost.

One proposed alternative is network function virtualization (NFV). In this approach, the network functions would be implemented in software and provided on top of cheap, generic computing and networking hardware. To support the elastic provision of the virtualized functions, private "telco clouds" are one option. Cloud computing has become the prevalent business model in IT owing to its on-demand flexibility. However, it might struggle with, for example, the latency and fault tolerance requirements of the mobile network operators accustomed to the performance of dedicated hardware.

Despite challenges, NFV and cloud computing hold promise for cost-effective provision of mobile network functions. Initially, virtualized functions could be deployed in parallel with the legacy infrastructure, for instance, to support the growing traffic of machine-to-machine communications. Meanwhile, part of the traffic could be supported by dedicated network hardware.

Software Defined Mobile Networks (SDMN): Beyond LTE Network Architecture, First Edition.
Edited by Madhusanka Liyanage, Andrei Gurtov, and Mika Ylianttila.
© 2015 John Wiley & Sons, Ltd. Published 2015 by John Wiley & Sons, Ltd.

It is typical of the high-tech business that possibilities for new technologies continuously open up even though no clear market need exists. Both failures in the commercialization process and unforeseen success stories happen unexpectedly. The technology evolution dynamics are very complicated, as many conditions on success have to be met simultaneously.

This chapter analyzes the factors influencing future evolution of telco clouds controlled by open-source platform software. The employed research methodology is a single-case study. In contrast to an instrumental study of multiple cases, it allows us to understand the market behavior in our specific case [1, 2]. The study itself is based on a review of existing literature and company Web sites. To deal with the volume of information, we define a framework that is based on the theory of generic technology evolution to structure the analysis.

The chapter is organized as follows: First, we describe the theoretical background and the employed framework. Second, we introduce the reader to the basics of cloud computing and its technology, followed by a description of an example open-source cloud platform, OpenStack. Next, we use the framework to analyze the case of open cloud platforms in a software defined mobile network (SDMN). Finally, we discuss the critical factors in this possible evolution and summarize.

9.2   Generic Technology Evolution

Technologies evolve through periods of incremental change interrupted by discontinuous innovations. Competence-enhancing discontinuities complement the existing competences and are initiated by incumbents, because they are unwilling to cannibalize existing products and services. In contrast, competence-destroying discontinuities, typically initiated by new entrants, make the previous competences obsolete [3].

Additionally, incumbents are likely to develop technological performance that finally exceeds even the most demanding customers' needs. Typically at the same time, new cheaper technologies start to gain market share among less demanding customers. These technologies, originally ignored by the incumbents, begin to gain share of the mainstream market. These technologies and the related innovations are called disruptive [4].

Technological discontinuities are likely to cause changes in the existing industry structures and especially in the competitiveness of the incumbents. Expectations of a growing market and high profits encourage new companies to enter the market and challenge the incumbents. The success of many new entrants has led to a phenomenon called the "attackers' advantage." This term refers to the new entrants who are better than the incumbents in developing and commercializing emerging technologies because of their smaller size, limited path-dependent history, and commitment to the value networks of the previous technology [5, 6].

Industries, however, have barriers to entry, which protect the existing profit levels of the incumbents and constrain new entrants from entering the market. Barriers to entry are unique to each industry, and these barriers include cost advantage, economies of scale, brand identity, switching costs, capital requirements, learning curve, regulation, access to inputs or distribution, and proprietary products [7].

In the beginning of the technology evolution, there is a phase called variation, where these technologies and their substitutes seek market acceptance. The speed of change in this phase is slow because the fundamentals of the technology and new market characteristics are still inadequately understood. During this phase, the companies experiment with different forms of technology and product features to get feedback from the market [8]. An important factor affecting technology evolution is the relative advantage and added value over older technologies. Experimentation then relates to the extent to which the technology can be experimented with a low threshold when seeking emerging sources of added value. Easy experimentation possibilities enhance the overall technology diffusion [9].

The standardization and related openness increase the overall market size and decrease uncertainty caused by variation. The competition between several incompatible technologies from the evolving new market is called a "standard war." New standards change the competition for the market into a more traditional market share battle and from systems into the component level. They also increase price competition and decrease feature variation. Companies can also differentiate their products by promoting an own de facto standard, which provides unique performance. Rival de facto standards have a negative impact on the success of the technologies developed in the formal standardization process. However, a trade-off between openness and control exists: proprietary technologies tend to decrease the overall market size, and the optimum solution lies in between these extremes [10].

Highly modularized standards will increase the flexibility to adapt to uncertain market needs by providing a larger field of options from which to select and by allowing experimentation and market selection of the best outcomes. Standards should be introduced in an evolutionary way by starting from one that is simple and building it up in complexity as the market uncertainty decreases thus allowing for a staged investment in creating and expanding the standard. Centralized architectures can, however, be used in the technological discontinuities where the market uncertainty related to the end-user needs is low [11].

An incumbent that has a large installed base and locked-in customers can gain a competitive advantage by a controlled migration strategy. The company can prevent backward compatibility for new entrants with its own legacy systems by influencing interface definitions of standards or by introducing an early new generation of equipment with the advantage of backward compatibility [10].

Evolution of compatibility and revolution of compelling performance are distinguishable, and their combinations are also possible. There is a trade-off between these extremes because improved performance decreases customers' switching costs, while in evolution existing customers can be better locked into the supplier. An ideal solution would be a significantly improved system or product that is also compatible with the existing installed base of the company [10].

In a virtual network of technologies that share a common platform, complementarities influence the value of individual parts of the system. The complementarities between interdependent technologies can have both negative and positive effects on the success of the technology evolution. In a virtual network of complementary goods that share a common technical platform, network externalities arise because a larger availability of the complementary components increases the value of their counterparts [10].

The technology that first creates a critical mass of users simultaneously benefits from the demand and supply sides of economies of scale simultaneously. The diffusion is also accelerated by network externalities, while the value of the subscriptions to the network is increasingly higher as the number of the users of the network increases. The associated process is also called the "bandwagon effect." These drivers of increasing returns lead to a situation where the winner technology faces an exponential growth of a virtuous cycle, while the loser technology gets increasingly weaker in a vicious cycle [10].

The variation phase is closed when the market selects a dominant design. Typically, the new technology and the related standards do not become the dominant design in their initial form, and the dominant design is not based on the leading edge of the technology. The dominant design does not embody the most advanced features, but a combination of the features that best meet the requirements of the early majority of the market [8]. The emergence of the dominant design leads to the further development of product platforms and related architectural innovations. This also leads to benefits of an increased offer of subsystem products as well as linking different technologies to a bigger system [12].

A dominant design emerges out of the competition between the alternative technological evolution paths driven by companies, alliance groups, and governmental regulators, each of them with their own goals [13]. Especially, regulation has a significant impact on the success of new technologies. Regulation defines the general boundaries of the business, while standardization provides a filtering impact that reduces the uncertainty by increasing predictability [14].

It is assumed that a harmonized market enables the economies of scale and lowers the price levels of telecommunication products and services. The telecommunication industry has been a sector with a strong and broad regulation of the wireless spectrum, technologies, services and competition, and several other aspects. The primary goal of regulation is to balance the sharing of social welfare among the market players, for example, vendors, operators, and consumers [15].

The dominant design tends to command the majority of the market until the next technological discontinuity. Companies now gain a deeper understanding of the technology, and its performance improvement starts to accelerate in an incremental manner [8]. The selection of the dominant design shifts the balance of innovation from product to process in order to decrease the production cost because of the increasing price competition. The product variation decreases and products develop on the basis of incremental evolution. As a result, the industry starts to consolidate because of the increasing number of acquisitions and mergers [7, 16]. At some point, diminishing returns begin to emerge as the technology starts to reach its limits and it is likely to be substituted by a new technology.

9.3 Study Framework

The theoretical background of technology evolution allows us to create a framework with the following 10 dimensions. Next, we describe the dimensions in more detail:

- Openness—the extent of availability of new technologies for all players in the industry
- Added value—the relative advantage over older technologies
- Experimentation—the threshold of end users to experiment with new technology
- Complementary technologies—the interdependence between complementary technologies
- Incumbent role—the product strategy of existing players
- Existing market leverage—the extent of redirection of existing customers to new technologies
- Competence change—the extent of required new competences
- Competing technologies—the role of technology competitors
- System architecture evolution—the extent of new technologies induced to the architecture
- Regulation—the influence of government regulation

Evidently, a number of dimensions correlate with each other. At least three examples emerge. Firstly, increased openness lowers the threshold to experiment. Secondly, contrary to the first example, the required system architecture evolution hinders experimentation via competence change. Finally, the incumbent players leverage the existing markets of their own, possibly to discourage moving to novel technologies.

To speculate the future evolution, the positively and the negatively affecting dimensions must be identified. However, this is highly dependent on the case where the framework is applied. Therefore, these assumptions must be addressed in the analysis.

9.4 Overview on Cloud Computing

Cloud computing is a concept in which a cloud provider offers computational resources remotely as a service. The computational work moves from dedicated servers into a data center of the provider, where a great number of servers perform computation for multiple customers. A cloud customer could, for instance, be a video-on-demand service provider that purchases the content storage and processing capacity as a service from the cloud provider.

Cloud providers offer services on three abstraction levels—**software as a service (SaaS)**, **platform as a service (PaaS)**, and **infrastructure as a service (IaaS)**:

- On the highest abstraction level (SaaS), the provider offers the customer applications that are, commonly, accessed via a Web interface. An example of such an application is Gmail, Google's Web-based email application.
- On the PaaS level, which is the middle abstraction level, a customer gets a platform on which they can run their own software. The platform is commonly restricted to one or few programming languages and provides its own services, such as platform-specific storage and databases. One such platform is Heroku.[1]
- The lowest level, that is, IaaS, offers customers access to virtual machines (VMs) that are logical abstractions of physical computers as well as other IT infrastructures such as storage and computer networks. As with physical machines, the customer can install its own software on the machine, beginning with the operating system. An example of an IaaS is Amazon's Elastic Compute Cloud (EC2) and its related services.

The three abstraction levels also require different levels of competence from the cloud customer. For example, a PaaS cloud customer must implement its own software, but the cloud platform manages the service scaling. On the other hand, an IaaS customer is required to handle the scaling features (increasing the amount of VM instances) and the distributed system communication by itself.

An important feature of cloud computing is that the provider typically offers the services on an "on-demand and pay-per-use" basis. Thus, the customer can acquire the required computing resources immediately and release the resources when they are no longer required.

Cloud computing benefits both the cloud customer and the cloud provider. It relieves the former of purchasing own computation hardware, therefore lowering upfront costs, together

---

[1] https://www.heroku.com

with removing the need to maintain and administer the own IT infrastructure. The latter, in turn, has economies of scale while providing services to many customers at the same time: hardware is cheaper in large quantities, and the computation tasks can be allocated to the physical machines more efficiently.

Furthermore, cloud computing has three different deployment models. The first model is a **public cloud**, where the customer buys the service from a separate company, for example, Amazon. The second model is a **private cloud**, where the cloud provider is, actually, the customer company itself. The last model is a hybrid cloud, which is a combination of the two previous models. In a **hybrid cloud**, the customer itself provides the base part of the capacity while the remaining required capacity is elastically acquired from a public cloud.

To provide cost-effectiveness, cloud computing bases itself on computer virtualization and extensive use of automation:

- In computer virtualization, the underlying hardware resources of a physical computer are shared by multiple VMs. A piece of software running on the physical machine, the hypervisor, is responsible for sharing access to the physical resources between the VMs and isolating the VMs from each other to provide security. In the context of clouds, an important feature of VMs is that they can be moved from one physical machine to another.
- Automated resource provisioning hastens decisions and execution of configuration changes and provides fault tolerance in the data center. For example, the selection of a physical machine for a new VM instance and VM migration to another physical machine must be automatized to reach sufficient efficiency. Automation is typically provided in the cloud service via its control software, the "cloud operating system." For example, launching a new VM instance does not require the user to select a physical machine for the VM, for the control software makes the selection.

Multiple cloud platform solutions, both proprietary and open, exist. However, proprietary solutions might lead to vendor lock-ins. To avoid lock-ins, an open solution might be preferred. Such solutions include Apache CloudStack[2] and OpenStack.[3]

9.5   Example Platform: OpenStack

OpenStack is a project that provides open-source software to create an IaaS cloud. Originally, it was created by Rackspace and NASA.

The project is a collection of subprojects that provide functionality for different areas of the cloud platform, such as compute and networking. Furthermore, similarly to other large-scale open-source projects, it is developed and led by a vast community that includes individual and corporate developers, cloud providers, and other project personnel.

This section describes OpenStack as an example of an open cloud platform. First, it details the general design and architecture of the platform. Second, it will describe the community of the project.

---

[2] http://cloudstack.apache.org
[3] https://www.openstack.org

9.5.1 OpenStack Design and Architecture

OpenStack is divided to subprojects that implement the different functional parts of a cloud platform. The main idea of the project is to provide well-defined application programming interfaces (APIs) that allow the user to access services providing the actual functionality required to control the hardware infrastructure. The primary APIs are based on Representational State Transfer (REST). Moreover, the project also bundles command line tools to interact with its services.

OpenStack supports numerous infrastructure services via its plug-in-based architecture. For example, support for a novel hypervisor requires implementing a driver plug-in and releasing it for others to use. This flexible nature allows OpenStack to interact with a number of different, proprietary or open, infrastructure technologies.

Figure 9.1 shows the main components of OpenStack. In more detail:

- Compute (code name Nova) controls the virtualized resources, such as virtual CPUs and memory and storage interfaces via interaction with the hypervisors. For instance, cloud users or a dashboard service accesses it to deploy a new VM. The list of supported hypervisors is extensive; Nova also supports other technologies, such as Linux Containers (LXC). Formerly, Nova also provided virtual networking interfaces for VMs. However, that responsibility is nowadays mostly transferred to the networking component Neutron.
- Dashboard (Horizon) provides a Web GUI to access other OpenStack services.
- Networking (Neutron) manages the virtual network connections between the virtual network interfaces of the VMs. It allows the user to create complex virtual network architectures that include virtual routers, load balancers, and firewalls. Its supported network back-end technologies range from Linux bridges to proprietary network control methods and software defined networking (SDN) controllers.

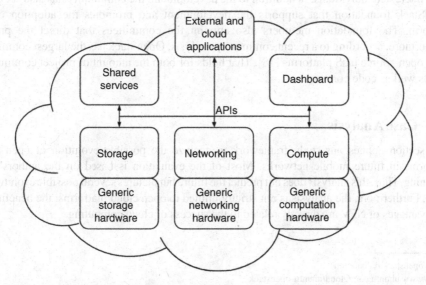

Figure 9.1 OpenStack architecture (Adapted from http://www.openstack.org/software/).

- Storage provides support for block storage, that is, disk volumes via Cinder and distributed object storage via Swift.
- Shared services include the identity service (Keystone), the image service (Glance), the orchestration service (Heat), and the telemetry service (Ceilometer). As the name implies, they serve other services, as well as human users. Keystone is responsible for user authentication and authorization, together with managing user credentials. Glance, in turn, manages the disk and server images. Heat, on the other hand, allows deploying cloud resources in predefined setups using templates. Finally, Ceilometer centrally collects and provides metering data from the cloud.

Additional subprojects have been constantly added to the OpenStack project in each release. The most recent release of April 2014, Icehouse, introduced four new capabilities: database service (Trove), bare-metal service (Ironic), queue service (Marconi), and Hadoop data processing (Sahara).

Today, OpenStack is available in multiple private cloud distributions and public cloud platforms. Private cloud platforms include Red Hat's RDO,[4] Ubuntu OpenStack,[5] and Rackspace Private Cloud.[6] Many cloud providers, such as Rackspace and HP, offer massive OpenStack-based public clouds, as well. In addition, it is already employed in telco systems as well. For example, Ericsson uses OpenStack to power its cloud system product [17].

9.5.2   OpenStack Community

An important part of an open-source project is the community behind it. It consists of people implementing and testing new features, people fixing bugs, and people deciding where the project should be headed.

The OpenStack community consists of the employees of the supporting companies and other interested individuals. In addition to the development, the community has also set up the OpenStack foundation that supports the development and promotes the adoption of the platform. The foundation members also appoint the committees that direct the project. Furthermore, according to a recent community analysis, OpenStack has the largest community of all open-source IaaS platforms [18]. That holds for both the amount of project communication as well as code commits.

9.6   Case Analysis

This section applies our study framework to analyze the possible evolution of open cloud platforms in future mobile networks. Most of the evaluation is based on the authors' own reasoning. Thus, this analysis does not predict the future but pictures several possible evolutionary paths. Furthermore, the analysis is not strictly limited to open cloud platforms: the benefits and disadvantages of NFV are directly related to the success of cloud computing.

---

[4] http://openstack.redhat.com
[5] http://www.ubuntu.com/cloud/ubuntu-openstack
[6] http://www.rackspace.com/cloud/private

9.6.1   Openness

We defined openness as the extent of availability of the new technology for all players in the industry. Naturally, open cloud platforms are open by definition: the source code is openly available; they are freely available; they can be deployed to generic hardware; and their development and user communities are open to join.

The open-source code of the platforms is generally distributed via public code repositories using different version control systems. In addition, it is released under a free software license, for example, GNU General Public License (GPL) or Apache License. The biggest differences between the licenses are whether they require changes to be released and whether the license must remain the same in future derivatives.

The platforms are, inherently, provided free of charge. The major Linux platforms also offer it via their packet managers. Furthermore, the platforms support generic commercial hardware. To support the platform deployment, many companies provide paid consultation and training.

Moreover, open cloud platforms are free of the threat of vendor lock-in.

Finally, open platforms of course allow the network functions and other services to be either open or closed solutions. Thus, the adoption of open IaaS cloud software does not limit the network operator's options to choose the solution providers.

9.6.2   Added Value

A new technology must provide a fair amount of added value to be accepted to the market. Open cloud computing platforms, together with virtualization, have several advantages and drawbacks compared to the current solutions.

In general, virtualization offers flexible allocation of computation capacity provided by generic, commercial off-the-shelf (COTS) hardware. Moreover, the automatized allocation introduced by the cloud platform offers increased utilization of the hardware. Increased utilization, in turn, allows greater energy efficiency. Therefore, this approach would also reduce overall expenses.

The benefits of open cloud platforms include modifiability and constant access to the novel features and updates on the platform together with lower cost. The first means that anyone can make changes to the platform code to support their own needs. The second fact promises that the progress the community makes is available to all members. Finally, the open platforms are free, thus lowering the total cost of the network.

Introduction of cloud computing in the mobile operator's core network also facilitates several novel business models. First, the operator could rent computing capacity from the telco cloud to third-party providers. Furthermore, the third-party services provided in the operator cloud could be offered similar control of spare capacity of the underlying infrastructure (i.e., reserve network tunnels with given quality of service). Therefore, this approach would transform the operator from a bit pipe into a computing and networking infrastructure provider.

In contrast to dedicated hardware devices, the software-based approach provides significant advantages. Compared to hardware development, software development allows a faster development cycle. The shorter development time also means that new services and technologies can be deployed more often.

However, there are a number of drawbacks, including decreased performance. A virtualized network function is supposedly slower than its dedicated counterpart. Thus, the same work must be distributed to multiple virtualized instances, which leads to modified architecture and functionality. For example, a distributed network function might require an additional aggregation layer solution.

The introduction of open cloud technologies to the mobile networks would open a new market for virtualized network function solutions. This allows new players to enter the networking business, which in turn could lead to competition resulting in faster development times and lower prices.

9.6.3   Experimentation

The threshold of experimenting with open cloud platforms is minimalistic. Usually, a cloud platform can be entirely installed to a single commodity computer. Since the platforms can be installed on top of free operating systems, usually with their package managers, beginning to experiment is easy, indeed.

Furthermore, the whole existing architecture is not required to change. The LTE core network, for example, is already fully IP based. Therefore, the network providers can start by virtualizing a single network function and replacing the corresponding dedicated machine with a virtual instance run on a cloud. Another approach could be to separate some traffic, such as machine-to-machine communication, to be served by virtual network functions [19].

Open cloud platforms also power numerous public cloud services, such as Rackspace's public cloud. An alternative approach to study a specific cloud platform is to try a publicly available instance in the beginning.

To stay up-to-date with current computing trends, the academia is also researching and experimenting on different, most likely open, cloud platforms. This benefits the whole community since a lot of innovation is also done at the universities and the results are published to the community.

9.6.4   Complementary Technologies

The complementary technologies of open cloud platforms include generic COTS computing hardware and SDN together with its related technologies.

Cloud platforms are typically run on standard hardware, that is, x86 servers, instead of proprietary dedicated hardware. Moreover, they might support other computing architectures. For example, OpenStack also runs on ARM-based hardware.

Currently, the idea of openness is also reaching hardware. For example, the Open Compute Project[7] aims to provide open schematics and designs of cloud computing infrastructure. In the future, open cloud platforms might be optimized to utilize open hardware or vice versa.

The other main complementary to cloud computing in mobile networks is SDN. Data center networking has been one of the main drivers for SDN. Therefore, it is logical to integrate the control of an SDMN to the cloud platforms to allow advanced network control, such as traffic

---

[7] http://www.opencompute.org/

engineering. Open cloud platforms provide drivers for both proprietary and open back-end technologies. Alternatives as the network controller include the OpenDaylight[8] or the OpenContrail.[9]

9.6.5 Incumbent Role

The incumbent players of the mobile network market comprise of hardware vendors, network providers, and network operators. In the center of the market are the network providers that manufacture the networks using devices from the hardware vendors and their own products, finally selling them to the mobile network operators.

In the past, network operators preferred multivendor solutions and possessed significant skills in system integration. However, modern networks are usually provided as a whole by a single provider since operators have reduced their effort in network construction and maintenance.

Massive increase in mobile data traffic and operator business demands highly developed solutions. Previously, network providers have based their solutions on top of hardware providers' proprietary products. The results have been expensive dedicated hardware devices that are designed to support the vendors' other products. However, the recent interest in more flexible solutions forces the network providers to consider their future strategy.

The open cloud platform approach would separate the hardware and the software business of mobile networks, thus allowing new players to enter the network business. The economies of scale would make the IT hardware providers the logical choice to provide the generic computing and networking infrastructure. In the software side, the development of network functions does not, anymore, require massive resources. Therefore, the network function market opens to both incumbent network providers and newly entering software companies.

To keep up with the competition, the existing mobile network providers have two options. On the one hand, they can continue with the dedicated device path and improve the existing solutions to support increased traffic and flexibility. On the other hand, they can seek for novel solutions from the virtualization domain via software development. Naturally, the approaches can be combined. This makes sense since virtualized and hardware solutions are interoperable. At least two examples of such a hybrid approach exist: First, Ericsson already provides the LTE core network in both virtualized and dedicated hardware solutions [20]. Second, Nokia Solutions and Networks (NSN) offers a telco cloud solution that supports both hardware and virtualized network functions together with multiple cloud platforms [21].

In general, we believe that the existing network vendors will offer virtualized solutions in parallel with dedicated devices. And, with increasing emphasis, the network solutions will become software based.

Another new market business opportunity to hardware and network providers is to design and provide the computing infrastructure for the mobile operators. The incumbent players are familiar with the requirements of mobile networks, which give them initial advantage.

---

[8] http://www.opendaylight.org/
[9] http://opencontrail.org/

9.6.6   Existing Market Leverage

The existing market is filled with dedicated offerings. Mobile operators have heavily invested to existing networks and devices. These investments must be amortized, which discourages moving totally to cloud-based network solutions.

The future of cloudified networks lies in future network investments. The technological disruption in LTE adoption requires the operators to invest in new networks. Fortunately for cloud approach, the interoperability of hardware and virtualized network functions allows operators to choose this approach.

Overcoming the existing leverage of hardware solutions requires positive experiences of virtualized solutions. Such experiences may be obtained via trials in test and production networks. Positive opinion on virtualized solutions could also affect the ongoing design of future, 5G and later, networks.

Another market leverage is the possible preference for proprietary solutions. However, the success of open approach in the IT business could provide an incentive to employ such technologies in mobile networks, as well.

9.6.7   Competence Change

Moving to cloud solutions requires significant change in both development and operating competences.

Developing virtualized solutions is, first and foremost, a software development effort. The main idea is to exploit generic hardware and differentiate with software products. Therefore, the network providers would have to integrate and implement software components instead of designing novel products from hardware components.

Another major change in cloudification is that network operators must become cloud service providers and administrators. Although cloud computing relies on automation, the operator must, nevertheless, have experienced cloud administrators configure, update, and troubleshoot the infrastructure and platform software. Furthermore, cloud computing bases itself to a different idea of fault tolerance: the massive amount of commodity hardware makes it bound to fail at some point, thus requiring the platform to tolerate failures instead of resisting them with custom high-availability hardware.

Moreover, the operators must be capable of selecting and integrating together different virtualized solutions. Otherwise, they must rely on providers for complete network solutions.

Hardware manufacturers typically provide training for the users of their proprietary technologies. In turn, training is also available for open cloud platforms. A number of networking and cloud computing companies offer consultation, training, and support for installing and using open cloud technologies.

9.6.8   Competing Technologies

Open cloud platforms face competition from three directions: traditional dedicated hardware, proprietary cloud platforms, and public clouds.

While virtualized products reach the market, the dedicated hardware solutions continue to evolve. Although the performance and benefits of these dedicated products may not increase significantly, the network operators are familiar with the technology and other aspects of the approach. Thus, continuing to invest in dedicated technologies might attract the operators unwilling to take risks.

Another competitor for open cloud platforms is the commercial, proprietary platforms, such as VMware vSphere[10] and Microsoft System Center.[11] These platforms might lock the user in certain technologies and providers. On the other hand, they might offer superior support because the provider controls the whole platform.

9.6.9   System Architecture Evolution

Cloudification of the mobile core networks requires the addition of generic computing capability. Two approaches exist: the first option is to add data centers, that is, facilities that house hundreds or thousands of servers to the core network infrastructure. Another possibility is to distribute the computation devices across the network.

The data center approach is the current de facto way in IT systems. Therefore, it could benefit from the experiences of administering IT cloud services. However, it only provides a small number of sites for the execution of network functions.

On the other hand, distributing computation across the core network would allow, for example, more flexible spatial allocation of network functions and other services. The idea has already been realized by NSN whose product portfolio includes NSN base stations that include computing capacity [22].

In the distributed approach, the latencies to the nearest network function instances could be lower. However, maintaining the hardware would be more difficult. For example, a server residing in a base station is harder to replace quickly than a server in a rack in a data center. Another question is network speed and latency from a remote location to supporting systems, such as database servers, and other virtualized instances.

Independent of the architecture choice, the existing legacy networks and their devices will not disappear. Therefore, the backward compatibility of novel solutions is also important. However, the virtualization of legacy network elements is also possible. Thus, the operators could, for example, in the case of device failure, replace the device with a virtualized solution.

Finally, it remains to be seen whether virtualized solutions and cloud computing affect the future 5G and later networks and become the dominant solution. Without question, the experiences of virtualization and cloudification of parts of 4G networks will affect their design.

9.6.10   Regulation

Rationally, regulation should not present obstacles for employing open cloud technologies in the mobile networks. Regulation might even encourage the opening of the network function market and, thus, require operators to seek multiprovider solutions, where multiple vendors provide the virtualized network elements.

---

[10] http://www.vmware.com/products/vsphere/
[11] http://www.microsoft.com/en-us/server-cloud/products/system-center-2012-r2/

On the other hand, regulation might affect the possible business cases of telco clouds and its complementary technologies. For example, net neutrality might become an issue when some over-the-top (OTT) services are offered dedicated network slices in the form of virtual private networks (VPN).

9.7 Discussion

As Section 9.2 pointed out, the path from a novel technology to a dominant design includes many surprising events. However, the possible evolutionary paths can be speculated.

This section will tie together the case-by-case analysis presented in the previous section and, based on that, discuss the future of open cloud platforms in mobile core networks. We separate the framework dimensions to enablers, neutral factors, and inhibitors based on the case analysis. Furthermore, we will point out the relations between the dimensions in this specific case.

As the enablers for the success of open cloud platforms in future networks, we identify the following dimensions: openness, added value, experimentation, and complementary technologies.

Firstly, high openness is a clear enabler for open cloud technologies. The technology and the communities are fully available to all, both existing and future, players in the industry. Thus, the mobile network industry would gain immediate access to the advances of the IT industry that are based on the success of cloud computing model. Furthermore, the virtualized network element market possibly attracts new players from the IT industry and open-source communities to develop products that compete with the new and the old solutions of the incumbent players. Thus, the development speed increases and solution prices would possibly decrease.

In the big picture, openness is clearly related to the threshold of experimentation and added value. Open platforms are available to everybody, making them easy to try and study. They also create a market for applications that employ the platform.

Secondly, open cloud platforms also offer added value. The value proposition of open platforms, including price, modifiability, and access to the whole development effort of the community, is attractive to the network operators and providers. Moreover, virtualization and the cloud computing approach would address many challenges present in modern networks.

Network providers have already realized the benefits of open cloud platforms and employ them in their products. Thus, the added value of such platforms already affects the incumbents and their role in the evolution of mobile networks. Furthermore, the value increase promotes the architecture evolution to integrate cloud computing to the networks.

The third enabler is the low threshold to experiment with the technology that is boosted by the increased openness. Different industry players can study the employment of the platforms to their benefit. Furthermore, openness attracts the academia to study the platforms. Altogether, the combined experimental efforts might lead to new value propositions via creative ways of using the technology.

Finally, the complementary technologies of open cloud platforms support the evolution toward cloudified mobile networks and open platforms as well. For instance, network providers see SDN as an important technology to expand the capabilities of modern mobile networks. In turn, the open cloud platforms quickly support novel technologies via the community development effort. Thus, the novel networking features would be promptly supported by the network platform.

The complementary technologies also affect the added value of open cloud platforms, the architectural evolution of mobile networks, and the required competence change. The complementary technologies introduce generic hardware to the core network and should thus lower the costs. However, the required competences of the network administrators and designers are significantly different.

Next, we identify the neutral dimensions, that is, the dimensions that do not seem to clearly resist nor promote the evolution toward the inclusion of open cloud platforms to the mobile networks. These three factors consist of the role of the incumbent players, system architecture evolution, and regulation.

Firstly, the incumbent players have a significant but uncertain role in the success or failure of the employment of open cloud platforms in mobile networks. To date, network providers have already included them to their product portfolios. On the other hand, they have not abandoned the dedicated solutions business either. Therefore, it is evident that network providers are unsure of the future dominant design. Thus, the success of open cloud depends on the reactions of the network operators in the live deployments. In summary, the effect the incumbents have on the evolution is uncertain.

Secondly, system architecture evolution is actually the end result of technology evolution. Therefore, we identify it as a neutral dimension. On the other hand, an extensive change in the system architecture leads to significant change in required competences.

Thirdly, regulation is also seen as a neutral factor. For instance, it might favor the use of open cloud platforms in the core networks. On the other hand, it might limit many of the possible new business models, such as dedicated network slices and quality of service differentiation. Regulation might also lead to multiple network function and service providers joining the market, thus affecting the future system architecture as well.

The final three dimensions, that is, existing market leverage, competing technologies, and competence change, seem to restrict the adoption of open cloud platforms.

Firstly, the competing technologies, including dedicated solutions and proprietary cloud platforms, could hinder the employment of open cloud technology. Some industry players will definitely resist change and continue to offer and employ dedicated solutions in their networks. However, dedicated and virtualized solution can and will coexist. In turn, the competition between open and closed cloud platforms will probably exist similar to the corresponding competition in the IT sector. Altogether, the market share between the competing solutions will also affect the future system architecture.

Secondly, the leverage of the existing market inhibits the future cloudification. Network operators have a large installed base of dedicated hardware solutions, and the dedicated products are proven to work. Furthermore, the existing network vendors are not likely to welcome new entrants to the network business.

Finally, we think the most notable resistance to open cloud technologies emerges from the required competence change. Cloud computing is a revolutionary approach to offer mobile connectivity. Therefore, the operators must reeducate the network administrators and technicians or recruit new employees with the required skills. The amount of change might also increase the attractiveness of continuing with existing approach. Moreover, it also affects the role of incumbent vendors: the failure of existing vendors to transform themselves to provide virtualized solutions and clouds will offer a chance for new players to enter the market.

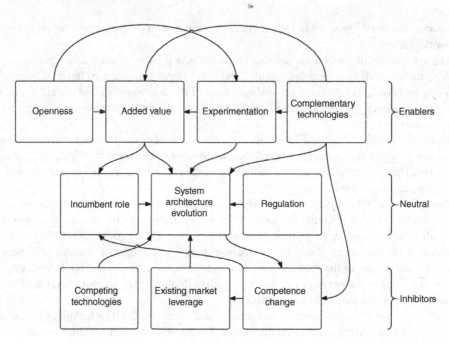

Figure 9.2 Identified roles and relations of the framework dimensions in the technology evolution of open cloud platforms in mobile networks.

Based on the discussion in this section, Figure 9.2 summarizes the roles and relations of the dimensions of the employed framework. It is evident that in this case the future system architecture evolution depends on success in a number of areas. Open cloud approach addresses many present challenges of mobile network business, such as vendor lock-ins, flexibility, and cost. It also supports the other key developments, such as NFV and SDN. On the other hand, cloudification is a major technological disruption that requires significant changes in competence and solution methods.

We believe that the cloud approach will be first trialed in some specific use cases, such as machine-to-machine communication. If the performance seems appropriate, new networks will be increasingly built from virtualized components running in operator clouds that are run on generic hardware. In that case, the virtual network function market opens and new players enter the business. However, the dominant cloud design may consist of either open or proprietary platforms. Differences in price and hardware support promote open solutions. In turn, the vendors have a long history of using proprietary technologies. Solutions of both types have room in the market, but the dominant type cannot be foreseen.

9.8 Summary

This chapter discussed the technology evolution in mobile networks in the specific case of open cloud platforms. We provided an analysis based on a framework drawn from the theory of generic technology evolution.

Generic technology evolution theory suggests that in a market of numerous players and solutions, a dominant design emerges as a product of legacy product evolution and novel disruptive solutions. The actual reasons for the success of one technology over others are difficult to identify exactly. However, common features in successful technology allow us to analyze possible evolutionary paths. In this case, we selected 10 dimensions for our analysis framework.

The chapter also briefly introduced the concepts of cloud computing, including the division to different service abstraction levels (SaaS, PaaS, and IaaS) and deployment models (public, private, and hybrid). We also presented OpenStack as an example of an open cloud platform project.

Based on the analysis, open cloud software addresses some problems present in the modern networks that are provided with dedicated hardware. It also supports the modern network developments, such as NFV and SDN. However, complete cloudification is not possible due to large installed base of dedicated hardware and sudden competence change. Therefore, we predict that cloud technology, open or proprietary, will be introduced gradually to the network. For example, it could be initially deployed to face the increasing machine-to-machine traffic. The possible positive experiences from the initial trials and development cloud technologies would lead to wider usage and the birth of the virtual network function market. The competition between the developers of such functions would drive the virtualization of mobile networks forward. However, the final selection between open and closed cloud platforms depends on the network provider or operator preferences.

Acknowledgments

This work has been performed in the framework of the CELTIC-Plus project C2012/2-5 SIGMONA. The authors would like to acknowledge the contributions of their colleagues. This information reflects the consortium's view, but the consortium is not liable for any use that may be made of any of the information contained therein.

References

[1] Stake RE. *The art of case study research*. Sage, Thousand Oaks; 1995.
[2] Yin RK. *Case study research: Design and methods*. Sage, Thousand Oaks; 2003.
[3] Tushman ML, Anderson P. Technological discontinuities and organizational environments. Adm Sci Q; 1986:439–65.
[4] Christensen CM. *The innovator's dilemma: When new technologies cause great firms to fail*. Harvard Business School Press/HBS Press Book, Boston; 1997.
[5] Foster RN. *Innovation: The attacker's advantage*. Summit Books, New York; 1986.
[6] Christensen CM, Rosenbloom RS. Explaining the attacker's advantage: Technological paradigms, organizational dynamics, and the value network. Res Pol; 1995;24(2):233–57.
[7] Porter ME. *Competitive advantage: Creating and sustaining competitive performance*. Free Press, New York; 1985.
[8] Anderson P, Tushman ML. Technological discontinuities and dominant designs: A cyclical model of technological change. Adm Sci Q; 1990;35(4):604–33.
[9] Gaynor M. *Network services investment guide: Maximizing ROI in uncertain times*. John Wiley & Sons, Hoboken; 2003.
[10] Shapiro C, Varian H. *Information rules*. Harvard Business Press, Boston; 1998.

[11] Gaynor M, Bradner S. The real options approach to standardization. Proceedings of the Hawaii International Conference on System Sciences, Outrigger Wailea Resort, Island of Maui, Hawaii; 2001.

[12] Henderson RM, Clark KB. Architectural innovation: The reconfiguration of existing product technologies and the failure of established firms. Adm Sci Q [Internet]. Sage Publications, Inc. on behalf of the Johnson Graduate School of Management, Cornell University; 1990;35(1):9–30. Available from: http://www.jstor.org/stable/2393549. Accessed February 18, 2015.

[13] Tushman ML, Anderson PC, O'Reilly C. Technology cycles, innovation streams, and ambidextrous organizations: Organization renewal through innovation streams and strategic change. Manag Strateg Innov Chang. Oxford University Press, New York; 1997:3–23.

[14] Longstaff PH. *The communications toolkit: How to build and regulate any communications business*. MIT Press, Cambridge; 2002.

[15] Courcoubetis C, Weber R. *Pricing communication networks, economics, technology and modelling*. John Wiley & Sons, Chichester; 2003.

[16] Abernathy WJ, Utterback JM. Patterns of industrial innovation. Technol Rev 2. 1978:40–7.

[17] Ericsson. Cloud system. 2014 [cited May 28, 2014]; Available from: http://www.ericsson.com/spotlight/cloud-evolution. Accessed February 18, 2015.

[18] Jian Q. CY14-Q1 community analysis—OpenStack vs OpenNebula vs Eucalyptus vs CloudStack [Internet]; 2014 [cited May 15, 2014]. Available from: http://www.qyjohn.net/?p=3522. Accessed February 18, 2015.

[19] ETSI. Network functions virtualization (NFV); Use Cases—White Paper [Internet]; 2013. Available from: http://www.etsi.org/deliver/etsi_gs/NFV/001_099/001/01.01.01_60/gs_NFV001v010101p.pdf. Accessed February 18, 2015.

[20] Ericsson. Launch: Evolved packet core provided in a virtualized mode industrializes NFV/Ericsson [Internet]; 2014 [cited May 27, 2014]. Available from: http://www.ericsson.com/news/1761217. Accessed February 18, 2015.

[21] Nokia Solutions and Networks. Nokia telco cloud is on the brink of live deployment [Internet]; 2013 [cited May 27, 2014]. Available from: http://nsn.com/file/28161/nsn-telco-cloud-is-on-the-brink-of-live-deployment-2013. Accessed February 18, 2015.

[22] Nokia Solutions and Networks. NSN intelligent base stations—white paper [Internet]; 2013 [cited May 27, 2014]. Available from: http://nsn.com/sites/default/files/document/nsn_intelligent_base_stations_white_paper.pdf. Accessed February 18, 2015.

Part III

Traffic Transport and Network Management

10

Mobile Network Function and Service Delivery Virtualization and Orchestration

Peter Bosch,[1] Alessandro Duminuco,[1] Jeff Napper,[1] Louis (Sam) Samuel,[2] and Paul Polakos[3]

[1] Cisco Systems, Aalsmeer, The Netherlands
[2] Cisco Systems, Middlesex, UK
[3] Cisco Systems, San Jose, CA, USA

10.1 Introduction

The concept of virtualization in telecommunications is not new. Ever since the need for the testing of telecommunication functions appeared, there had to be some means of testing a function either in a simulated environment or simulating the function itself to see if modifications and evolutions of the functions were viable prior to deployment. In either case, virtualization has gained ground as the potency of computation and storage has increased with every evolution of processors and storage. This has led to the point in telecommunications where network functions that were once considered only suitable to be run on bespoke hardware because of various limitations are now potential applications that can be run on common off-the-shelf processing. The idea of orchestration in telecoms is also not a new one. Orchestration in essence is the intelligent automation of repetitive processes whether for business or engineering.

Modern telecommunication networks have evolved over many decades. The evolution has usually meant that newer portions of the network have to coexist with older portions of the network. The resulting heterogeneity naturally led to increased network complexity. What this also meant is that the mechanisms by which networks and their elements are provisioned

Software Defined Mobile Networks (SDMN): Beyond LTE Network Architecture, First Edition.
Edited by Madhusanka Liyanage, Andrei Gurtov, and Mika Ylianttila.
© 2015 John Wiley & Sons, Ltd. Published 2015 by John Wiley & Sons, Ltd.

have become tied to vendor-specific systems. The implication of both of these things is that to deploy new services and simply manage the existing network has become a relatively expensive task.

The confluence of improving off-the-shelf processing, leading to the enablement of virtualization, and of the application of more engineering processes can lead to an advantageous situation where not only large portions of the network can be virtualized onto inexpensive commoditized hardware but also the ensemble can be orchestrated into new solutions. The telecommunication industry has realized this, and there are many initiatives underway that seek to exploit this confluence. One such initiative is the ETSI network function virtualization (NFV). This initiative seeks to standardize the interfaces between the virtualized functions and the overall management of these functions. NFV relies heavily on an underlying programmable networking substrate, commonly referred to as software defined networking (SDN), to allow dynamic deployment, isolation, and control of a multiplicity of network services (NS) in a multitenant data center environment. In this chapter, we describe in detail and in the mobile network context the ETSI NFV architecture and underlying support provided by SDN.

10.2   NFV

The ETSI NFV architecture is set apart from other cloud management approaches through its aim to manage and operate virtualized critical network functions in a private data center (such as packet cores, IMS, etc.). In this architecture, guarantees for high availability and reliability (greater or equal to the current "5–9 s"[1]) are to be met while simultaneously achieving lower capital and operational expenses for the same services provided via traditional means. At the time of writing, work is not complete and there are still debates on how such data centers will be operated. However, there is enough information to report on their architecture and likely modes of operation. The basic premise of NFV is that network functions have their software implementation decoupled from the computation, storage, and network resources that they use. This means that for the telecom service provider (SP), new ways of operating, administering, maintaining, and provisioning such network functions will emerge.

10.2.1   The Functionality of the Architecture

Figure 10.1 shows the current NFV Management Architecture, and each of its functional blocks is described in more detail in the following text (for greater detail, see Ref. [1]).

10.2.1.1   Network Function Virtualization Orchestrator

The top-layer orchestrator of the NFV system is the Network Function Virtualization Orchestrator (NFVO). It instantiates and manages services provided through virtual network functions (VNFs) of a VNF 3GPP virtual Evolved Packet Core (vEPC) or 3GPP IMS. The NFVO endeavors to accomplish three objectives: (i) instantiate VNFs, (ii) manage the

---

[1] In short, they have service outages of less than 315 seconds per year.

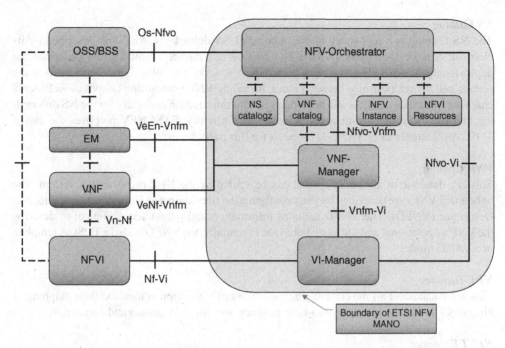

Figure 10.1 ETSI NFV Management and Orchestration (MANO) architecture.

service-level agreements (SLAs) of the VNFs, and (iii) perform NS orchestration. In order to achieve these objectives, the NFVO uses a set of descriptor files (NS, network function, virtual link, and Virtual Network Function Forwarding Graph Descriptors (VNFFGD)) to drive the instantiation of the VNFs by orchestration system. The descriptor files are templates or models that describe a VNF or an NS in terms of its required resources, configurations, and associated SLA descriptions. The descriptor files are covered in more detail in a later section.

In order to manage the SLA of the NS and VNFs under its control, the NFVO has the end-to-end view of the resources being allocated to the NS and VNFs. To accomplish this, the NFVO possesses the current snapshot of the distribution of in-use and available NFV data center resources retrieved from the Virtualized Infrastructure Manager (VIM) via the NFVO-Vi interface. The NFVO uses this information along with application-specific information contained in the NFV descriptors (such as scaling rules for a given VNF) and the overall projected resource consumption to make placement decisions where the NFV should be instantiated into the NFV data center. To assist in this activity, the NFVO notionally uses a number of databases. There are essentially two categories of databases:

- Databases that hold information (model definitions and descriptions) regarding NS and their components—the NS Catalog and VNF Catalog (see Fig. 10.1)
- Databases that describe what NS are currently deployed and state and availability of infrastructural resources—the NFV Instance and Network Function Virtual Infrastructure (NFVI) Resource databases (see Fig. 10.1)

NS Catalog

The NS Catalog is a repository of the onboarded NS definitions. An NS is described by its Network Service Descriptor (NSD). The NSD is a template or model that describes the deployment of an NS. The description includes the service topology (the VNFs used in the service and the relationships between them, including VNF Forwarding Graph) as well as NS characteristics such as SLAs and any other relevant information necessary for the NS onboarding and life cycle management of its instances. Currently, ETSI NFV proposes the use of TOSCA [2] templates or YANG [3] models for this purpose.

VNF Catalog

This is a database of all the VNFs that can be started by the NFV orchestration system. The onboarded VNFs are described by their configuration files known as Virtual Network Function Descriptor (VNFD). The VNFD contains information and parameters sufficient to describe the VNF's operational and life cycle behavior. Potentially, the VNFD is also a TOSCA template or a YANG model.

NFV Instance

This is a database of all the currently active (running) VNF applications and their mapping to virtual NS as well as additional run-time instance-specific information and constraints.

NFVI Resource

This is effectively a database that an inventory of all the available, reserved, and allocated resources across the entirety of SP domains. This database is kept consistent with the state of the NFV system so that it can be interrogated to reserve, allocate, or monitor the state of resources in the system. This is the entry point to the process by which SLAs in the system are maintained.

Although the NFVO in principle possesses broad information on a VNF, it in fact does not and should not need to understand the functional role or operation of the VNF itself. The information it does have is there to manage a service and enforce its SLA.

10.2.1.2 VNF Manager

The middle layer of the orchestration system is provided by the VNF Manager (VNFM). The VNFM is responsible for the life cycle management of VNF instances. Life cycle management in this context means:

- Handling VNF instantiation, that is, the allocation and configuration of virtual machines (VMs)[2] for the VNF.
- Monitoring the VNF—This can mean one of two things:
  - Either direct monitoring of the VMs in terms of CPU load, memory consumption, etc.
  - Or the presentation of application-specific data from the VNF VMs.
  In both cases, the VNFM collects the VNF application-specific data and NFVI performance measurements and events. This data is passed up to the assurance function of the NFVO for analysis or to the VNF Element Management System (EMS).

---

[2] In ETSI NFV terminology, an NFV VM is termed as Virtual Deployment Unit (VDU).

- Elastic control of the VNF VMs, that is, if the VNF application allows it, given a set of triggering events spin up new VMs or remove existing VMs.
- Assisted or automatic healing of the VNF, that is, restarted, stalled, or stopped VMs (assuming the VNF is capable of such management).
- Terminating the VNF VMs, that is, withdraw the VNF from service when requested by the OSS via the NFVO.

A VNFM can potentially manage single or multiple VNF instances of the same or different types. Moreover, as NFV matures, the possibility of a generic VNFM becomes likely. However, for reasons of pragmatism (ETSI NFV may mature slowly), the NFV Management and Orchestration (MANO) architecture also supports cases where VNF instances need specific functionality for their life cycle management, and this may be delivered as part of the VNF package. In this context, a VNF package is deemed to be the collection of the VNFD and its software image(s) and any additional information used to prove the validity of the package. This means that the VNFM can come bundled with the application. The advantages and disadvantages of this approach are covered in a later section.

10.2.1.3   VIM and NFVI

The lower layer of the orchestration system is the VIM. It is responsible for controlling and managing the compute, storage, and network resources within one operator's subdomain (i.e., a physical data center). In ETSI NFV, the physical resources (the compute, storage, and network) and software resources (such as the hypervisor) are collectively termed the NFVI. The VIM may be capable of handling multiple types of NFVI resources (via, e.g., Openstack cloud management system), or it may only be capable of handling a specific type of NFVI resources (an example of this would be VMware's vSphere).

In either case, the VIM is responsible for:

- The management of allocation/upgrade/deallocation and reclamation of NFVI resources
- The association of the virtualized resources to the compute, storage, and networking resources
- The management of hardware (compute, storage, and networking) and software (hypervisor) resources
- The collection and forwarding of performance measurements and events from the hardware and software to either the VNFM or the NFVO

At the time of writing, there is some debate as to whether network resources are explicitly managed by the VIM, for example, through open-source SDN controllers as provided by OpenDaylight [4], in which Openstack Neutron can interface with OpenDaylight [5].

10.2.1.4   Traditional Elements: Element Manager and Operation and Business Subsystems

Figure 10.1 shows some traditional elements included within. These are the Element Management (EM) and Operation and Business Subsystems (OSS/BSS) responsible for the conventional means of managing and assuring network functions. There are a couple of implications associated with the inclusion of these elements. Firstly, the inclusion of both elements means that a VNF can be managed via conventional means, and secondly, their inclusion provides a starting point for operational migration. We note that at the time of writing, ETSI NFV has not resolved how functions such as the EM would evolve given that the combination of NFVO and descriptor files has the nucleus of similar functionality.

Figure 10.1 shows the current NFV Management Architecture, and each of its functional blocks is described in more detail in the following text (for greater detail, see Ref. [1]). The NFV MANO architecture is a three-layer orchestration system. The top layer (the NFVO) orchestrates the NS, the middle layer (the VNFM) orchestrates the life cycle of virtual functions, and the bottom layer orchestrates data center resources and infrastructure. There is cardinality associated between these layers. There is a single NFVO that controls one or more NFV data centers. Therefore, one NFVO controls one or more VIMs. There is a one-to-one relationship between the VIM and a data center or data center partition. An NFVO can be associated with multiple VNFM, and a VNFM can have an association with one or more VNFs.

10.2.2 Operation of the ETSI NFV System

The entire ETSI NFV system can work in one of two main modes:

(i) Traditional, that is, driven directly via the OSS through an existing NMS/EMS that manages the VNF. In this mode, essentially all that has changed is that the network function has been virtualized.
(ii) Orchestrated, that is, driven via the ETSI NFV MANO system. In this mode, the entire system is driven through a collection of descriptor files that are held in the NFVO catalogs.

10.2.2.1 ETSI NFV Descriptor Hierarchy

The ETSI NFV descriptor files have a hierarchy to them (Fig. 10.2). At the top level are the NSDs: the NSD and VNFFGD. The NSD and the VNFFGD describe the service in terms of the VNFs that compose the service, its required network topology, and its KPIs. The middle level of the hierarchy includes the Virtual Link Descriptor (VLD) and VNFD. The VLD describes the inter-VNF link requirements (bandwidth, QoS, Hypervisor, vSwitch, NIC, etc.), while the VNFD describes the VNF. The VNF is further broken down to form the lower level of the hierarchy and is composed of three subcomponents; these are the VNFD_element, VNF_element, and VDU_element. The VNFD_element essentially names the VNFD. The VNF_element describes the composition of the VNF in terms of the number of separate VMs and their overall management, whereas the VDU_element describes the individual VMs that make up the VNF. The NSD and VNFD are described in more detail in the following subsections.

The NSD
The NSD describes the service in terms of a list of VNFs that go to compose the service, a list of VNF Forwarding Graphs that essentially describe how the various NVFs are connected together to effect the service, and a list of the associated dependencies for each of the NFV. The dependencies can be as simple as describing the order in which the NFV are brought up to more complicated relationships, for example, predication by other NS. The NSD also

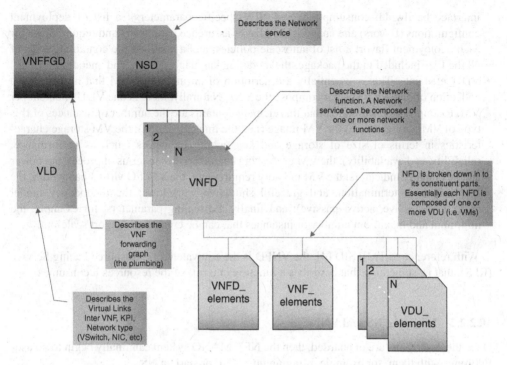

Figure 10.2   Hierarchy of NFV descriptors.

includes information that describes the SLA for the service. This is achieved by disclosing a list of parameters to monitor associated with a given configuration of the service and a list of autoscaling policies described in terms of metadata for that service.

With reference to TMF SID (see Ref. [4] and references contained therein), the NSD loosely approximates a Customer Facing Service (CFS), that is, something that can be ordered by a customer of the system. In this respect, examples of an NSD would be voice services, Internet access, video optimization, or a content delivery network—depending on who or what the end user is (e.g., a subscriber or an enterprise).

The VNFD
The VNFD describes the VNF components down to the VM level. The VNFD is composed of three elements:

- VNFD_element—Basically the name of the VNF and who the vendor of the VNF is.
- VNF_element—This is a list of elements that make up the VNF. Examples of information contained in this element are the number of VMs that make up the VNF; the set of workflows that describe initiation, termination, graceful shutdown, and other life cycle events; a list of the type of network connectivity required by the NFV components; a list of the external interfaces exposed by the VMs enabling connection to other virtual and physical network functions; a list of the internal connectivity/interfaces between the VNF VMs; a list of dependencies of the VNF components; a list of monitoring parameters (e.g., CPU utilization,

interface bandwidth consumption, and NFV-specific parameters); a list of deployment configurations (flavors) and an associated list of assurance parameters and requirements for each deployment flavor; a list of autoscale policies; and a manifest file containing a list of all the files included in the package, the modeling language version, and encoding format.
- VDU_element—This is essentially a description of an individual VM that makes up the collection of distinct VMs that compose the VNF. Naturally, there is one VDU_element per VM. Examples of information that this element contains are the number of instances of **this** type of VM; a reference to the VM image (i.e., the image location); the VM storage characteristics in terms of size of storage and key quality indicators (such as performance, reliability, and availability); the VM processing characteristics in terms of processing power and key quality indicators; the VM memory requirement; the VM I/O virtual bandwidth; the VM initiation, termination, and graceful shutdown workflows; the redundancy model (e.g., active–active, active–passive); and finally the scaling parameters, for example, the minimum and maximum number of instances that can be created to support scale-out.

With reference to TMF SID [4], the VNFD is the equivalent of a Resource Facing Service (RFS), that is, something that describes a function in terms of the resources it consumes.

10.2.2.2 Initiating NS and VNFs

Once the descriptors are onboarded, then the NFV MANO system can finally begin to do useful things with them, for example, bringing up a VNF or start an NS.

NS Instantiation
Exactly how the OSS gets the stimulus for initiating the NFV NS is out of scope of this document; suffice it to say that an initiation command is passed between the OSS and NFVO. Currently, this is the only way that an NFV service can be brought into existence. However, there are proposals for a more dynamic operation of NS instantiation, where the stimulus can arrive from more than just the OSS, for example, a self-service portal that the SP may expose to its business partners.

This instantiation command carries the NSID information. This information includes:

- NSD reference that is the NS identifier of an already onboarded NS
- Optionally, a list of already running VNFs from which the service can be constructed
- The method of scaling for the network work service (manual or automatic)
- The flavor (the particular configuration) for the service
- A list of threshold descriptors for the service
- A list of autoscale policy descriptors for the service if autoscaling is permitted

The logical implication of the last three items in the NSID is that the:

- VNFs for this particular service must be tagged in a way that the VNFM can pass monitoring data relevant to the service to an assurance system capable of grouping the data for monitoring how the service is performing.

- Thresholds and autoscale descriptors override the existing thresholds and autoscale descriptors contained in the NSD if they are valid (consistent and within accepted permitted range of values).

Service instantiation can proceed along one of several alternative paths. The NFVO determines the appropriate path by comparing the service VNF dependencies (as described in the NSD) against the list of active VNFs maintained in the NFV instance database. The paths are categorized as follows:

- *None of the service components are running.* If the VNFs that compose the NS already exist in the VNF Catalog and none of the VNFs are already instantiated, then the NFVO instantiates the creation of the necessary VNFs using the information already stored to in the VNF Catalog, and the service is stitched together using the information contained in the VNFFG references of the NSD.
- *All of the service's VNFs are already running.* If this is the case and the necessary VNFs have no restriction on them being used for multiple services, then all that may be required is for the service to be plumed using the VNFFG references of the NSD.
- *Some of the service's VNFs are running.* In this case, the NFVO then initiates only those that are missing and once again stitches together the VNFs (new and existing) using the information contained in the VNFFG references of the NSD.

The flow captured in Figure 10.3 shows this activity.

VNF Instantiation

VNF instantiation refers to the process of identifying and reserving the virtualized resources required for a VNF, instantiating the VNF, and starting the VMs associated with each VNF, in short the initial phases of VNF life cycle management.

Unlike the NS instantiation request, the VNF instantiation request can come from a number of sources. It can be as part of an NS instantiation (see previous section); it can be as part of commissioning a new VNF as a direct request from the OSS; it can be as a result of a scaling request from a VNFM; or it can be as a request from the EMS of the VNF. In any event, the NFVO receives a Virtual Network Function Initiation Descriptor (VNFID) that includes (i) a reference to the VNFD to instantiate the VNF, (ii) a list of network attachment points for the VNF, (iii) the scaling methodology and threshold descriptors (i.e., the parameters on which to trigger scaling if automatic scaling has been selected), and (iv) the flavor (configuration) to use.

We note that much of the information presented here is still under discussion by ETSI NFV and that at the time of writing, the operational flows and procedures have yet to be defined normatively and may therefore change.

10.2.3   Potential Migration and Deployment Paths

Given the architectural blocks described in a previous section, it can be seen that the VNF, EMS, and VNFM can be packaged in a number of configurations that can give rise to a number of potential deployment migration paths from current to future modes of operation. These configurations range from having individual VNFs mostly orchestrating themselves to fully

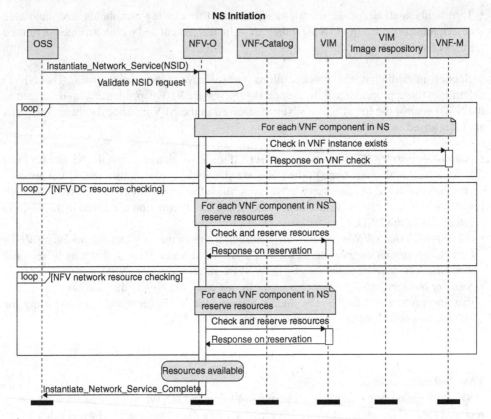

Figure 10.3 Initiation of network service.

automated operation of the SP network based on ETSI NFV MANO. The next four subsections indicate those migration configurations for a VNF, from the simplest and closest to current mode of network operation to the configurations closest to the future mode of operation end state.

10.2.3.1   VNFM Packaged with EMS and VNF

The simplest configuration is where the VNFM is packaged with the application (the VNF) and its EMS/NMS (see Fig. 10.4). The advantages of this configuration are that it:

- Hides the complexity of running the application from the NFVO:
  - The VNF is self-contained. In effect, the NFVO system brings up the very first VM of the application—the VNFM. The VNFM then brings up the rest of the application.
  - The assurance for the application is provided by traditional means. The VNFM may provide an aggregation point for collecting monitoring and performance information before passing it to the application EMS.

- Scaling events (for instance, to satisfy SLA requirements) can be handled either as manual interventions into the EMS via the OSS or as events triggered by scaling rules in the VNFM that subsequently lead to automatic action or manual intervention from the OSS. Note that all requests for additional resources (VMs and network bandwidth) still have to be made via the NFVO and authorized before the resources can be used.
- Allows the application to be run via more traditional means (OSS/NMS/EMS/VNF). In other words, operation of the VNF is possible in the absence of an NFVO.

There are possible long-term disadvantages to this configuration. Future SP operation may require composite end-to-end solutions requiring tighter control over the way the NFV data center operates as an NFV forms part of a larger solution. Therefore, assurance, control, and apportionment of resources may be better satisfied via actions taken at the NFVO and generic VNFM level rather than as depicted in Figure 10.4.

To support such cases, the VNFM-specific functionality needs to be integrated with the VNFM generic common functionality or, more broadly, integrated into the NFVO. How such integration is achieved is out of scope for NFV MANO but is a much relevant operational problem for any data center's NFV orchestration system.

Pragmatically, in the early stages of NFV, it is likely that many virtualized applications (VNFs) will be capable of running independent of the ETSI NFV system.

10.2.3.2 VNFM Packaged with VNF (Separate or Existing EMS/NMS)

This configuration (see Fig. 10.5) is much the same as the previous one, with the exception that the VNF and VNFM come as a single package and may integrate with an existing EMS and NMS. The advantages and disadvantages are much the same as stated previously.

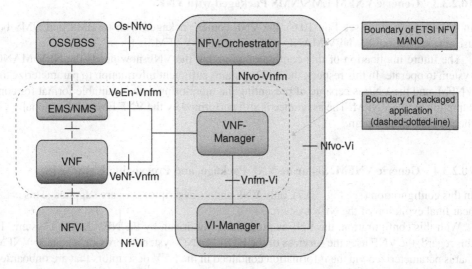

Figure 10.4 NFV Management Architecture where the VNFM is supplied with the application and its EMS/NMS.

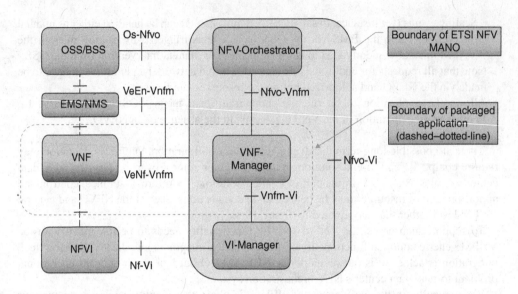

Figure 10.5 NFV Management Architecture where the VNFM is supplied with application.

This configuration is the next step along the operational migration path. In this configuration, the assurance of the VNF may not necessarily be presented to the existing EMS/NMS system, and assurance information in the future may be reported back through the NFVO via the VNFM. This situation may occur in the future when composite end-to-end solutions may be composed out of the NFV Catalog.

10.2.3.3   Generic VNFM EMS/NMS Packaged with VNF

In this configuration (see Fig. 10.6), the VNF comes packaged with its EMS and NMS but uses the services of the NFV MANO system's generic VNFM.

The future implication of this configuration is that the VNF now needs the NFV MANO system to operate. In this respect, the NFD contains sufficient information to parameterize the VNFM, and the VNF is capable of presenting the monitoring data in suitable format for consumption by the VNFM. The assurance is still performed by the VNF EMS and optionally via the NFV MANO system.

10.2.3.4   Generic VNFM, Separate VNF Package, and Preexisting EMS/NMS

In this configuration (see Fig. 10.7), only VNF and its descriptor files are supplied. This is the near final evolution of the NFV system.

With this configuration, the VNF is predominantly driven by the NFV MANO system. In this regard, the VNF uses the services of the ETSI MANO system to provide a generic VNFM that is parameterized via the information contained in the NFV descriptors that are onboarded to the system (see later section for more information). The VNF has a monitoring agent within it (e.g., Ganglia [5]) and provides all the necessary monitoring data to the generic VNFM.

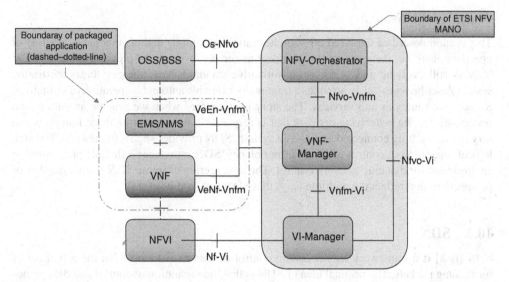

Figure 10.6 NFV Management Architecture where the VNF is supplied with its EMS/NMS. The VNFM is generic to the ETSI NFV MANO architecture.

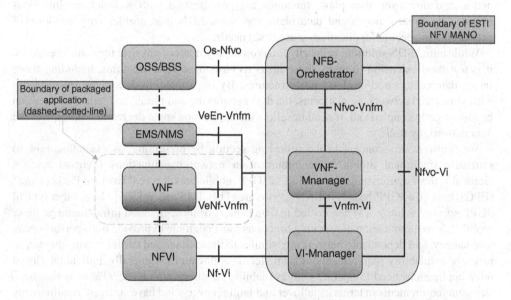

Figure 10.7 NFV Management Architecture where the VNF is supplied as an executable.

The assurance of the VNF may also be provided by the NFV MANO system. In theory, the traditional OSS system can drive the NFV, but then if this continues to happen, there is little point to the NFV MANO system. In effect, in this final configuration, there is also an implied OSS transformation from an OSS that manages the network via direct stimulation of the NMS and EMS for given network functions to one where the network is managed through configuration scripts (automatically) driven by the NFV MANO system.

10.2.4   NFV Summary

This section described the ETSI NFV orchestration system, its main components, method of operation, and the likely groupings of functions for deployment. As previously noted, ETSI NFV is still evolving and is not yet a normative standard. Nevertheless, the orchestration system described previously provides a framework to enable automatic operation of virtualized SP network functions and services. The story is completed when we provide an automation framework for the *underlying network* that is responsible for integrating these functions and services into a fully connected end-to-end system. SDN provides such a framework. Through logical separation of control and forwarding planes, SDN introduces a flexible programming environment for dynamic network control. The next section examine SDN from a number of perspectives that reflect the multiple ways that SDN can be used

10.3   SDN

SDN [6–8] is a framework for separating control of data networking from the actual act of forwarding packets. The original idea of SDN is that the separation of control and data planes allow optimization, evolution, and innovation to proceed independently for these two domains. With SDN, the control of network forwarding engines can be aggregated by common control nodes, and moreover, data plane functions are programmed with well-defined interfaces between the control nodes and data plane functions. SDN thus enables easy adoption of networking to specific (application or use case) needs.

While initial SDN solutions primarily focus on deployments of physical forwarding engines, it is a natural extension to apply these ideas to virtualized infrastructures, including those inside data centers and/or cloud infrastructures. By implementing data plane functionality with virtualized networking platforms, the data networking and virtual compute instances can be more tightly coupled and, if need be, allow the application some degree of control over the data networking itself.

We illustrate this concept in the following section by presenting, as examples, how to virtualize traditional mobile telecommunication networking functions ("virtual telco"). Generally, these applications may include a 3GPP mobile packet core ("Evolved Packet Core" (EPC)) [9–11], a 3GPP-based Gi-LAN service area [12], IMS and VoLTE [13], or other typical 3GPP-specific solutions. When hosted in data centers or through cloud infrastructures, these mobile telecommunication networking functions must continue to provide high-performance, low-latency, and dependable networking, similar to those described earlier in this chapter, to meet the availability requirements of such "telco" applications. Generally, individual virtual telco applications need to operate with availability requirements of 99.999% or better, have stringent requirements in terms of failover and fault recovery, and have stringent requirements for packet loss and packet delivery times. FCAPS lists a series of requirements specifically for such applications. In other words, these are critical network applications that require a high level of quality assurance.

When deploying "virtual telco" services in data centers, some of these applications require a "service chain" of functions, that is, a construct where one or more packet flows are steered through one or more virtualized or service functions that adapt the packet flows with service functionality as these packets traverse between ingress and egress. An example of this is a Gi-LAN service area that is comprised of, for example, a packet classification function, an

HTTP application function, a DPI application function, and an NA(P)T application function. To support such functions in data centers and/or cloud infrastructures, each of the application function's output (virtual) Ethernet needs to be connected to the next service function's input (virtual) Ethernet to form a service chain of services. This procedure is generally called service chaining [14]. Steering and classification principles can then be used to selectively assign packet flows to service chains of varying types.

Given that mobile solutions oftentimes produce relatively modest amounts of bandwidth (typically up to a maximum of a few hundred gigabits/second), managing all of the data flows for mobile service delivery is well suited for data center and/or cloud infrastructures. The prime benefits of hosting such functions in the data center and/or cloud infrastructures then involve the speed of service delivery and speed with which capacity can be provided. In the following text, we explore how such virtual service delivery can be provided for, specifically for a combined vEPC and Gi-LAN service area solution.

10.4   The Mobility Use Case

As an example of SDN applied to mobile networks, we explore the use case of a combined vEPC and Gi-LAN service gateway as depicted in Figure 10.8. In this example, the vEPC provides mobile subscriber line termination and charging functionality, while a Gi-LAN service gateway provides an infrastructure to host value-added services toward mobile service delivery. Here, we only describe the high-level functionality of a vEPC, and we refer to external material for more in-depth descriptions of the functions of an EPC [10, 11].

An EPC function is composed of a control and data plane. The control plane ("Mobility Management Entity" (MME)) addresses mobility, authentication, and paging operations, while the data plane is best characterized as a series of GPRS Tunneling Protocol (GTP) routers: the serving gateway (router) ("S-GW") providing for local mobility operations, while

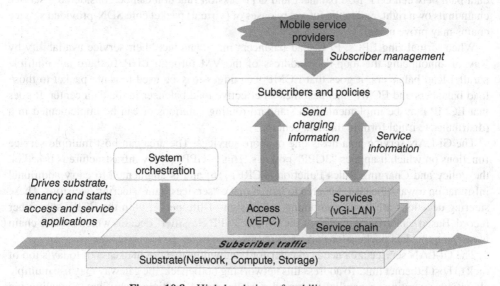

Figure 10.8   High-level view of mobility use cases.

the Packet Data Network (PDN) Gateway ("P-GW") providing for GTP line termination, roaming, and charging functions. Typically, if a subscriber is "at home," the S-GW and P-GW are collapsed into a single entity and termed as "SAE-GW."

An S-GW connects to base stations (eNBs) southbound, while a P-GW connects to the Internet or Gi-LAN area northbound. Southbound connections from MME/S-GW are 3GPP specific (S1-U/S1-MME reference points) and carry data packets in an IP-in-GTP tunnel or SCTP [15] sessions between base station and S-GW and/or MME. Northbound traffic is routed by way of standard routing protocols and may involve MPLS, virtual local area network (VLANs), or other network connectivity. BGP is oftentimes used for signaling connectivity, and ECMP may be used for load distribution.

A typical EPC is internally composed of two main components—GTP-based load balancers and GTP session engines. The load balancers terminate external sessions, for example, the GTP session originated by the base stations and the PDN sessions toward the Internet, while the session engines address the actual 3GPP-specific functions. Communication between the EPC load balancers and session engines is oftentimes a proprietary function and uses 3GPP session state for appropriate distribution of load across various servers. The reason for such 3GPP-specific solutions is primarily legacy—when 3GPP systems were first designed, "SDN" as a concept simply did not exist, virtual operation of said systems was not considered or deemed inappropriate, and application-specific "SDN"-like functionality was devised for load distribution of 3GPP calls in fixed network elements.

While "SDN" techniques may be applicable to operating EPCs in the cloud, immediately transforming today's EPCs to be more "data center- and/or cloud-like" takes time and investment. A fine balance needs to be found between efficiently using existing EPC assets while transitioning to new "cloud-like" environments. One approach to this transformation is to host the 3GPP load balancers and session functions as VMs in a data center while leveraging existing 3GPP assets to implement the business end of the service. Over time, these legacy solutions can be replaced by virtualized implementations. In this sense, the networking data path between GTP load balancer and GTP session function can be considered a service chain in its own right, and in the future, releases of (virtual) packet core SDN-provided service chains may prove useful.

When virtualizing EPCs, EPC load balancers may announce their service availability by way of announcing the loop-back address of the VM through BGP. If there are multiple parallel load balancers, it goes that ECMP techniques may be used to route packet to those load balancers and ECMP thus becomes an effective load balancer in the data center. It goes that ECMP may be implemented on standard routing solutions or can be implemented in a (distributed) virtual form in the data centers.

The Gi-LAN service area hosts one or more services. The area can host multiple service functions on which it applies "3GPP" policies. This "3GPP" policy infrastructure is based on the Policy and Charging Rules Function (PCRF) [16] and is augmented to carry additional information toward the Gi-LAN area to help it make "service chain" steering decisions. These steering decisions are made combining a first-sign-of-life packet with the PCRF subscriber record. Based on the combination of the two, a "3GPP classifier" decides which service chain is carrying the data packets for that specific flow.

The Gi-LAN service area needs to cater for bandwidths that are in excess of today's top of rack to host Ethernet link. To address this networking bottleneck, the gateway may use multiple classifiers operating in parallel to spread packets across multiple classifiers. Signaling via

BGP may provide upstream nodes the necessary information on how to distribute load across the classifiers, for example, by way of ECMP [17]. When BGP is used to announce service availability, loop-back addresses signaled are equivalent to the IP addresses of the VMs implementing the classification function.

The business end of the Gi-LAN service gateway can be defined by static service chains that carry packet flows between the P-GW function and the Internet. In this configuration, multiple logical service chains can share services, services can off-load flows to the fast path, services may participate in further classification of the services, and packet flows can be completely agnostic; these are being used as part of a Gi-LAN service area. A thorough description of such service chaining can be found separately [14].

10.5   Virtual Networking in Data Centers

In today's data centers, all hosts are connected to an Ethernet-based local area network (LAN) that supports VLAN tagging. In a VLAN, every Ethernet frame carries a VLAN tag that identifies tenancy. A VLAN in a data center operates as a standard Ethernet-based layer 2 network, with all associated helper functions such as the Address Resolution Protocol (ARP) [18], potentially augmented with the Dynamic Host Configuration Protocol (DHCP) [19] combined with all other IP- and Ethernet-based tools to operate such a virtualized or isolated layer 2 network.

In IaaS, the notion of a "tenant" exists. A tenant is defined as a project, group, enterprise, or any other grouping for which joint operations exist. Groups of hosts can be assigned to a tenant, and IaaS uses tenant information to allocate VMs to hosts, and networking constructs such as VLANs can be assigned specifically to tenants.

Allocation of VLANs to tenants is a function defined within the broad context of "system orchestration." Such system orchestration tool needs to maintain a VLAN allocation table, and it needs the ability to manage the various hardware-based routers and switches to connect the VLANs to the appropriate hosts. Various modes for such orchestration exist, ranging from manual configuration to fully automated orchestration. Management of such VLAN ranges enables a data center operator to segregate traffic between the various VLAN domains and thus tenants.

Many data centers are managed by way of an IaaS/Openstack [20] orchestration system. The role of Openstack is to start, manage, and stop VMs on Openstack-managed "host" computers and to establish a host networking infrastructure to connect tenant VMs to the appropriate VLANs. Moreover, the host networking infrastructure (Quantum/Neutron) for Openstack based on Open vSwitch (OVS) is elaborate and feature rich: through a series of software bridges and IP rule tables, packets are steered from a host-resident Ethernet controller card by way of the VLAN tagging to the appropriate tenant VM.

While most virtual telco applications can be considered single tenant, it is not uncommon that control over such telecommunication's applications and the "data plane" for such applications are separated in multiple segregated domains, that is, the "virtual telco" application as a whole becomes multitenant. Inherently, the control and data portion of such applications needs to communicate, and thus, bridging capabilities are needed between various tenants. This bridging may potentially involve IP address translation and firewalling and may include functions to detect intrusions. Given that all assets in a data center are to be virtualized, all these tenant bridging functions need to operate on data center resources themselves.

10.6 Summary

NFV promises the SP greater operational flexibility in the future leading to improved OPEX and CAPEX. The chapter explored two tools that are used by NFV to help achieve these aims, namely, orchestration and SDN. As this chapter has indicated, while the technology used for both SDN and orchestration is still maturing, the standardization of NFV is not quite at the same level of maturity. The implication is that NFV is likely to continue to evolve as the technology and the application models are refined further.

References

[1] ETSI, "GS NFV MAN 001 V0.3.3 (2014–02) Network Function Virtualization (NFV) Management and Orchestration," ETSI, Sophia-Antipolis Cedex, March 2014.

[2] OASIS, "Topology and Orchestration Specification for Cloud Applications Version 1.0," OASIS Standard, 25 November 2013 [Online]. Available: http://docs.oasis-open.org/tosca/TOSCA/v1.0/os/TOSCA-v1.0-os.html (accessed January 19, 2015).

[3] M. Bjorklund, "RFC 6020—YANG-A Data Modeling Language for the Network Configuration Protocol," October 2010 [Online]. Available: http:/tools.ietf.org/html/rfc6020 (accessed January 19, 2015).

[4] tmforum.org, "tmforum Information Framework (SID)," 2014 [Online]. Available: http://www.tmforum.org/DownloadRelease14/16168/home.html (accessed January 19, 2015).

[5] sourceforge, "Ganglia Monitoring System," 2014 [Online]. Available: http://ganglia.sourceforge.net (accessed January 19, 2015).

[6] T. Lakshman, N. Nandagopal, R. Ramjee, K. Sabnani, and T. Woo, "The SoftRouter Architecture," in *HotNets-III*, San Diego, CA, 2004.

[7] Open Networking Foundation, "Software-Defined Networking: The New Norm for Networks," April 13, 2013 [Online]. Available: https://www.opennetworking.org/images/stories/downloads/sdn-resources/white-papers/wp-sdn-newnorm.pdf (accessed February 18, 2015).

[8] N. McKeown, T. Anderson, H. Balakrishnan, G. Parulkar, L. Peterson, J. Rexford, S. Shenker, and J. Turner, "OpenFlow: Enabling Innovation in Campus Networks," ACM SIGCOMM Computer Communication Review, vol. 38, no. 2, p. 6, 2008.

[9] 3rd Generation Partnership Project, "The Evolved Packet Core," 2014 [Online]. Available: http://www.3gpp.org/technologies/keywords-acronyms/100-the-evolved-packet-core (accessed January 19, 2015).

[10] 3rd Generation Partnership Project, "General Packet Radio Service (GPRS) Enhancements for Evolved Universal Terrestrial Radio Access Network (E-UTRAN) Access, 3GPP TS23.401, v12.4.0," March 2014 [Online]. Available: http://www.3gpp.org/DynaReport/23401.htm (accessed January 19, 2015).

[11] 3rd Generation Partnership Project, "Architecture Enhancements for Non-3GPP Accesses, 3GPP TS23.402, v12.4.0," March 2014 [Online]. Available: http://www.3gpp.org/DynaReport/23402.htm (accessed January 19, 2015).

[12] H. La Roche and P. Suthar, "GiLAN and Service Chaining," Cisco Live, May 14, 2014. [Online]. Available: https://www.ciscolive2014.com/connect/sessionDetail.ww?SESSION_ID=3138 (accessed January 19, 2015).

[13] 3rd Generation Partnership Project, "IP Multimedia Subsystem (IMS); Stage 2, 3GPP TS 23.228," 24 June 2014 [Online]. Available: http://www.3gpp.org/DynaReport/23228.htm (accessed January 19, 2015).

[14] W. Haeffner, J. Napper, N. Stiemerling, D. Lopez and J. Uttaro, "IETF draft—Service Function Chaining Use Cases in Mobile Networks," July 4, 2014 [Online]. Available: https://datatracker.ietf.org/doc/draft-ietf-sfc-use-case-mobility/ (accessed January 19, 2015).

[15] R. Stewart, "RFC 4960—Stream Control Transmission Protocol," September 2007 [Online]. Available: http://tools.ietf.org/html/rfc4960 (accessed January 19, 2015).

[16] 3rd Generation Partnership Project, "Policy and Charging Control Architecture, 3GPP TS23.203, v12.4.0," March 2014 [Online]. Available: http://www.3gpp.org/DynaReport/23203.htm (accessed January 19, 2015).

[17] C. Hopps, "RFC 2992—Analysis of an Equal-Cost Multi-Path Algorithm," November 2000 [Online]. Available: http://tools.ietf.org/html/rfc2992 (accessed January 19, 2015).

[18] D. Plummer, "RFC 826—Ethernet Address Resolution Protocol," November 1982 [Online]. Available: http://tools.ietf.org/html/rfc826 (accessed January 19, 2015).

[19] R. Droms, "RFC 2131—Dynamic Host Configuration Protocol," March 1997 [Online]. Available: http://tools.ietf.org/html/rfc2131 (accessed January 19, 2015).

[20] Openstack Foundation, "Openstack Cloud Software," [Online]. Available: http://www.openstack.org (accessed January 19, 2015).

11

Survey of Traffic Management in Software Defined Mobile Networks

Zoltán Faigl[1] and László Bokor[2]

[1] Mobile Innovation Centre, Budapest University of Technology and Economics, Budapest, Hungary
[2] Department of Networked Systems and Services, Budapest University of Technology and Economics, Budapest, Hungary

11.1 Overview

Due to the evolution of mobile technologies, in these days, high-speed data services are dominating in mobile networks both in uplink and in downlink. In function of the characteristics of data services, the usage patterns, and the user and network mobility patterns, the utilization of network resources is varying in time and location. As the volume of traffic demands increases, the amplitude of their variations grows as well. Existing network, resource, traffic, and mobility management mechanisms are too inflexible to adapt to these demands. Software defined mobile networks (SDMNs) aim at improving the scalability and adaptability of the mobile network architectures to varying traffic demands by applying host and network virtualization concepts, restructuring the network functions into parts that are running in data centers in virtualized environment and parts, which cannot be virtualized, for example, base transceiver stations.

This chapter first defines the scope of traffic management in mobile networks in Section 11.2, including microscopic, macroscopic, improved content resource selection, and application-supported traffic management. Section 11.3 gives an overview of QoS enforcement and policy control in 3G/4G networks, which should be kept also in SDMNs. Section 11.4 surveys new research problem areas in software defined networks (SDNs) for traffic and resource management. Following that, an example of traffic engineering mechanism will be discussed in Section 11.5, that is, application-layer traffic optimization (ALTO) applied in SDN environments. ALTO–SDN provides improved resource selection and ALTO transparently for the users. This example shows the feasibility of SDN-based techniques for traffic management in SDMNs.

Software Defined Mobile Networks (SDMN): Beyond LTE Network Architecture, First Edition.
Edited by Madhusanka Liyanage, Andrei Gurtov, and Mika Ylianttila.
© 2015 John Wiley & Sons, Ltd. Published 2015 by John Wiley & Sons, Ltd.

11.2   Traffic Management in Mobile Networks

Traffic management methods may be both necessary and warranted in the operation of broadband networks because of overbooking, that is, the network capacity requirement of the services sold generally far exceeds the available network capacity. Traffic management methods can mitigate the negative effects of congestion and can contribute to a more fair distribution of scarce network resources among users. Moreover, traffic management allows service providers to define service features.

Regulation, for example, in many European countries and in the United States, requires transparency of the network, no blocking of content, and no unreasonable discrimination of content. However, some users or applications, especially in content delivery, require quality of service (QoS) guarantees and data discrimination. Therefore, the regulation of such countries requires from network providers the definition of QoS criteria in the QoS Decree in a detailed manner based on the establishment of the service; the error ratio, availability, troubleshooting, etc.; and the specification of various quality target values depending on the nature of the service. Other QoS target values may also exist, which are required by a specific service but not included in the QoS Decree.

Modern traffic management possesses a very rich toolset of interventions, which may influence the traffic demands arriving in the network of the operator, the load distribution in the network, the priorities of traffic classes, etc. Traffic management consists of the following six different building blocks, as defined in the Celtic-Plus MEVICO project [1].

Microscopic-level traffic management is associated with all mechanisms with the primary objective to improve performance of individual flows based on application type, user profile, and other policy-related information. For example, multipath transport control protocol, congestion control, and QoS differentiation of service dataflows are such areas.

Macroscopic-level traffic management includes all mechanisms with the primary objective to improve efficient usage of network resources. Parameters for optimization in the latter case describe traffic patterns without detailed knowledge of individual flow attributes. Sample mechanisms for macroscopic traffic management are (re)selection of core network elements and IP flow mobility, energy-efficient and QoS-aware routing, load balancing, and technologies enabling the improvement of the usage of multiple interfaces and enforcing breakout of part of data services from the mobile network operator's network toward other networks.

The third category of traffic management technologies is called *improved resource selection*. The mechanisms associated to improved resource selection address the selection of the best service endpoint in the case of distributed services, such as Web-based content delivery by peer-to-peer networks, content distribution networks, or in-network caches. ALTO is a good example from this category, since it provides better-than-random endpoint selection for applications, considering both of the aspects of network operator and content provider (or distributor).

The previous technologies are associated with mechanisms, which may require support from lower layers (below application) and which may require support from each other. For example, improved resource selection may require support from macroscopic traffic management for finding the best path toward the optimal endpoint and from microscopic traffic management for enforcement of QoS policies.

The next three building blocks may require only little or no support at all from the previous categories. *Application-supported traffic management* aims at optimizing performance from end user perspective without getting support from network elements. Many traffic management applications of CDNs, multimedia streaming optimization techniques, P2P services, and even Provider Portal for P2P Applications (P4P) fall into this category.

Mainly network operators, but possibly also other stakeholders, may influence user behavior by defining certain constraints for usage of networks/services and certain incentives to comply with the usage constraints. Such procedures are called *traffic steering usage models*. They do not have too much technical aspects but influence traffic demands in the network.

Extension of network resources, or *overprovisioning*, is the sixth category of traffic management. When the network is regularly in high load conditions, network capacities need to be increased. It is a challenge to apply an intelligent planning process for extending the available resources.

11.3 QoS Enforcement and Policy Control in 3G/4G Networks

Connectivity to Packet Data Networks (PDN) is provided by *PDN connections* in 2G/3G/4G packet core of the 3rd Generation Partnership Project (3GPP) networks. A PDN connection comprises several aspects, that is, IP access, in-band QoS provisioning, mobility, and charging.

PDN connections are provided by *Packet Data Protocol (PDP) contexts* in the 2G/3G core between the User Equipment (UE) and Gateway GPRS Support Node (GGSN) and *Evolved Packet System (EPS) bearers* between the UE and P-GW in Evolved Packet Core (EPC) (4G core network) when UEs attach to evolved UMTS terrestrial radio access network (E-UTRAN).

Several options are available to provide PDN connection between 2G/3G access and the PDN GW or E-UTRAN and GGSN. For example, a UE can access from a 2G/3G radio access network (RAN) the Serving GPRS Support Node (SGSN) through PDP context, have a one-to-one mapping between PDP contexts and EPS bearers in the SGSN, and reach the S-GW and P-GW with EPS bearers.

2G/3G core supports two types of PDP contexts related to IP connectivity: IPv4 and IPv6. A PDN connection in EPC supports three options: the allocation of one IPv4, one IPv6, or both an IPv4 and an IPv6 address to the UE within the same PDN connection. 3GPP Release 9 introduced support for dual-stack PDP context also in 2G/3G GPRS core network.

IP address is allocated during the attach (PDP context activation) procedure to the UE. Another option is the usage of DHCPv4 after the attach procedure or PDP context activation. Stateless IPv6 address autoconfiguration is also supported by sending routing advertisements through the PDN connection advertising a 64-bit prefix allocated to the specific PDN connection.

In the case of E-UTRAN access, multiple EPS bearers can belong to the same PDN connection: a default bearer and optionally other dedicated bearers provide PDN connectivity. During the attach procedure, a default bearer is established to provide always-on connectivity for the UE. In 2G/3G GPRS core, PDP contexts are only activated when an application requests IP connection.

Each EPS bearer is associated with a set of QoS parameters and traffic flow templates (TFTs). TFTs specify the traffic filters related to the IP flows that are mapped to the specific EPS bearer. TFTs may contain traffic filters for downlink and uplink traffic (denoted by DL TFT and UL TFT, respectively). All traffic flows matching the traffic filters of an EPS bearer will get the same QoS treatment.

The filter information is typically the five-tuple of source and destination IP addresses, transport protocol, and source and destination ports. Wild cards can be used to define a range

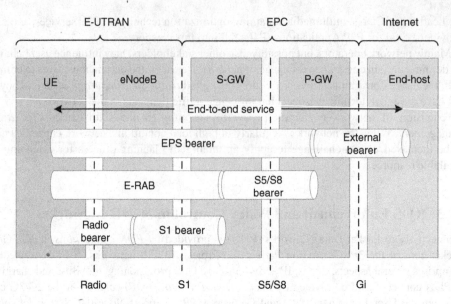

Figure 11.1   Hierarchy of bearers in LTE–EPC.

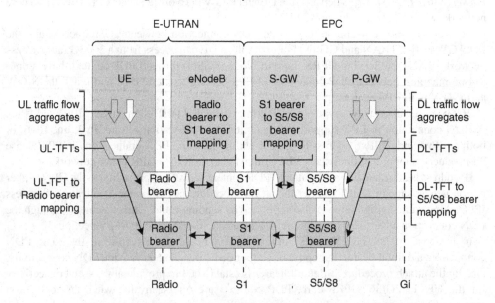

Figure 11.2   EPS bearer in E-UTRAN access and GTP-based S5/S8.

of addresses or ports. Other parameters of traffic filters can be the IPsec security parameter index, type of service (IPv4)/traffic class (IPv6), or flow label (IPv6).

EPS has adopted network-centric QoS control approach, that is, it is basically only the P-GW that can activate, deactivate, and modify EPS bearers and decide flow mapping to EPS bearers. That is different in pre-EPS systems. Originally, in 2G/3G GPRS, it was only the UE that could initiate new PDP context activation and decide about flow mapping to PDP contexts. Then 3GPP Release 7 introduced network-requested secondary PDP context activation where the GGSN initiates the creation of a new "bearer" (PDP context) and assigns IP flows to the bearer. This change is due to the introduction of policy control within the 2G/3G GPRS core and in EPC.

The GPRS Tunneling Protocol (GTP) is responsible for the control of PDP contexts in 2G/3G GGSN core (GTP-C) and the tunneling of IP packets of the user (GTP-U). In EPC, a new version for the GTP-C has been developed to manage EPS bearers over the S1 and S5/S8 interfaces, but the tunneling of user IP traffic remains the same as it was. It is called GTPv2.

Depending on the tunneling option, EPS bearers are implemented in different ways. Figure 11.1 represents the hierarchy and terminology of bearers for E-UTRAN access. For end-to-end (E–E) QoS provision, an EPS bearer and an external bearer are required. An external bearer is not under the control of mobile network operator. An EPS bearer consists of an evolved radio access bearer (E-RAB) and an S5/S8 bearer. An E-RAB includes a radio bearer and an S1 bearer. Figure 11.2 presents the realization of EPS bearers in the user plane when E-UTRAN access and GTP-based S5/S8 interfaces are deployed.

11.3.1   QoS for EPS Bearers

EPS differentiates two types of EPS bearers. Guaranteed bit rate (GBR) bearers are typically used for those services where it is better to block a service rather than degrade already admitted services. For example, VoIP, video streaming benefit from a constant bandwidth, hence GBR is needed to provide satisfactory user experience. An important characteristic of GBR bearer is that it is associated with a certain amount of bandwidth, independently of being utilized or not. The GBR always takes up resources over the radio link, even if no traffic is sent. Hence, in normal cases, the GBR bearer should not experience any packet loss.

Non-GBR bearers are used for those services, which normally do not require a constant fixed bandwidth, such as Web browsing, email, and chat. No transmission resources are reserved for non-GBR bearers.

An EPS bearer QoS profile however is broader than this categorization. It includes the parameters QoS class identifier (QCI), allocation and retention priority (ARP), GBR, and maximum bit rate (MBR), explained in the following.

For both non-GBR and GBR services, QoS parameters are the following:

- QCI: QCI is just a pointer to node-specific parameters, which define what packet forwarding treatment a particular bearer should receive (i.e., scheduling weights, admission thresholds, queue management thresholds, link layer protocol configuration, etc.). On the radio interfaces and the S1 interface, each protocol data unit is indirectly associated with one QCI via the bearer identifier carried in the header. The same applies to S5/S8 if GTP-based option is used. In GTP-U, the identifier is the tunnel endpoint identifier (TEID) conveyed in the GTP header. Table 11.1 summarizes the QoS requirements for different traffic types. Further details on standardized QCI characteristics can be found in TS 23.203 [2].
- ARP: ARP is used to indicate the priority for the allocation and retention of bearers. It includes:

  Priority level: Higher priority establishment and modification requests are preferred in situations where resources are scarce.
  Preemption capability: If true, then this bearer request could drop away another lower priority bearer.

Table 11.1 QoS requirements for different traffic types

Traffic type	Priority	Maximum delay (ms)	Maximum packet loss	Guaranteed bit rate
Control, signaling	1	100	10^{-6}	No
Voice call	2	100	10^{-2}	Yes
Real-time games	3	50	10^{-3}	Yes
Video call	4	150	10^{-3}	Yes
Premium video	5	300	10^{-6}	Yes
Interactive games	7	100	10^{-3}	No
Video, WWW, email, file transfer	6, 8, 9	300	10^{-6}	No

Preemption vulnerability: If true, then this bearer can be dropped by a higher priority bearer establishment/modification.

QoS parameters for GBR bearer are as follows:

• GBR: Is the minimum bit rate that an EPS bearer should get.
• MBR: The MBR limits the bit rate that can be expected to be provided by a GBR bearer (e.g., excess traffic may get discarded by a rate shaping function). Currently, MBR is set to the same value as GBR in EPC, that is, the instantaneous rate can never be greater than the GBR for GBR bearers.

Aggregate QoS parameters for nonguaranteed bearers (aggregate values) include:

• Per APN aggregate maximum bit rate (APN-AMBR): It defines the total bit rate that is allowed to be used by the user for all non-GBR bearers associated with a specific APN. It is enforced by P-GW in DL and the UE and P-GW in UL.
• Per UE aggregate maximum bit rate (UE-AMBR): The UE-AMBR limits the aggregate bit rate of all non-GBR bearers of the user. It is enforced by the eNodeB in UL and DL. The actually enforced rate is the minimum of the sum of all active APN's APN-AMBR and the subscribed UE-AMBR value.

The HSS defines, for each PDN subscription context, the "EPS-subscribed QoS profile," which contains the bearer-level QoS parameter values for the default bearer (QCI and ARP) and the subscribed APN-AMBR value.

The subscribed ARP shall be used to set the priority level of the EPS bearer parameter ARP for the default bearer. In addition, the subscribed ARP shall be applied by the P-GW for setting the ARP priority level of all dedicated EPS bearers of the same PDN connection unless a specific ARP priority-level setting is required (due to P-GW configuration or interaction with the Policy and Charging Rules Function (PCRF)). The preemption capability and the preemption vulnerability information for the default bearer are set based on Mobility Management Entity (MME) operator policy.

The mapping of services to GBR and non-GBR bearers is the choice of the operator and can be controlled with static rules in the Policy and Charging Enforcement Function (PCEF) or dynamic Policy and charging control (PCC) and QoS rules by the PCC framework.

11.3.2   QoS for Non-3GPP Access

In 2G/3G RANs, a more complicated QoS concept is used; hence, operators are not using many of the parameters in practice. That concept is referred to as the release 99 QoS. Its main characteristics are the following: 4 traffic classes, one mapped at the same time to a PDP context, and 13 attributes, such as bit rate, priority, error rate, max. delay, etc. For 2G/3G radio access to EPS via SGSN, the QoS attributes must be translated from release 99 QoS to EPS QoS parameters, when one-to-one mapping of PDP contexts to EPS bearers is performed. Mapping is described in Annex E of TS 23.401 [3].

11.3.3   QoS Enforcement in EPS

The following QoS treatment functions are deployed in the user plane of E-UTRAN and EPC. The maximum granularity of QoS control achieved by these functions is the EPS bearer granularity.

PCEF enforces traffic gating control for UL and DL based on policies. The mapping of packets to actual EPS bearer using TFTs is realized by UE for UL, P-GW (or S-GW if GTP is not deployed between the S-GW and P-GW) for DL.

Admission control (bearer establishment, modification) and preemption handling (congestion control, bearer drop) when resources are scarce, using the ARP to differentiate the handling of bearers, are executed by the eNodeB and P-GW (or S-GW).

Rate policing is enforced in the following way. eNodeB enforces the maximum rate for the aggregate of non-GBR bearers of the UE in UL and DL, based on the UE-AMBR UL and DL values. P-GW enforces the maximum rate for the aggregate of non-GBR bearers of the UE in UL and DL, using APN-AMBR values for UL and DL. eNodeB enforces GBR/MBR for GBR bearers in UL. P-GW (or S-GW) enforces GBR/MBR for GBR bearers in DL.

Queue management, scheduling, and configuration of L1/L2 protocols to enforce QCI characteristics, such as packet delay budget and packet loss in E-UTRAN, are enforced by eNodeB in UL and DL.

Mapping of QCI values to DSCP values in IP transport network between EPC elements is deployed in eNodeBs and S-GWs for IP transport between eNodeB and S-GW and/or S-GWs and P-GWs, for IP transport between S-GW and P-GW.

Finally, in order to enforce QoS on the path of EPS bearers in the transport network layer, routers and switches may deploy queue management and UL and DL scheduling.

11.3.4   Policy and Charging Control in 3GPP

Policy and charging control (PCC) provides QoS and charging control for operators. It provides a general, centralized framework to control the QoS procedures of heterogeneous access networks. It supports control of the user plane for IP Multimedia Subsystem (IMS) and non-IMS services. It solves the problem of lacking on-path QoS control in the case of non-GTP-based tunneling options. PCC can provide off-path control using the diameter protocol to any access network, which provides QoS bearers.

The "bearer" in PCC denotes an IP data path with desired QoS characteristics; hence, it is more generic than the EPS bearer and PDP context concept and is access network agnostic. Multiple service sessions can be transported over the same bearer. PCC enables service-aware

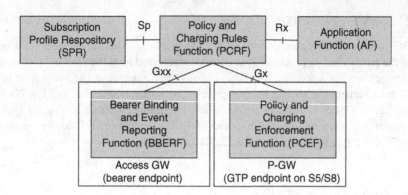

Figure 11.3   Policy control part of the PCC architecture (nonroaming case).

QoS control, having higher granularity than the bearer-level QoS control provided by EPS bearers. PCC allows QoS control over wireless non-3GPP access networks, such as High-Rate Packet Data Services (HRPD) and Worldwide Interoperability for Microwave Access (WiMAX). For the fixed access, interworking with policy control has not come as far as for the wireless access. It supports policy control in roaming scenarios as well.

11.3.5   Policy Control Architecture

Figure 11.3 presents the policy control part of the PCC architecture. The elements related to policy control are the following.

The application function (AF) interacts with services that require dynamic PCC. For example, in the case of IMS, AF is the proxy-call session control function. The AF extracts session information (e.g., from service description protocol field) and sends the information to PCRF over the Rx interface. Such information includes, but is not limited to, IP filter information to identify the service dataflow for policy control and/or differentiated charging and media/application bandwidth requirements for QoS control.

The AF can also subscribe at the PCRF to the notification of events in the network, such as IP session termination or access technology-type change.

The subscription profile repository provides user-specific policies and data over the Sp interface.

The PCRF receives session information on Rx, subscriber-specific policies over Sp, and access network information over Gx or, if Bearer Binding and Event Reporting Function (BBERF) is used, then over Gxa/Gxc. Operators can configure policies in the PCRF, which must be applied to given services. Based on that information, it brings service session-level policy decisions and provides them to PCEF and optionally to BBERF. PCRF also sends event reports from PCEF and optionally the BBERF to the AF, for example, for video/audio codec adaptation.

The Policy and Charging Enforcement Function (PCEF) enforces policy decisions based on the PCC rules provided by PCRF over the Gx interface. It may perform measurements of user plane traffic (e.g., data volume, session duration). It reports the usage of resources for offline charging and interacts with online charging. PCEF is part of the P-GW in EPC.

The BBERF is required if no on-path QoS negotiation is available (by GTPv2-C), and DSMIPv6/IPsec or PMIP/IP GRE tunnels are used between the P-GW and the access GW of the UE, not capable of implementing QoS bearers for the services of the UEs. BBERF is responsible for bearer binding and QoS enforcement based on QoS rules provided by the PCRF over the Gxa/Gxc interface. Furthermore, it is responsible for event reporting toward the PCRF, about access network type, bearer state, and other information.

Policy control comprises gating control and QoS control. Gating control is applied by the PCEF on a per service dataflow basis.

11.3.5.1 PCC Rule and QoS Rule

The Policy and charging control rule (PCC rule) comprises the information that is required to enable the user plane detection of the policy control and proper charging for a service dataflow. The packets detected by applying the service dataflow template of a PCC rule are designated a service dataflow.

Two different types of PCC rules exist: dynamic PCC rules and predefined PCC rules. The dynamic PCC rules are provisioned by the PCRF via the Gx reference point. Predefined PCC rules are configured in the PCEF, and the PCRF only refers to them. While packet filters in a dynamic PCC rule are limited to the five-tuple of source and destination IP, source destination port, transport protocol, and some more header fields, the predefined PCC rules may use DPI filters for more fine-grained flow detection, charging control. Those filters are not standardized by 3GPP. TS 23.203 [2] contains more details on PCC rules.

In the case of off-path QoS control, PCRF needs to provide QoS information to the BBERF via the Gxa/Gxc reference points. QoS rule includes only a subset of PCC rule but with the same service-level granularity. It includes hence typically the filter information (SDF template, precedence) and QoS parameters (e.g., QCI, bit rates), but not charging-related information.

11.3.5.2 Network-Initiated and UE-Initiated QoS Control

For services provided by the access provider, such as IMS voice, mobile TV, etc., the network-initiated QoS control procedure is preferable. For services that are not known by the operator, UE-initiated QoS control is possible.

A network-initiated QoS control procedure may have the following steps:

1. Application-level signaling between the UE and the AF (e.g., SIP, SDP).
2. Session information provision from the AF to the PCRF (over the Rx reference point). In the case of IMS services, the SDP information is mapped to QoS information, such as bit rate and service type.
3. The PCRF may request subscriber-related information from the SPR.
4. PCRF makes policy decision based on session information, operator-defined service policies, and subscription information and generates PCC/QoS rules.
5. PCC rules are pushed by the PCRF to the PCEF and PCEF enforces the policy and charging rules, and conditionally, if BBERF is required, then QoS rules are pushed to the BBERF and installed.

A UE-initiated QoS control procedure may have the following steps:

1. Application-level signaling between the UE and the AF (e.g., SIP, SDP),
2. Session information provision from the AF to the PCRF (over the Rx reference point). In the case of IMS services, the SDP information is mapped to QoS information.
3. The PCRF may request subscriber-related information from the SPR.
4. The application on the UE side makes request through vendor-specific APIs for the access interface to request the needed QoS resources.
5. UE sends resource request, including QoS class and packet filters for the service. In E-UTRAN, that is called UE-requested bearer resource modification. In 2G/3G RAN, it is realized by secondary PDP context activation/modification.
6. If BBERF exists, it initiates PCRF interaction over Gxa/Gxc interface. If there is no BBERF, the PCEF initiates PCRF interaction over Gx interface.
7. The same as step 4 in network-initiated case.
8. The same as step 5 in network-initiated case.

11.4 Traffic Management in SDMNs

Dynamic, service dataflow-based policy control will be more and more needed by mobile network operators due to the increasing diversity of services and the related policy rules. Hence, in general, the QoS provisioning mechanisms specified by 3GPP, such as EPS bearers or PDP contexts and policy control by PCRF, should be kept also in case of virtualization of mobile core and transport network.

It is still uncertain whether GTP tunneling will be kept in SDN-based transport network segments. PCRF supports both on-path (GTP based) and off-path QoS configuration. Off-path QoS configuration is applicable over any transport network technology, which supports some sort of QoS bearers. Therefore, for the application of dynamic QoS enforcement in SDNs, two main challenges should be solved:

- The SDN transport should be able to provide QoS enforcement.
- Gx and Gxa/Gxc interfaces must be adapted for communicating PCC/QoS rules to the SDN controller, and the SDN controller shall be able to signal application-specific information to the PCRF through the Rx interface.

The service-chaining concept requires network function forwarding graphs both through virtual and traditional transport network segments. Operators need to be able to control logical and physical interconnections, configure traffic class conditioning and forwarding behaviors (capacity, priority, packet loss, delay, shaping, dropping, etc.), and map traffic flows to appropriate forwarding behaviors.

11.4.1 Open Networking Foundation

Open Networking Foundation (ONF) is a nonprofit industry alliance in charge of supporting the researches of software defined networking and of the standardization activities having OpenFlow (OF) in the main focus. OF is a completely open protocol that was originally published by Stanford

University researchers in [4] aiming to enable network developers to run experimental protocols in the university campus network. According to the Open Networking Foundation, SDN is an emerging network architecture that decouples the network control and forwarding functions.

ONF-based SDN architectures inherit a number of benefits for traffic management-related challenges of mobile and wireless environments, including their wireless access, mobile backhaul, and core networking segments. These benefits and potentials are listed in the following.

The paradigm of flow-based communication in SDN architectures fits well to provide efficient E–E communications in multiaccess environments, when different radio technologies, like 3G, 4G, WiMAX, Wi-Fi, etc., are simultaneously available for users. SDN is able to provide fine-grained user flow management aiming to improve traffic isolation, QoS/QoE provision, and service chaining.

In current networks, the decision logic and organization of network functions and protocols are distributed and multilayered, enabling the evolution of each layer, separately. That makes very complex the understanding and management of networks, when network providers want to fulfill E–E connectivity and QoS requirements over different access networks for different services. SDN tries to hide this complexity and introduces centralized control of network. Centralized control plane allows for efficient resource coordination of wireless access nodes, which makes possible to implement efficient intercell interference management techniques.

The fine-grained path management in SDN networks provides various optimization possibilities based on the individual service needs and independently from the configuration of the underlying routing infrastructure. In mobile and wireless environments, it is useful as users are frequently changing their network points of access, the used applications and services vary in bandwidth demands depending on the nature of the content to be transmitted, and considering that wireless coverages are providing a naturally changing environment.

Virtualization of network functions efficiently abstracts services from the physical infrastructure. Multitenancy permits each network slice to possess its own policy, independently of whether that slice is managed by a mobile virtual network operator, over-the-top service provider, virtual private enterprise network, governmental public network, traditional mobile operator, or any other business entity.

11.4.2 The OF Protocol

In SDN networking, the network operating system (NOS) is in charge of controlling the SDN-capable networking elements (SDN switches) in a centralized way. The NOS has southbound and northbound APIs that allow SDN switches and network applications to communicate over the common control plane provided by the NOS. In order to support multivendor environments for SDN switches and controllers, the southbound APIs must be standardized. OF protocol is one of the most known standards for the southbound API.

An OF switch contains multiple flow tables, which implement pipeline processing for incoming packets. Each table may contain flow entries. A flow entry contains a set of match fields for matching the packets, priority for matching precedence, a set of counters to track packets, and a set of instructions to apply. Furthermore, it includes timeouts to determine the maximum amount of time or idle time before flow is expired by the switch and cookie set and used by the controller as a group identifier of flow entries, enabling filtering queries for flow statistics, flow modification, or flow deletion.

An instruction either modifies pipeline processing by sending the packet to another (higher number) flow table or contains a list of a set of actions. The action set includes all actions accumulated while the packet is processed by the flow tables. The actions are executed when the packet exits the processing pipeline. Possible actions are the following: output a packet on a given port; enqueue the packet to a given queue; drop packet; rewrite packet fields, such as time to live, virtual local area network ID, and multiprotocol label switching label.

In regard to QoS provisioning, the enqueue action is the most relevant action. The "enqueue" action in OF version 1.0 was renamed to "set_queue" in version 1.3 [5]. Its main purpose is to map a flow to a queue; it also sets up simple queues.

QoS provisioning of OF-capable switches is still not enough developed. Currently, both OF version 1.4 and OpenFlow Management and Configuration Protocol (OF-Config 1.1.1) [6, 7] can set up queues using only two input parameters:

- Minimum rate: It specifies the guaranteed rate provided for the aggregate of flows mapped to a queue. The minimum rate is relevant when the incoming data rate of an egress port is higher than the maximum rate of the port.
- Maximum or peak rate: It is relevant when there is available bandwidth on the output port.

OF-config and OF protocols do not support hierarchical queueing disciplines, which are necessary to implement standard or other per hob behaviors (PHB) specified for DiffServ architecture.

The OF protocol supports two queueing disciplines, that is, hierarchical token bucket (HTB) and hierarchical fair-service queue (HFSC). These queueing disciplines have much more configuration possibilities than minimum rate and maximum rate, such as the maximum queue size for HTB or delay curves for real-time traffic in HFSC.

The advantages of queueing disciplines could be more leveraged if more queuing disciplines were available, the establishment of more than one level of QoS class hierarchies was possible, and more parameters of the queuing disciplines were allowed by the OF and OF-config specifications.

It is possible to build hierarchical queueing disciplines in switches using their administration tools and map flows to queues based on traffic control filters. For example, the DSCP value in IPv4/IPv6 headers or other packet headers and fields can be used to map packets to more complex queues.

The OF 1.4 has specified requirements for counters that could be set for flow tables, flow entries, ports, queues, etc. [5]. An OF controller may set meters in an OF switch to measure performance metrics related to flows, ports, queues, etc. It can set meter bands and appropriate actions if the actual measured metric falls into the meter band. Such actions could be dropped, realizing rate limiting or DSCP remarking for assigning the packet to a new behavior aggregate. However, it depends on the implementation of the OF switch, whether these functionalities are available.

11.4.3   Traffic Management and Offloading in Mobile Networks

One of the most straightforward use cases of ONF is traffic steering and path management that have received tremendous attention within the SDN community. Tools of smart traffic steering can be applied for advanced load balancing, load sharing, content filtering, policy control and enforcement, error recovery and redundancy, and, in general, any application that involves traffic flow operations and control. Putting all of these potential SDN applications into the context of mobile and wireless networks, we gather another set of potential use cases like traffic offloading and roaming support, content adaptation (e.g., adaptive streaming solutions), and mobile traffic optimization.

OF enables mobile Internet traffic to be dynamically and adaptively moved and removed in the mobile network based on a number of possible trigger criteria, such as individual or aggregate flow rate (such as per application or per user aggregation), aggregate flow number on a particular port or link, flow duration, number of UEs per cell, available bandwidth, IP address, type of application, device utilization rate, etc. All of these criteria can be defined either by the user or by the mobile operator. For example, the operator could measure network conditions and decide to offload mobile traffic in case of need. As a user-centric alternative, subscribers could opt in based on their preferred parameters and predefined policies, like (i) voice calls should never be offloaded and (ii) FTP download traffic should always be offloaded to Wi-Fi. In a more advanced use case, it could be envisioned that users travel in a multiaccess radio environment simultaneously connecting to multiple base stations. Network parameters such as congestion, QoS, and quality of experience (QoE) are measured, and triggering factors (e.g., a flow rate threshold) are set and changed dynamically by the mobile operator. For example, "If the flow is an FTP download, and the flow rate exceeds 100 kbps, hand over the flow from LTE to Wi-Fi." As the example shows, distinct criteria and thresholds could be applied for different applications and therefore different flow types running on the same UE or on the terminals of different subscribers. Of course, thresholds could be based on the widest range of possible criteria like user/flow profile, location, service plan, etc.

11.5 ALTO in SDMNs

We call ALTO problem when someone is concerned with better-than-random peer selection, optimization of rendezvous service for applications fetching distributed content. Typical fields where the ALTO problem occurs are peer-to-peer networks, content distribution networks, and data centers.

In peer-to-peer networks, peers can exchange pieces of information in an incremental way until they obtain the entire content. When a peer has not a global view on the network, it may pick randomly a candidate peer, which may result in lower QoE.

CDNs distribute content and may cover large geographical areas. With the increasing demand for streaming video services, CDN servers/caches are deployed deeper in the network of Internet service providers, including mobile network operators. CDN operators elaborated different technologies to direct the end users to the best CDN server or in-network cache of operators for appropriate level of QoE for the users.

A third area for ALTO problem is related to cloud services. Cloud services run on top of data centers. Users should be served by the closest data center by an enough lightly loaded server. In case of virtual private clouds, the obtainment of proximity measures is more complicated because the service is provided through overlay networks; servers in the same virtual network may be located at different geographical locations.

Gurbani et al. [8] provide a good survey on existing solutions for the ALTO problem. ALTO solutions can be divided into two categories: (1) application-level techniques to estimate parameters of the underlying network topology and (2) layer cooperation. Techniques in (1) can be further divided into (i) end-system mechanisms for topology estimation, such as coordinates-based systems, path selection services, and link layer Internet maps, and (ii) operator-provided topological information services, such as P4P [9], oracle-based ISP–P2P collaboration [10], or ISP-driven informed path selection [11].

The authors of [8] argue that these techniques have limitations in terms of abstraction of network topology using application-layer techniques, for example, unable to detect overlay paths shorter than the direct path or accurately estimate multipath topologies, or do not measure all the relevant metrics for appropriate selection of the best endpoint. For example, round-trip times do not reveal information on throughput and packet loss. Furthermore, topology estimations may converge slowly to the result. Moreover, application-layer measurements induce additional network resource utilization.

Hence, there is need of cooperation between the application and network layer, where network operators should be able to provide network maps and cost maps representing distance-, performance-, and charging-related criteria.

11.5.1   The ALTO Protocol

A new protocol, that is, ALTO protocol, is on track to become a proposed IETF RFC specified by Alimi et al. [12] for interoperability between ALTO solutions of different vendors.

The two main information elements provided by ALTO service are the network map and the related cost maps. A network map consists of the definition of host groups but not the connectivity of host groups. The identifier of host groups is called provider-defined identifier (PID). A PID may denote, for example, a subnet, a set of subnets, a metropolitan area, a PoP, an autonomous system, or a set of autonomous systems.

A cost map defines one-way connections between the PIDs and assigns a cost value to each one-way connection. It also determines the metric type (e.g., routing cost) and the unit type (numerical or ordinal), furthermore the network map name and version, where the PIDs are defined.

ALTO protocol is based on HTTP and uses a RESTful interface between the ALTO client and server. The protocol encodes message bodies in JSON [13]. Several JSON media types are proposed in [12], which realize required and optional functions. Required functions are the information resource directory and network and cost map request and responses. Optional functions of ALTO service are, for example, filtered network and cost map queries, endpoint property queries, etc.

11.5.2   ALTO–SDN Use Case

Gurbani et al. proposed in [14] the application of ALTO service in the SDN application layer. They argue that the ALTO protocol is a well-defined and mature solution that provides powerful abstraction of network map and network state that can be leveraged by distributed services in SDNs. ALTO hides unnecessary detail of the underlying networks without unnecessarily constraining applications; hence, privacy of network information of network operators and content providers can be kept.

An important limitation of ALTO protocol is that it does not specify network information provision service. Creation of network and cost maps in the ALTO server should be automated and policy driven. There is ongoing work for distribution of link-state and TE information from BGP routers [15–17]. A similar approach should be used in the case of SDN networks,

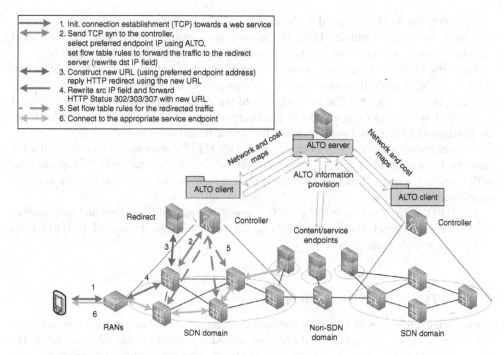

1. Init. connection establishment (TCP) towards a web service
2. Send TCP syn to the controller,
   select preferred endpoint IP using ALTO,
   set flow table rules to forward the traffic to the redirect
   server (rewrite dst IP field)
3. Construct new URL (using preferred endpoint address)
   reply HTTP redirect using the new URL
4. Rewrite src IP field and forward
   HTTP Status 302/303/307 with new URL
5. Set flow table rules for the redirected traffic
6. Connect to the appropriate service endpoint

Figure 11.4 ALTO–SDN use case; HTTP-based video streaming scenario.

that is, the SDN controllers should be able to provide network information from which the ALTO server derives network and cost maps.

Xie et al. [18] prepared an IETF draft discussing possible use cases for the integration of ALTO service in SDNs. The benefits of the integration of ALTO network information service into SDNs are the following. ALTO becomes transparent for the end users or the service claimant entity (no deployment cost in the UE). Due to ALTO information, the ALTO client in the SDN controller can overwrite the initial peer selection decision of the service claimant entity (e.g., UE). Any flow can be dynamically selected for getting ALTO guidance, and SDN controller provides built-in redirection mechanisms with flow rewrite rules. Furthermore, SDN controllers are aware of the topology and state of served network areas and hence can provide abstract network and cost maps to the ALTO server.

Figure 11.4 illustrates the use case where ALTO guidance is used for better-than-random endpoint selection for HTTP-based video streaming service.

The SDN controller shall be notified about a new TCP connection establishment request by the edge SDN switch of the SDN domain. Since ALTO network and cost maps basically apply IP addresses, and not, for example, HTTP URIs, the IP and TCP header of TCP SYN message shall be used to decide whether this connection should get ALTO guidance or not. If yes, then the ALTO client shall find the appropriate network and cost maps for the service and shall determine the candidate IP addresses/PIDs for the service. If not already cached, it may request the appropriate maps from the server with target-independent query. Next, the ALTO client may, for example, calculate k-shortest paths for each cost type. That is followed by a multiattribute ranking procedure, to calculate the aggregated ranking of the endpoints.

After that, the ALTO client and SDN controller shall check resource availability for candidate paths to the best endpoint. If the E–E path crosses multiple SDN domains, this would require communication over the west–east interfaces interconnecting SDN controllers.

Then, the SDN controller shall install the necessary flow entries in its SDN domain and notify other SDN controllers on the path to do the same for this flow.

If the procedure does not find any path toward one of the endpoints, the TCP SYN should be dropped. If the service can support IP address rewriting, the controller should install rewrite the destination IP address downstream and the source IP address upstream.

Another option is that the TCP connection (and the HTTP communication on top of that) is redirected to a local HTTP redirect server. The related flow entries must only be kept until the HTTP redirect server redirects the source to the appropriate endpoint; hence, these are very short-lived flow entries.

The HTTP redirect server must be notified about the selected IP address and may resolve the DNS name to generate the new HTTP URI for the client. Then it can send the HTTP redirect message back to the client.

11.5.3   The ALTO–SDN Architecture

An important change in ALTO–SDN architecture compared to the original SDN architecture is that the selection of the preferred endpoint (decision making) is moved from the ALTO server to the ALTO client. Consequently, the ALTO server mainly is utilized as a pure network and cost map information service. From the proposed functions of ALTO client-to-server API [12], we realized information resource directory, network map, and cost map query services. The change in the concept was made due to the fact that it is better to implement communication-intensive SDN applications as an application module in the controller.

Another important functionality of the ALTO server is the automatic merging of network and cost map information coming from different sources (over ALTO server-to-network APIs), such as CDNs, BGP speakers, and SDN controllers. Currently, we are only focusing on network and cost map provision from one SDN controller. Figure 11.5 illustrates the main components of ALTO server.

The ALTO client is implemented as an application module in the SDN controller, as depicted in Figure 11.6. Its basic functionality is the query of network and cost maps from the ALTO service during the connection establishment phase of distributed services requiring ALTO guidance. It also stores locally in its cache the maps in order to reduce signaling. Additionally, it also provides ranking of endpoints based on the cost maps obtained from the ALTO server.

The SDN controller must know which service classes require ALTO service and ALTO-related policies of the operator must be provided. The proposed configuration XML schema includes the definition of service classes, which have a name (id); reachability information of servers (network addresses, port numbers), which specify the name of the related ALTO network map; the cost types to be considered for the service; and the main direction of the service (downlink, uplink, or both). If cost type is missing, then all cost maps should be considered in the ranking of service endpoints. Additionally, the reachability of ALTO server and redirect servers must be given. There is an additional field called PID mask. It is currently an IPv4 network mask, which defines the boundaries between subnets assigned to the same PIDs and represents the policies of the operator regarding the level of abstraction of the SDN network areas.

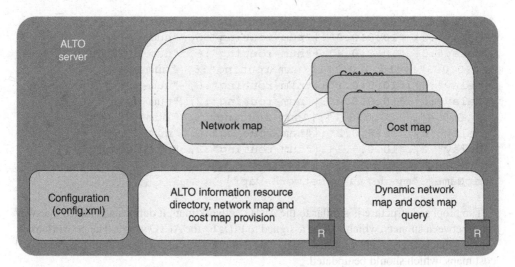

Figure 11.5 ALTO server.

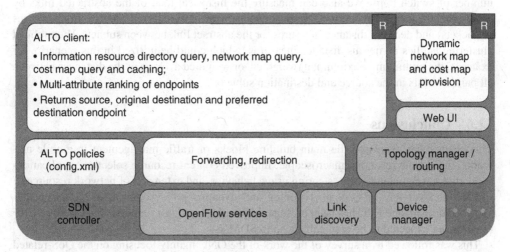

Figure 11.6 ALTO client in SDN controller.

11.5.4   Dynamic Network Information Provision

In our proposal, the ALTO server can request dynamically network information from the SDN controller. The SDN controller provides an up-to-date single-node network view over RESTful interface with JSON media type. An example for the JSON message is given in the following:

```
{"topology":{"10.0.0.1":{"10.0.0.2":{"num-routing":2, "num-
    delay":0}, "10.0.0.3":{"num-routing":6, "num-delay":0},
    "10.0.0.4":{"num-routing":6, "num-delay":0}},
```

```
"10.0.0.2":{"10.0.0.1":{"num-routing":2,  "num-
   delay":0},"10.0.0.3":{"num-routing":6,  "num-
   delay":0},"10.0.0.4":{"num-routing":6,  "num-delay":0}},
"10.0.0.3":{"10.0.0.1":{"num-routing":6,  "num-
   delay":0},"10.0.0.2":{"num-routing":6,  "num-
   delay":0},"10.0.0.4":{"num-routing":2,  "num-delay":0}},
"10.0.0.4":{"10.0.0.1":{"num-routing":6,  "num-
   delay":0},"10.0.0.2":{"num-routing":6,  "num-
   delay":0},"10.0.0.3":{"num-routing":2,  "num-delay":0}}},
"pidMask":"255.255.255.255",
"mapName":"my-default-network-map"}
```

The proposed structure is similar to the ALTO network map; it defines abstracted one-way links between subnets, which will be assigned to PIDs by the ALTO server. The network mask for the subnets is given by "pidMask." The "mapName" gives the network map and associated cost maps, which should be updated.

The cost maps are created using the different distance metrics given for each one-way abstract link in the topology structure. Num-routing cost is a metric proportional with the number of switch hops. We also can measure the historical load of the abstracted links by monitoring the increments in switch port statistics (in terms of received, sent, or dropped bytes or packets) and deriving distance measures for the abstract link between subnets. Hierarchical clustering applies numerous distance measures, which could be utilized in this scenario, for example, the minimum, maximum, unweighted, or weighted average of the distances between all pairs of hosts in the source and destination subnets.

11.6 Conclusions

This chapter has discussed the main building blocks of traffic management in mobile networks, that is, microscopic, macroscopic, improved content resource selection, application-supported traffic management, steering usage behavior, and extension of network resources.

Then an overview of QoS provisioning and dynamic policy control in 2G/3G packet-switched domain and EPC has been presented. The policy control functions realized by PCRF function are expected to be applicable also in SDMNs.

This was followed by a survey of the work of the ONF, mainly focusing on the QoS-related features of OF protocol, and an important traffic management-related use case defined by ONF.

Following that, an ALTO–SDN solution has been presented, showing the feasibility of SDN-based traffic management.

References

[1] Bokor, L., Faigl, Z., Eisl, J., Windisch, G. (2011) Components for Integrated Traffic Management—The MEVICO Approach, Infocommunications Journal, vol. 3, no. 4, pp. 38–49.
[2] 3GPP (2013) Policy and Charging Control Architecture (Release 12), TS 23.203. http://www.3gpp.org/DynaReport/23203.htm. Accessed February 16, 2015.
[3] 3GPP (2013) General Packet Radio Service (GPRS) Enhancements for Evolved Universal Terrestrial Radio Access Network (E-UTRAN) Access (Release 12), TS 23.401. http://www.3gpp.org/DynaReport/23401.htm. Accessed February 16, 2015.

[4] McKeown, N., Anderson, T., Balakrishnan, H., Parulkar, G., Peterson, L., Rexford, J., Shenker, S., and Turner, J. (2008) OpenFlow: Enabling Innovation in Campus Networks, SIGCOMM Computer Communication Review, vol. 38, no. 2, pp. 69–74.

[5] Open Networking Foundation (2013) OpenFlow Switch Specification, version 1.3.2. https://www.opennetwork ing.org/images/stories/downloads/sdn-resources/onf-specifications/openflow/openflow-spec-v1.3.2.pdf. Accessed February 16, 2015.

[6] Open Networking Foundation (2013) OpenFlow Management and Configuration Protocol (OF-Config 1.1.1), version 1.1.1. https://www.opennetworking.org/images/stories/downloads/sdn-resources/onf-specifications/ openflow-config/of-config-1-1-1.pdf. Accessed February 16, 2015.

[7] Open Networking Foundation (2013) *Solution Brief: OpenFlow™-Enabled Mobile and Wireless Networks*, Wireless & Mobile Working Group. https://www.opennetworking.org/images/stories/downloads/sdn-resources/ solution-briefs/sb-wireless-mobile.pdf. Accessed February 16, 2015.

[8] Gurbani, V., Hilt, V., Rimac, I., Tomsu, M., and Marocco, E. (2009) A survey of research on the application-layer traffic optimization problem and the need for layer cooperation, IEEE Communications Magazine, vol. 47, no. 8, pp. 107–112.

[9] Xie, H., Yang, Y. R., Krishnamurthy, A., Liu, Y. G., and Silberschatz, A. (2008) P4P: Provider Portal for Applications, in Proceedings of the ACM SIGCOMM 2008 Conference on Data Communication (SIGCOMM '08), Seattle, WA, USA, August 17–22, 2008, pp. 351–362.

[10] Aggarwal, V., Feldmann, A., and Scheideler, C. (2007) Can ISPS and P2P Users Cooperate for Improved Performance?, SIGCOMM Computer Communication Review, vol. 37, no. 3, pp. 29–40.

[11] Saucez, D., Donnet, B., and Bonaventure, O. (2007) Implementation and Preliminary Evaluation of an ISP-driven Informed Path Selection, in Proceedings of the 2007 ACM CoNEXT Conference, New York, USA, pp. 45:1–45:2.

[12] Alimi, R., Penno, R., and Yang, Y. (Eds.) (2014) ALTO Protocol, IETF Draft, draft-ietf-alto-protocol-27, March 5, 2014. https://tools.ietf.org/html/draft-ietf-alto-protocol-27. Accessed February 16, 2015.

[13] Crockford, D. (2006) The Application/JSON Media Type for JavaScript Object Notation (JSON), IETF RFC 4627, July 2006. http://www.ietf.org/rfc/rfc4627.txt. Accessed February 16, 2015.

[14] Gurbani, V., Scharf, M., Lakshman, T. V., Hilt, V., and Marocco, E. (2012) Abstracting Network State in Software Defined Networks (SDN) for Rendezvous Services, in Proceedings of the IEEE International Conference on Communications (ICC), 2012, Ottawa, Canada, pp. 6627–6632.

[15] Medved, J., Ward, D., Peterson, J., Woundy, R., and McDysan, D. (2011) ALTO Network-Server and Server-Server APIs, IETF Draft, draft-medvedalto-svr-apis-00, March 2011. https://tools.ietf.org/html/draft-medved-alto-svr-apis-00. Accessed February 16, 2015.

[16] Racz, P., and Despotovic, Z. (2009) An ALTO Service Based on BGP Routing Information, IETF Draft, draft-racz-bgp-based-alto-service-00, June 2009. http://www.ietf.org/archive/id/draft-racz-bgp-based-alto-service-00.txt. Accessed February 16, 2015.

[17] Gredler, H., Medved, J., Previdi, S., Farrel, A., and Ray, S. (2013) North-Bound Distribution of Link-State and TE Information Using BGP, IETF Draft, draft-ietf-idr-ls-distribution-04, November 2013. https://tools.ietf.org/ html/draft-ietf-idr-ls-distribution-04. Accessed February 16, 2015.

[18] Xie, H., Tsou, T., Lopez, D., Yin, H. (2012) Use Cases for ALTO with Software Defined Networks, IETF Draft, draft-xie-alto-sdn-use-cases-01, June 27, 2012. https://tools.ietf.org/html/draft-xie-alto-sdn-use-cases-00. Accessed February 16, 2015.

12

Software Defined Networks for Mobile Application Services

Ram Gopal Lakshmi Narayanan

Verizon, San Jose, CA, USA

12.1 Overview

Cloud, virtualization, and software defined networking (SDN) are emerging IT technologies, which revolutionizes business model and technical realization. Enterprise and network operators are consolidating their network and data center resources using virtualization technologies into service architecture. First, it is better to understand essential requirements that are driving these technologies so that we can realize how these technologies are achieving the end goals. The driving requirements are:

- Virtualization: Implement network function in software, and decouple the hardware dependency. Then run the network functions anywhere without the need to know physical location and how it is organized.
- Programmability: Topology should be flexible and able to change the behavior of the network on demand.
- Orchestration: Ability to manage and control different devices and software uniformly with simple and fewer operations.
- Scaling: System should be scalable up or down based on the usage of the network.
- Automation: System should provide automatic operations to lower operational expense. It must support troubleshooting, reduced downtime, easy life cycle management of infrastructure resource, and load usage.
- Performance: System must provide features to understand the network insights and take actions to optimize network device utilization such as capacity optimization, load balancing, etc.

Software Defined Mobile Networks (SDMN): Beyond LTE Network Architecture, First Edition.
Edited by Madhusanka Liyanage, Andrei Gurtov, and Mika Ylianttila.

- Multitenancy: Tenants need complete control over their addresses, topology, routing, and security.
- Service integration: Various middleboxes such as firewall, security gateway, load balancers, video optimizer, TCP optimizer, intrusion detection systems (IDS), and application-level optimizer must be provisioned on demand and placed appropriately on the traffic path.
- Open interface: Allow multiple equipment suppliers to be part of the topology and open control functions to control them.

The SDN goals are being envisioned by providing (i) separation of control and user plane traffic, (ii) centralized control of network functions and policies, (iii) open interfaces to hardware and software that need control, and (iv) control of traffic flows and programmability from external application. As SDN is a bigger concept, there are various organizations working toward standard approach including network function virtualization (NFV) from the European Telecommunication Standards Institute (ETSI), OpenFlow from the Open Network Forum (ONF), Interface to the Routing Systems (I2RS) from the Internet Engineering Task Force (IETF) [1–5].

Mobile network is expanding in all directions including the number of base station to accommodate traffic growth, increase in number of connected devices to provide machine-to-machine (M2M) services, and number of applications being developed for users. Operators are challenged on how can they manage and control such sudden growth in mobile network and, in parallel, allow growth in mobile networks. IT infrastructure has already consolidated their data center using cloud and SDN-based architecture when they faced with similar challenges. Therefore, telecommunication operators are following similar evolution on wireless networks. In this regard, 3GPP standard organization has started study group activity on SDN for Long-Term Evolution (LTE) wireless architectures. The goal of this chapter is to provide an overview of mobile network and then describe how NFV and SDN-based mechanisms are applied to wireless architecture. Next, we describe various application-level use cases and how SDN can be applied to improve the operation of network. Finally, we conclude with list of open research problems for future study.

12.2 Overview of 3GPP Network Architecture

Mobile broadband access network consists of packet core network (CN), radio access network (RAN), and transport backhaul network (TN). Simplified 3G and 4G mobile broadband network architecture is shown in Figure 12.1. Third-generation RAN shown in Figure 12.1a consists of NodeBs and radio network controller (RNC). The functions of RAN include radio resource management (RRM), radio transmission and reception, channel coding and decoding, and multiplexing and demultiplexing. Layer 2 radio network protocol messages are used to carry both control and user plane traffic from RAN to user terminal. RNC identifies signaling and user plan messages and forwards layer 3 mobility management messages toward CN HSS for authentication and authorization of users. Uplink user plane traffic from UE is received at RNC, and RNC performs GPRS Tunneling Protocol (GTP) operation and forwards the IP packet toward GGSN. Similarly, downlink packets for UE are received at RNC, and GTP packets are terminated and inner IP packet is forwarded to UE.

3G packet core consists of Gateway GPRS Support Node (GGSN) and Serving GPRS Support Node (SGSN). The functions of packet core include IP session management,

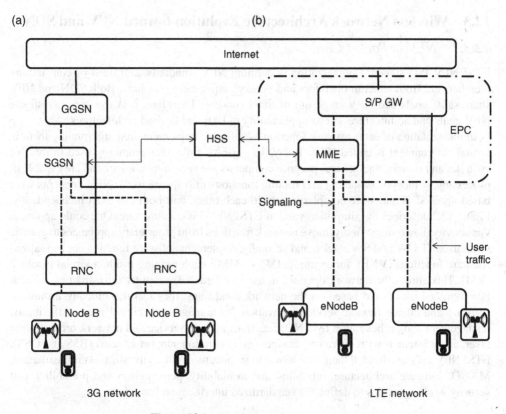

Figure 12.1 Mobile broadband networks.

legal interception functions, policy-based routing and charging functions, etc. User mobility and its associated sessions are handled by SGSN node. SGSN also acts as anchor point for ciphering and authentication of session between UE and wireless access networks.

Fourth-generation LTE radio network is shown in Figure 12.1b and it consists of eNodeB network element [6]. LTE design is based on flat architecture with reduced number of network elements. The RAN functions are consolidated into single network element eNodeB. 3G and 4G RAN networks are not backward compatible as their physical layer technologies and radio network protocols are different. 4G uses orthogonal frequency division multiplexing (OFDM), whereas 3G uses wideband code division multiple access (WCDMA)-based technologies for wireless physical layer processing.

Similar to 3G, there is a logical separation of networks such as RAN, packet core, and transport network in 4G networks. Mobility management entity (MME) and serving and packet gateway (S/P-GW) are part of Evolved Packet Core (EPC) in LTE. The functions of MME include radio signaling functions and mobility management sessions maintenance. The EPC functions are similar to that of 3G GGSN and contain improved IP mobility management functions.

12.3   Wireless Network Architecture Evolution toward NFV and SDN

12.3.1   NFV in Packet Core

ETSI NFV ISG standard organization is defining NFV standards, and most of contributing members are from telecom operators and network equipment suppliers. Both SDN and NFV share same goals and NFV came out of SDN concept. Therefore, it is worth to investigate NFV goals and architecture and its applicability to LTE and beyond architectures.

Implementation of each network function as software implementation and running them in virtual environment is called NFV. Today, 10 gigabit/s links are commonly used in most of switches and routers, and general-purpose computers are becoming cheaper and are capable of processing of most of switching and routing functions in software itself. NFV concepts were based upon SDN and were complementary to each other, and both can exist independently. Figure 12.2 describes the simplified view of ETSI NFV ISG architecture. One could approach this activity in two steps: firstly, move network functions from proprietary appliances to generic virtualized IT HW and SW stacks, and secondly, implement software functions as virtualized software functions (VNF), for example, IMS or MME each running inside a virtual machine (VM). This allows the network operator to use standardized computer infrastructure without HW vendor lock-in, and based on the network conditions, they can dynamically instantiate software and enable flexible service innovation. To manage with high degree of flexibility, management and orchestration (MANO) functions include service and network orchestration layer, and it interacts with operators' business and operation support systems (BSS and OSS). ETSI NFV ISG has divided their activities into architecture of the virtualization infrastructure, MANO, software architecture, reliability and availability, performance and portability, and security working groups to define the standardized interfaces and solutions.

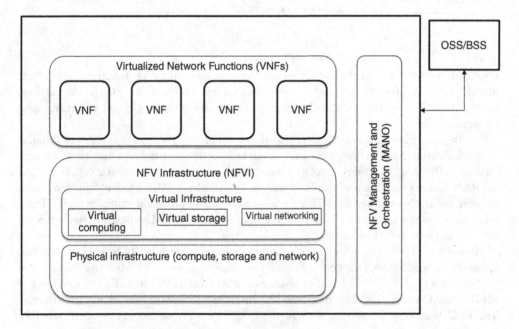

Figure 12.2   Simplified ETSI NFV ISG architecture.

The relationship between NFV and SDN is as follows:

- NFV came out from SDN, and SDN and NFV are complementary as both serve same goals.
- NFV can be implemented without SDN.
- According to NFV, virtualization of network function alone is sufficient to solve most of the problems, and this is used in current data centers.
- SDN relies on separation of control and data plane forwarding, defining additional control and interfaces.
- NFV and SDN can be combined to create potentially greater value.

12.3.2 SDN in Packet Core

SDN has become apparent as it provides simple interface to vendor equipment and allows dynamic network topology changes and easy integration of new services and reduces operating cost. The goal of SDN is to provide (i) separation of control and user plane traffic, (ii) centralized control of network functions and policies, (iii) open interfaces to hardware and software that needs control, and (iv) control of traffic flows and programmability from external application.

In 3GPP architecture, there exists separation of user plane and control plane in interface and in protocols. However, all these interfaces are implemented in proprietary hardware and it is difficult to decouple and introduce new services. Processing elements such as MME and policy charging and routing function (PCRF) process only control plane messages, and it contributes toward proper handling of user sessions. Elements such as eNodeB and EPC P/S gateway process both control plane and user plane traffic, and for performance reasons, each vendor implements it in dedicated hardware. To have clear separation between control plane and user plane, the following could be one of the possible approaches. First, virtualize the network function so that they can be made available on any hardware or on the same hardware. Secondly, centralize the control functions via separation of the control plane processing and user plane processing via appropriate SDN protocols. Both these steps are described in the following sections.

12.3.2.1 Virtualized CN Elements

Figure 12.3 describes the virtualized network functions of existing network interfaces. This step is similar to the proposal proposed by NFV standards. But centralize all network functions in single or less hardware so that higher degree of control and simplification to the internal communication can be achieved. As we could see that SDN, virtualization, and cloud are related and essential concepts:

- Removing hardware and software dependency: As the network element is no longer a collection of integrated hardware and software entities, the evolution of both is independent of each other. This enables the software to progress separately from the hardware and vice versa.
- Flexible network function deployment: As network function is virtualized, it improves the resource utilization of the hardware. For example, MME and PCRF could be running on one

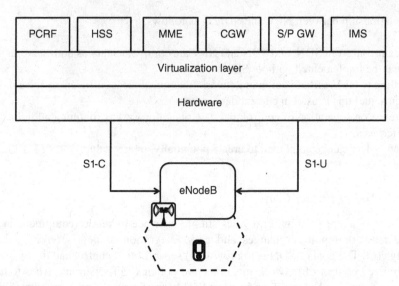

Figure 12.3   Virtualized 3GPP network functions.

hardware and EPC core on different target hardware, or if the network is serving smaller number of users, all of them can be made to run on the same hardware platform. We bring the concept of hardware pools and via virtualization any of the network function can be made to run on that hardware. Such dynamic start, stop, and movement of network functions bring different cloud and network topologies.

• Dynamic operation: When network functions can be virtualized and moved within the hardware pool, we can further extend and allow instantiated software to run with certain configuration and do dynamic operations.

There are several approaches to LTE wireless network element functions. Today, high-end commercial off-the-shelf (COTS) hardware and operating system are capable of handling higher traffic and can be configured as layer 2 switch or router by software configuration. Given this to the advantage, virtualize the 3GPP network elements into virtual environment without changing the 3GPP interfaces. This is one of form separation and gives more flexibility to operator who currently runs dedicated hardware for each network function. Virtualization gives improved resource utilization and ability to consolidate or distribute the processing functions at any time based on load and enables to optimize the resources. Figure 12.3 describes a scenario wherein all 3GPP control plane applications and EPC core functions are moved into virtualized network environment. For next-generation wireless architecture, virtualization becomes prerequisite for cloud; therefore, we must consider SDN on virtualized network environment. Pure control plane functions like MME, SGSN can run as applications in the cloud without impacts to other nodes as long as 3GPP-defined interfaces do not change. The initial step in the evolution process could involve virtualization of EPC nodes. The second step could involve moving whole P-GW/S-GW, other EPC functions like MME and PCRF to cloud or virtualized environment.

12.3.2.2 SDN Mechanism on Virtualized CN Elements

To achieve separation between control and user plane traffic, network element functions such as P-GW, S-GW must be logically separated and be able to communicate via open interface protocols. Identification of forwarding and control functions requires careful investigation of how and where each feature of S-GW and P-GW resides. The goal of this exercise is to achieve total separation of control and user plane functions. Figure 12.4 describes the control and user plane separated EPC core and running of SDN protocol as open interface to control the network functions. There are many protocol candidates such as OpenFlow and I2RS proposed by various standard organizations as SDN protocol. The purpose of this protocol is to support mechanism to configure, control, and manage the network functions and sessions seamlessly.

When control plane function of the gateway is virtualized (running on a VM) as shown in Figure 12.3 and the user plane function of the gateway application protocol (e.g., GTP-U) is not virtualized (running on dedicated hardware) as shown in Figure 12.4, the 3GPP specific user plane control and reporting functionalities shall be supported by the control protocol between the virtualized S/P-GW-C and the nonvirtualized S/P-GW-U.

12.4 NFV/SDN Service Chaining

12.4.1 Service Chaining at Packet Core

Network provides diverse functions. User gets service based on subscriber or type of application traffic types. Depending upon type of service, traffic passes through one or more middlebox functions deployed in the networks. Operator typically deploys several middlebox functions such as carrier-grade network address translation (CG-NAT), firewall, video optimizer, TCP optimizer, caching server, etc. Figure 12.5 shows deployment scenario of middlebox functions and how different packets go through series of network interfaces. Middleboxes are deployed

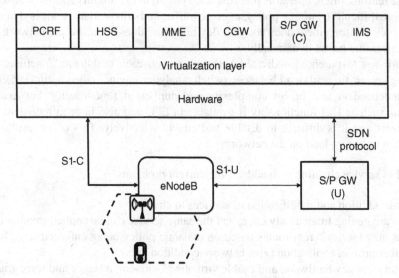

Figure 12.4 Control and user plane split architecture.

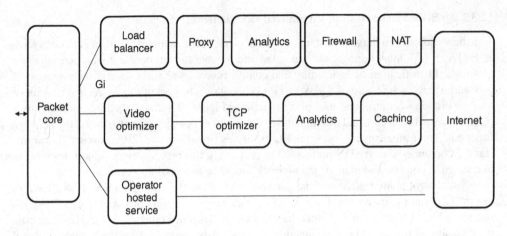

Figure 12.5 Traditional Gi LAN service chaining.

at SGi interface that connects data network and P-GW. When packet traverse through different path and treated by middleboxes on the path, it is called service chaining [7, 8].

Service chaining is currently operated via static policy and is not flexible. It has the following shortcomings:

- Network equipment supplier provides middlebox functions on dedicated hardware and configuration in SGi. LAN interface is hardwired to network interfaces and it is not easy to add or change the topology.
- Network operator does not have sufficient insights about how the traffic on their network is flowing and how it is impacting their service. Most of the services are offered by third-party applications and it is difficult to achieve standardized way of monitoring and control. Due to these reasons, often, operators preprovision systems to maximum capacity. Consequently, it is more difficult to enforce policy, manage traffic, and differentiate services dynamically.
- Static policy enforcement does not give flexibility and does not allow the network to scale when the traffic grows dynamically.
- Operator has to purchase dedicated hardware to run each middlebox functions such as caching server, firewall, load balancer switch, analytics engine, video optimizer, etc. Each such independent unit brings complexity and duplicated functionality. For example, a function such as DPI functionality is available in EPC and also in certain content filter or optimizers and it is difficult to disable and enable selectively. This often results in poor treatment and extra load on the network.

The goal of service chaining is to address the current problems:

- Dynamic addition and modification of services to chains.
- Packets are getting treated only once, and the same service can be applied to other chains.
- Packets may take different routes based on dynamic policy being enforced (e.g., based on subscriber profile, application type, network condition, etc.).
- Avoid unnecessary hardware, and enable virtualized software instance and form graphs and connect links to treat different packet flows.

This nature of dynamic service changing has been in discussion with various standardization bodies including IETF and NFV [7, 8].

Combining SDN and service chaining addresses dynamic control of each middlebox with central policies. SDN enables open interfaces and centralized control point for the middlebox and allows service creation functions such as bandwidth management, traffic steering, etc.

Virtualized network element function is a key enabler for dynamic service chaining. When network functions are virtualized and are running in same hardware, it is easy to perform different chain structure and create services easily. Each packet could be treated differently based on the service by making the packet to go through different next hop network functions and form a graph. As the packet has to go through series of middleboxes as part of service, the next hop vectors can be easily provisioned via flow label protocol such as OpenFlow. Figure 12.6 describes the service chain mechanism, wherein only necessary chains are selected based on the dynamic policy from orchestration layer. The orchestration entity has SDN controller functions and aggregates network state information and delivers service aware policy information dynamically. There is no preprovisioning involved network scale dynamically.

12.4.2   Traffic Optimization inside Mobile Networks

In 3GPP architecture, all user plane control and policy functions are centralized and reside at packet core. This creates difficulty to introduce any new service easily. To illustrate this, consider two users, say, U1 and U2 as shown in Figure 12.7a, are using voice over IP (VoIP) or P2P-type applications. All uplink traffic from user U1 goes through radio, transports, and reaches packet CN. At the packet core, GTP tunnel functions, charging, and NAT functions

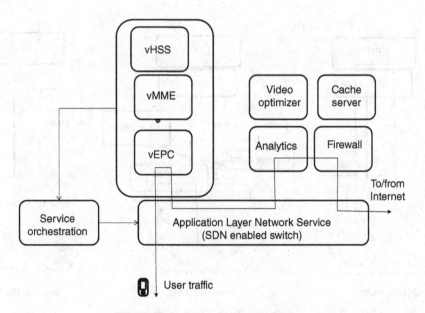

Figure 12.6 Dynamic service chaining.

are applied, and packet is sent back to U2 as downlink traffic. The traffic flow is north–south as shown in Figure 12.7a. It would be optimal to keep the traffic between the two users inside the RAN network (east–west traffic flow), and this will save bandwidth in transport network and reduce processing in EPC.

Concept of separation of signaling and switching is there in 2G mobile switching systems (MSC) and in plain old telephone systems (POTS) architectures. Traditionally, MSC is capable of handling both call flow signaling and switching function. Later, signaling and switching functions got separated into two different processing entities and are called soft switch. Operator will have one signaling server and deploy many switching servers closer to BTS or BSC in 2G architecture, thus allowing only signaling to be handled in the CN, and user traffic stays within the radio access networks.

Using SDN, it is possible to covert north–east to east–west traffic inside mobile networks. Design behind this is to have dynamic control of traffic flows, and traffic flow must be controlled based on the network state and internal network functions. Two key steps while designing SDN enabled network function are identification and exposure of internal state of network function and when to configure, monitor, and manage those states based on the traffic or network conditions. By having such network function, we could achieve traffic optimization for each application or flow level.

Figure 12.8 describes how to achieve traffic steering functions using SDN architecture. In this architecture, eNodeB is integrated with SDN layer 3 switch, and all traditional GTP tunnel endpoint functions are performed in this switch. Also, at the packet core, EPC is virtualized; control plane and user plane functions are separated. To illustrate the concept, we have shown only charging gateway (CGW), carrier-grade network address translation (CG-NAT or NAT) network functions. The following are the sequence of message that happen during a VoIP traffic stay within RAN network:

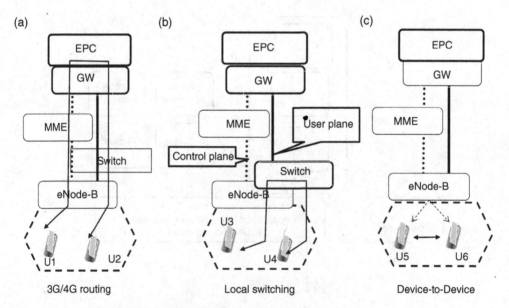

Figure 12.7 Traffic trend inside mobile networks.

1. SDN controller could be part of service and network orchestration layer function, and it provides initial configuration to CG-NAT. The CG-NAT is SDN enabled, and it will expose all its internal NAT table entries for management. Apart from traditional NAT, it includes extra control so that it will send a trigger message when two users behind CG-NAT are communicating. For example, when U1 traffic is seen first time on the NAT, it creates an outbound NAT table entry with U1 source IP address, destination address (public IP address of NAT), protocol, and port numbers. When NAT sees the traffic of U2, it will create an outbound traffic similar to U1. It compares internally when an entry is created with all other entries in the table to make sure that two users are not behind the NAT. SDN controller configures the CG-NAT, and CG-NAT will act on the user plane traffic and wait for the trigger conditions.
2. Either U1 or U2 is initiating VoIP session. VoIP application performs sequence of steps that includes peer endpoint IP address detection using simple traversal of UDP over NAT (STUN) or other suitable protocols and then exchange of signaling messages and starts application data transfer. As of now, IPv4 is still dominating and operators are using CG-NAT and deploy NAT function for all traffic toward the Internet. CG-NAT maintains mapping table in each direction.
3. CG-NAT detects the match on traffic flow between U1 and U2, and sends the traffic flow information trigger to SDN controller.
4. SDN controller instructs respective SDN-enabled eNodeB to apply traffic policies locally. Typically, policy would be to forward packet between U1 and U2 and start counting bytes in each direction. To have this functionality to be separately programmed, eNodeB's internal state processing must be identified and separated for external control. EnodeB typically performs GTP processing, and IP QoS functions toward packet core. Careful state identification and exposure is required so that higher degree of control is achieved. Timers such as dead-peer detection logics are part of OpenFlow protocols and those are also supplied along with the control messages. eNodeB SDSN switch function will hairpin the traffic locally for that session, thus saving the transport backhaul traffic. When traffic is being routed locally with eNodeB, the CG-NAT timer could expire; to avoid timer expiry, either SDN controller or port control protocol mechanism can be used explicitly to keep the NAT association active till the session is completed.
5. When either U1 or U2 or both terminate their sessions, the eNodeB SDN interface function will send IP flow statistics information pertaining to that flow including the duration and number of bytes transferred to SDN controller.
6. SDN controller could then send explicit command to purge the NAT table entry and also send the appropriate byte count information to CGW server.

To see the benefits of SDN, each network function when they are introducing control plane and user plane separation, they must identify each network internal state and expose them carefully. When control and user plane are separated properly along with internal state operations, SDN adds higher degree of freedom to service chaining and traffic steering.

Similarly, we can apply SDN to device-to-device (D2D) communication. 3GPP is standardizing D2D communication as part of Release 12 and Release 13 LTE features. These are going to be part of LTE-Advanced implementations. The goal of D2D is to combine location and proximity to enable direct D2D communication. There are two ways to use D2D, namely, infrastructure mode and ad hoc mode communication. Operator-assisted D2D communications are useful under following scenarios:

- Operators have limited spectrum available; infrastructure-based D2D communication will give relief for operator. By using D2D, operator could accommodate more devices to communicate and scale the network.
- When two users who wish to communicate among themselves are in cell edge of the network, they may not get sufficient bandwidth for their communication. In that situation, by using D2D network, they will be able to establish communications.
- A group of users want to share images or files among themselves. They share common profiles, and when D2D communication is present, operator could establish and manage their wireless connectivity to have proximity-based communications.

Without putting additional requirements on the user and application, we need to establish D2D communication when possible. To achieve this, SDN-based signaling can be used. For example, we described earlier a scenario of VoIP wherein two users U1 and U2 are behind the same cell or the same BTS. Now, BTS in the D2D communication is more flexible to lease frequencies to couple of mobile devices for D2D communication. If we extend that concept further, if two users are behind the same BTS (or cell) and are in close proximity, we could apply SDN principle to have D2D communication. Combining application endpoint information and location proximity information, we could create configuration parameters that will then eventually be used to establish D2D communication. Following are the possible additional procedures that must be incorporated in Figure 12.7.

Referring to sequence of message described earlier for VoIP application, there is no change to steps 1–3 of Figure 12.8. Additional functions are implemented in eNodeB. At step 3, eNodeB needs to check whether two devices are in close proximity, and they are allowed to communicate directly via D2D channel. eNodeB can verify the location and proximity information either with the help of MME or location-based service network elements. Assume that two UEs are in close proximity; then they will establish network-assisted D2D session, and control information will go to eNodeB.

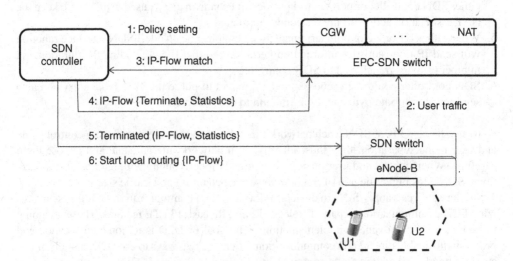

Figure 12.8   Traffic steering inside mobile networks.

12.4.3   Metadata Export from RAN to Packet CN

In the previous section, we described using SDN how to extract and transfer packet CN state information to radio network and achieve traffic optimization. There are situations wherein radio network states information needs to be extracted and transferred to packet core or to SGi LAN to achieve service differentiation. We will illustrate this use case by considering video streaming application.

Video traffic contributes toward a major portion of Internet traffic. Due to increase in penetration of smartphones and tablets, video traffic will grow by many folds. User-generated content (UGC) such as YouTube and premium traffic such as Netflix are major contributors in the United States [9, 10]. Such a sudden explosion of video traffic makes network unmanageable. Lessons learned from deploying fourth-generation mobile network such as LTE and third-generation mobile network such as WCDMA technologies have shown that video streaming does not perform well over mobile broadband networks. Video streaming session has a long duration and demands latency and bandwidth throughout the video play. Hence, network resource must be made available throughout the video play. To satisfy these requirements, mobile operators have increased their server and network capacity. However, wireless signal strength varies with respect to location, time, and environment, and delivering guaranteed bandwidth to video streaming application in such nonuniform wireless network condition becomes a challenge.

Three popular mechanisms exist in the Internet to deliver video content using standardized protocols. They are as follows: (i) user could download a video file from server using FTP service and watch video locally in his device at later time, (ii) video could be streamed as video on demand (VOD) from a server, and (iii) video could be delivered in real time. VOD and live video streaming are most popularly used today. Pseudo streaming and adaptive bit rate (ABR) streaming are two popular streaming mechanisms and both run on HTTP [11]. As of today, ABR is adopted by most of the content distributors. In ABR streaming, a video data may be coded into different quality levels such as high, medium, or low resolution. Each of those coded video data is further divided into chunks and kept as small files in the video content distribution server. Each chunk contains 3–8 s of video data. Based on the available bandwidth, client selects video quality and then requests respective chunk file for play. When the requested chunk is being downloaded by client, client computes several parameters including round-trip time, total download time for the chunk, etc. and keeps this history information. The stored information is then used to decide next chunk request. As the wireless network is time varying, TCP congestion control algorithms are not well suited for such burst video transmission and results in retransmission and video stalls. Application service providers are exploring the possibility to get additional information about radio conditions so that they can adjust their video transmission accordingly. As 3GPP standards suggest full inspection of control and data packets to solve RAN congestion [12], it is not possible when application service providers are enabling SSL-based encryption on user plane data. Therefore, one of the possible approaches is to have cooperative solution between application service provider and network operator. Figure 12.9 describes a solution wherein SDN controller is part of PCRF function and interacts with video service function such as caching server or video optimizer or TCP optimizer hosted on SGi LAN. The traffic from the UE is encrypted and is not possible to apply DPI and perform RAN traffic management functions.

There are two approaches to trigger the information export from eNodeB. (i) In the first approach, SDN controller receives periodic flow information that requires radio state information from eNodeB. (ii) In the second method, video server IP address is known to SDN

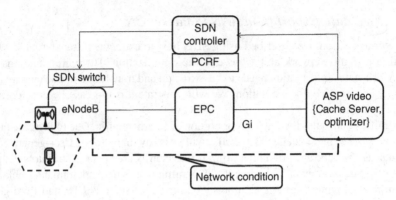

Figure 12.9   User plane RAN optimization.

controller, and it could instruct eNodeB to provide radio information when uplink traffic matches the video server IP address. In both approaches, eNodeB will export its internal radio state information such as cell load condition, traffic congestion, etc. EnodeB can piggyback this information on uplink traffic from UE toward video server, or alternatively, it could send at IP layer option information. This metainformation is then passed through EPC core and reaches video server on SGi LAN.

The role of SDN controller is to communicate the required flow configuration information to eNodeB. Then, eNodeB will act upon the IP traffic flow; eNodeB will generate metadata information when a match is found in IP traffic against the configuration. For example, whenever UE sends uplink traffic toward video server, as the server's IP address is put as part of eNodeB configuration, the eNodeB will provide the metadata and sends along with the UE uplink packet. The metadata may include radio load information, cell-level information, UE location information, etc. Existing eNodeB does not export this information at this moment. This is not yet defined in standards. When we are enabling SDN for each network function, there must be flexibility in configuring each network function, and hence, respective actions can be managed.

12.5   Open Research and Further Study

We described application scenarios and how it can be improved by combing service chaining and SDN principles. We emphasize that SDN provides separation of control and user plane and enhance higher degree of control and give the notion of programmable networks. As part of SDN design, careful design choice must be made on how to expose internal network and network element states. What we presented is just a beginning; as part of this exercise in wireless network, we need to consider several aspects. We believe that the following are still exploratory activities that are still open to both research and innovation:

- Legal interception gateway is performed in packet CN. Lawful agencies apply policies on user traffic and collect the user data [13]. When traffic optimizations as described in Figure 12.8 are performed, the traffic will not reach LIG. Therefore, we must distribute LI functions carefully to RAN network or disable the localized traffic routing for sessions requiring LI.
- Separation and exposing network state on each network function is a complex task. As EPC supports more than 100s of protocols, we need to ensure that state information can be manipulated or exported to outside world as programmable API.

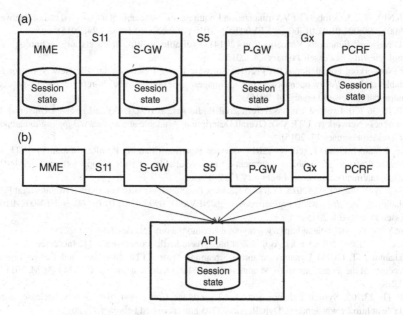

Figure 12.10 State consolidation and SDN API exposure.

- MME, S-GW, and P-GW maintain UE state information internally. Network elements such as MME process control plane information, S-GW acts as mobility anchor point, and PDN terminates traffic at SGi interfaces and acts as policy enforcement point. As the functionality is different, they need to maintain redundant information of each UE. As described in Figure 12.9, when a UE-attached procedure is performed, MME, S-GW, and P-GW creates state information for each UE and has lot of redundant information. When network functions are virtualized and going to run on a single hardware platform, we can revisit dedicated box approach to share virtual environment and improve the network design by making network element stateless as much as possible. Figure 12.10b describes one of the approaches to consolidate the state information in the CN elements and allow the database to be exposed with an API.

Acknowledgments

This work started while the author was working at Nokia. At the time of publication, the author works at Verizon. This work is supported both by Nokia and Verizon. Opinions, findings or recommendations expressed in this chapter is from the author and does not reflect the views of Nokia and Verizon.

References

[1] Atlas, A., Nadeau, T.D., and Ward, D. (editors) (2014) Interface to the Routing System Problem Statement, IETF, (work-in-progress) draft-ietf-i2rs-problem-statement-03, June 2014. Available at http://tools.ietf.org/id/draft-ietf-i2rs-problem-statement-03.txt (accessed February 17, 2015).
[2] ETSI, NFV (2013, October) Network Functions Virtualisation—Update White Paper. Available at http://portal.etsi.org/NFV/NFV_White_Paper2.pdf (accessed January 24, 2015).

[3] ETSI, NFV (2013, October) NFV Virtualization Requirements. Available at http://www.etsi.org/deliver/etsi_gs/NFV/001_099/004/01.01.01_60/gs_NFV004v010101p.pdf (accessed January 24, 2015).

[4] ETSI, Network Functions Virtualization (2014). Available at http://www.etsi.org/technologies-clusters/technologies/nfv (accessed January 24, 2015).

[5] Open Networking Foundation (ONF) (2012) Software-Defined Networking: The New Norm for Networks. Available at https://www.opennetworking.org/images/stories/downloads/sdn-resources/white-papers/wp-sdn-newnorm.pdf (accessed January 24, 2015).

[6] 3GPP TS 36.300, Evolved Universal Terrestrial Radio Access (E-UTRA) and Evolved Universal Terrestrial Radio Access Network (E-UTRAN); Overall Description. Available at http://www.3gpp.org/dynareport/36300.htm (accessed February 17, 2015).

[7] Quinn, P., and Nadeau, T. (editor) (2014) Service Function Chaining Problem Statement, IETF, (work-in-progress) draft-ietf-sfc-problem-statement-05.txt. Available at https://tools.ietf.org/html/draft-ietf-sfc-problem-statement-05 (accessed February 17, 2015).

[8] ETSI GS NFV 002 V1.1.1 (2013, October) Network Functions Virtualisation (NFV); Architectural Framework. Available at http://www.etsi.org/deliver/etsi_gs/NFV/001_099/002/01.01.01_60/gs_NFV002v010101p.pdf (accessed January 24, 2015).

[9] About YouTube, Available at https://www.youtube.com/yt/about/ (accessed April 9, 2015).

[10] Netflix. How does Netflix work?, Available at https://help.netflix.com/en/node/412 (accessed April 9, 2015).

[11] Stockhammer, T. (2011) Dynamic Adaptive Streaming Over HTTP: Standards and Design Principles. In: Proceedings of the second annual ACM conference on Multimedia systems, pp. 133–144. ACM, 2011. San Jose, CA, USA.

[12] 3GPP TR 23.705, System Enhancements for User Plane Congestion Management, Release, draft 0.11.0. Available at http://www.3gpp.org/DynaReport/23705.htm (accessed February 17, 2015).

[13] 3GPP TS 33.107, Lawful Interception Architecture and Functions. Available at http://www.3gpp.org/ftp/Specs/html-info/33107.htm (accessed January 24, 2015).

13

Load Balancing in Software Defined Mobile Networks

Ijaz Ahmad,[1] Suneth Namal Karunarathna,[1] Mika Ylianttila,[1] and
Andrei Gurtov[2]
[1] *Center for Wireless Communications (CWC), University of Oulu, Oulu, Finland*
[2] *Department of Computer Science, Aalto University, Espoo, Finland*

13.1   Introduction

Load balancing is a composition of methods to distribute workload among multiple networks or network components such as links, processing units, storage devices, and users to achieve optimality in respect to resource utilization, maximum throughput, and minimum response time. It also helps to avoid overload and provide quality of service (QoS). In a situation where multiple resources are available for a particular functionality, load balancing can be used to maximize network efficiency and increase fairness in network resource usage while keeping a balance between QoS and resource usage.

Network load balancing started in the form of load balancing hardware, which were application neutral and resided outside of the application servers. These network-based appliances could load balance using simple networking techniques. For example, virtual servers were used to forward connections to the real server deploying bidirectional network address translation (NAT) to load balance among multiple servers. A simple load balancing scenario is shown in Figure 13.1, where virtual servers balance the load among multiple real servers to ensure high availability and QoS.

Today, load balancing technologies are used mostly in the IP layer and application layer having layer-specific load balancing and distribution mechanisms. In this chapter, the basics of load balancing are introduced with the commonly used load balancing technologies in legacy wireless networks and state their problems and challenges. Moving forward to SDN, how SDMN-based load balancing technologies can solve the challenges existing in the

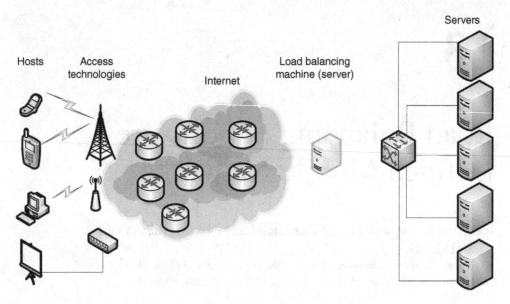

Figure 13.1 Load balancing among multiple servers.

currently used load balancing technologies in legacy wireless networks is elaborated. Toward the end of the chapter, future directions and research areas in load balancing in SDMN are discussed.

13.1.1   Load Balancing in Wireless Networks

In wireless networks, load balancing mechanisms are used to distribute traffic evenly among cells, nodes, and frequency bands to utilize resources of the network more efficiently. Highly loaded cells can be off-loaded to less heavily loaded neighboring cells, traffic on various backhaul or core network nodes can be shared among multiple nodes, and bandwidth can be dynamically shared to ensure QoS to subscribers. Since the aim of next-generation wireless networks is to provide high data rate services to mobile users in large coverage areas, bandwidth is a major consideration for operators to provide efficient services in dense and congested areas.

In order to use the available radio spectrum efficiently and effectively with high QoS, operators install small cells that significantly improve coverage and network capacity. These cells may use different technologies, such as cellular, WLAN, CDMA, or E-Band, to maintain the required QoS and quality of experience (QoE) through novel load balancing mechanisms. Technology convergence is an interesting approach to achieve high availability, fulfill QoS requirements, offer differentiated services, and provide network redundancy for network resilience. These goals of technology convergence can be achieved with the help of novel load balancing mechanisms for traffic and workload balancing.

Load balancing is a matured research topic that has been investigated for more than a decade in the context of mobile communications. However, due to vastly differing network architectures, intertechnology load balancing is mostly limited to researches because of the complexity in interoperability.

13.1.2   Mobility Load Balancing

In wireless networks, a user has the privilege to move around and still use network services. A mobile device can start or terminate connections randomly while traversing various cells or networks. Therefore, it is highly probable that a cell gets load of traffic that is beyond its capability with respect to its resources. Hence, mobility load balancing (MLB) is very important in cellular networks in particular and other wireless networks in general. MLB balances the load among available cells in certain geographical locations by means of controlling mobility parameters and configurations including UE measurement thresholds. MLB modifies handover (HO) regions to redistribute load between neighboring cells. The principle of MLB is adjusting the HO regions by biasing HO measurements, causing users in a cell edge to migrate from highly loaded cells to less heavily loaded neighboring cells to improve efficiency of resource utilization. Since the load redistribution is carried out automatically between neighboring cells, this MLB is an important feature of Self-Organizing Networks (SON).

13.1.3   Traffic Steering

Traffic steering is the capability of a network to control and direct voice and data traffic to the best suitable cell or radio technology within a network. It can be deployed in multiple layers such as frequency layers or hierarchical layers of cells (macro-, pico-, or femtocells) to provide resources to an end user in a certain geographical area. Traffic steering could optimize the network capacity and user experience through efficient utilization of the available pool of resources from a multitude of coexisting networking technologies in the core and edge. Traffic steering can be used to help MLB in a network. It can also be used to off-load macrocells toward low-power cells, HeNB, or Wi-Fi to accommodate large part of the traffic demand and minimize eNB power consumption. The primary challenge for traffic steering is coordinating mobility configurations in multiple overlaid cells.

13.1.4   Load Balancing in Heterogeneous Networks

Today, a typical smartphone can connect to the Internet via several different radio access technologies including 3GPP-standardized and non-3GPP technologies such as Wi-Fi (802.11x). Cellular base stations are getting diverse to satisfy user experience. Macrocells are shrunken to microcells, and picocells, distributed antennas, and femtocells are added continuously in cellular networks. Since the currently deployed networks are already dense in terms of nodes or base station installations, cell splitting is not a viable solution due to high intercell interference and costly capital expenditures (CAPEXs). Hence, the solution inclines toward overlaid structures where different varying architectures are overlaid to cowork and cooperate. These heterogeneous architectures would essentially use separate spectrum and different network architectures and topologies. With the introduction of heterogeneity from many directions in wireless networks, load balancing is crucial for the end user experience and overall system performance.

13.1.5   Shortcomings in Current Load Balancing Technologies

Load balancing is a critical requirement in large commercial networks, traditionally achieved with load balancers, which are expensive and independent entities in most of the cases. Generally, commercial load balancers sit on the path of incoming requests and then spread requests over several other servers. Current load balancing algorithms assume that the requests are entering to the network through a single gate where the load balancer is placed though thère can be several such choke points in a large network. On the other hand, servers and data centers may dynamically move across the network by means of virtualization introduced with programmability. Furthermore, different network sections may need totally different load balancing or optimization techniques to achieve the expected results.

It is clear that the traditional load technologies are not capable of meeting the requirements in today's large commercial networks. Thus, a different approach of load balancing is needed where the functions could come out of the box and deploy on top of traditional network elements to enable load balancing based on network and server congestion in an intelligent manner. Therefore, next-generation load balancers must have the following characteristics: (i) load balancing as a property of a network over traditional network elements, such as switches and routers; (ii) flexibility in application and service level (load balancing in application and service level that enhance ability to experiment new algorithms); (iii) dynamicity in terms of the ability to adapt to the changing conditions of the network where server congestion and route remapping are required; and (iv) dynamic configuration management to automatically adapt and scale with changes in network capacity, such as virtual machine (VM) mobility and data center mobility. Thus, load balancing in modern and future networks needs dynamism, which can be brought about by SDN through global visibility of the network state and open interfaces for programmability.

Current load balancing methods make a number of assumptions about the services that are not valid in the current requirements of higher data rates, need of seamless mobility, high availability, and expected and offered QoS. These assumptions [1] are as follows:

- Requests enter the network through a single point where load balancing devices can be placed at a choke point through which all the traffic must pass. This condition might not work for all networks, and hence, operators end up with congestion while using these expensive devices. In enterprise networks, there can be many choke points such as egress connections to the WAN, campus backbones, and remote servers.
- The network and servers are static, which can be true for a data center but not for wireless networks and enterprise networks. For example, in wireless networks, a base station can get congested at any time due to user movements, changes in channel conditions, etc. Similarly, operators of data centers move VMs of virtualized data centers to efficiently use their servers. With these changes, load balancers need to track changes in the network and movements in data centers to direct requests to the right places.
- Congestion is at the servers but not in the network, which might be true for data center hosting only one service, whereas, in cloud data centers, the network may be congested differently at different places.
- The network load is static, and hence, the load balancers spread traffic using static schemes such as equal-cost multipath (ECMP) routing. Such load balancing is suboptimal since some parts of the network might be heavily loaded and hence can be congested.
- All services require same load balancing algorithms, meaning that HTTP and video request can be served with same load balancing schemes. This is not feasible due to varying nature of requirements of different services such as bandwidth, mobility, and link capacity requirements. It is also difficult to provide each service its own type of load balancing, since in virtualized data centers, more and more services will be deployed by different users and they will be moving around.

Since the current networking technologies have no centralized control and lack global visibility, MLB is yet a challenge. In cellular networks, handoff is initiated by eNB with the help of measurements from the UE. These eNBs are weakly coordinated in terms of loose centralized control and visibility of near-cell traffic load or resource usage. Similarly, intratechnology mobility is not yet in practice, and hence, load balancing in heterogeneous networks (HetNet) cannot be materialized for better user satisfaction and efficient resource usage.

13.2 Load Balancing in SDMN

SDMN drives the motivation toward load balancing with the logically centralized intelligence or network operating system (NOS), which is capable of interoperability between different systems or network. SDMN enablers such as OpenFlow [2] introduce common programmable interfaces over which various network entities can talk regardless of the underlying technology. Added to that, replacement of network entities with software applications can substantially reduce the network cost and improve flexibility. In SDMN, load balancing mechanisms would enable to harvest the benefits of low-cost heterogeneous networking technologies to work in parallel with cellular networks. Even though spectrum scarcity is a major issue faced by the cellular network operators, cellular networks are still not capable to utilize the locally available wireless networks due to lack of efficient load balancing technologies.

The common centralized control plane in SDMN would enable redirecting network traffic through lower load middleboxes, links, and nodes. A common control plane in SDMN would be like the one shown in Figure 13.2. All the logical control plane entities such as Mobility

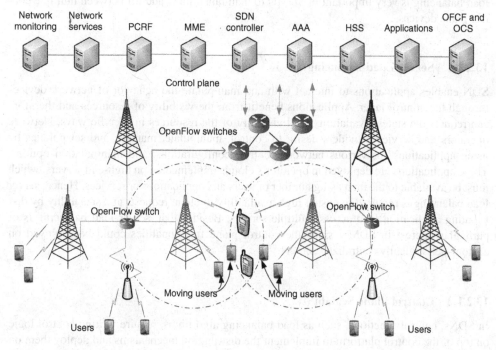

Figure 13.2 Software defined mobile network.

Management Entity (MME), AAA, PCRF, HSS, etc. are logically centralized and placed in high-end servers. These entities would (re)direct the data plane in real time. SDMN enables load balancing algorithms to be installed in the application server as a load balancing application in the control plane. The SDMN controller will fetch network load statistics from the data path and provide these statistics such as packet and byte counter values to the load balancing application. Similarly, UE mobility reports from MME would be provided to the load balancing application. Hence, centralized load balancing decisions based on the global view of the real network load would be taken.

OpenFlow enables flow-based routing and network virtualization with its extensions. The legacy network elements could be programmed with OpenFlow to enable flexible forwarding and management of commercial networks. This novel packet forwarding mechanism could be utilized for load distribution among technologically different systems as far as they are managed by a single controller or internetworked set of distributed controllers. The ability to move forwarding intelligence out of legacy network elements to a logically centralized control plane with an efficient forwarding approach is the significance behind OpenFlow. On the other hand, efficient load balancing is a counterpart of intelligent HO between the stations that are managed by the same NOS though they can be in technologically isolated domains.

13.2.1   The Need of Load Balancing in SDMN

Software defined networks are about a centralized control plane controlling and manipulating the forwarding behavior of the data plane from a logically centralized network view. Hence, load balancing is very important in SDMN to maintain a fair trade-off between multiple control plane devices.

13.2.1.1   Server Load Balancing

SDN enables applications to interact with and manipulate the behavior of network devices through the control layer. Applications benefit from the visibility of resources and therefore can request the states, availability, and visibility of the resources in specific ways. Network operators and service providers desire to control, manipulate, manage, and set policies by using applications for various network control, configurations, and manipulation options. These applications are deployed in operator's clouds implemented on high-end servers, which must be available to the increasing number of users and applications or services. Hence, server load balancing is required to ensure high availability to client requests and scalability by distributing application load across multiple servers. Besides that, server load balancing is of particular interest in SDMN since the control plane functionalities could be deployed on custom-belt logically centralized servers.

13.2.1.2   Control Plane Scalability

In SDNs, control functions, such as load balancing algorithms, require writing control logic on top of the control platform to implement the distribution mechanisms and deploy them on the forwarding elements such as switches and routers as shown in Figure 13.3. The control

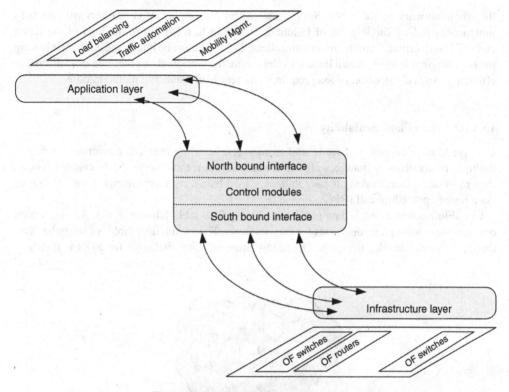

Figure 13.3   SDN-abstracted control plane.

platform of SDN can be either distributed or centralized, and the entity that implements the control plane functionalities is referred to as the SDN controller. The SDN controller is responsible for managing and controlling the whole network through an NOS from a central vantage point having global view of all the network resources.

However, centralization of the control logic of networks opens up its own type of challenges. Control plane scalability is one such challenge, which can be solved through efficient load balancing technologies. In SDN, most of the complexity is pushed toward the controller where forwarding decisions are taken in a logically centralized manner. A challenge for the currently available controller implementations is specifying the number of forwarding devices to be managed by a single controller to cope with the delay constraints. If the number of flows on the controller increases, there is a high probability that the sojourn time will increase, which is deeply dependent on the processing power of the controller. Today's controller implementations are not capable to handle the huge number of new flows when using OpenFlow in high-speed networks with 10 Gbps links [3]. This also makes the controller a favorite choice for denial of service (DoS) and distributed DoS attacks by targeting its scalability limitation. Therefore, controller efficiency has been the focus of many researches to enhance its performance through control platform distribution, devolving and delegating controller responsibilities, increasing memory and processing power of the controller, and architecture-specific controller designs. Proper load balancing mechanisms will enable the control plane platform to cost-effectively handle situations where scalability failure could be detrimental to the whole network performance. Simply increasing the number of controllers will not help mitigate the risk of single point of failure as shown in Ref. [4] where the load of the failed controller is distributed among other controllers. The load must be put on the controller having the least original load so that all the controllers share the workload and increase overall system efficiency. Such distribution of load requires efficient load balancing methodologies.

13.2.1.3 Data Plane Scalability

Data plane enables data transfer to and from users, handling multiple conversations across multiple protocols, and manages conversations to/from remote peers. SDN enables remote control of data plane, making it easy to deploy load balancing mechanisms in the data plane via a remote procedure call (RPC).

OpenFlow abstracts each data plane switch as a flow table (shown in Fig. 13.4b), which contains the control plane decisions for various flows. The switch flow table is manipulated by the OpenFlow controller using the OpenFlow protocol. One challenge for SDN is that how

(a)

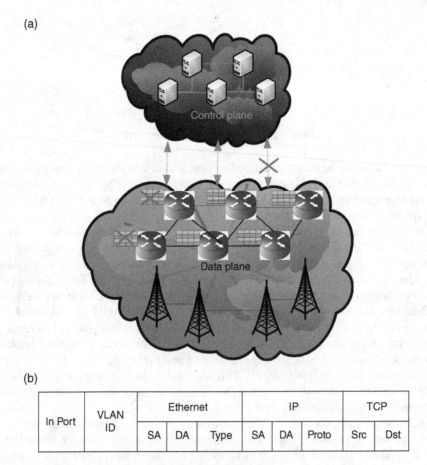

(b)

In Port	VLAN ID	Ethernet			IP			TCP	
		SA	DA	Type	SA	DA	Proto	Src	Dst

Figure 13.4   (a) The SDN data plane. (b) OpenFlow switch flow table.

efficiently the forwarding policies can be set from a logically centralized control plane on the forwarding devices. A scenario is shown in Figure 13.4a, where the SDN switches acquire flow rules from the controller. If the controller–switch path has higher delay, resources in the switches can be exhausted. For example, a switch has limited memory to buffer TCP/UDP packets for flow initiation until the controller issues the flow rules. Similarly, if a link to the controller is congested or the controller is slow in installing flow rules due to any reason (e.g., fault), the switch resources might be already occupied to entertain new flows. Besides that, recovering from link failure can take longer than the required time due to the centralized control plane. These challenges necessitate using OpenFlow switches according to their capacities through novel load balancing technologies.

13.2.2   SDN-Enabled Load Balancing

13.2.2.1   The Basis of Load Balancing in OpenFlow

As we know that SDN separates the control and data planes in a network, the OpenFlow variant of SDN defines an application programming interface (API) on the data path to enable the control plane interact with the underlying data path. The controller uses packet header fields such as MAC addresses, IP address, and TCP/UDP port numbers to install flow rules and perform actions on the matching packets. The action set comprises of, for example, forward to a port, drop, rewrite, or send to the controller. Flow rules can be set for either microflow that matches on all fields or wildcard rules that have empty (don't care bits) fields. A typical switch can support larger number of microflow rules than wildcard rules since wildcard rules often rely on expensive TCAM memory, while microflow rules use the SRAM, which is larger than TCAM. The rules are installed either with a fixed timeout that triggers the switch to delete (called the hard timeout) or with specified time of inactivity after which they are deleted (called the soft timeout). The switch also counts the number of bytes and packets for each rule, and the controller can fetch these counter values as shown in Figure 13.5.

The most basic mechanisms of load balancing in OpenFlow can use these counter values from the switches to determine how much load a switch is handling. Thus, the traffic load on various switches can be easily seen in the control plane, which can enable the controllers to load balance among the switches by using various coordination mechanisms. Load balancing in OpenFlow can also use the choice of using either the wildcard rules matching mechanism or microflows matching. Microflows matching would require the controller to be involved in small flows rather than aggregated flows and hence use more resources of the control plane. There can be a trade-off among the wildcard matching and microflows matching based on the controller load and availability. If matching on microflows is necessary, other mechanisms such as distributed control plane architectures can be used. That would require load balancing mechanisms among the controllers.

In the OpenFlow standard of SDN, a controller installs separate rule for each client connection, also called "microflow," leading to installation of a huge number of flows in the switches and a heavy load on the controller. Therefore, various approaches are suggested to minimize the load on the controller. These include using the wildcard support in the OpenFlow switches so that the controller directs an aggregate of client requests to server replicas. The wildcard mechanisms exploit the switch support for wildcard rules to achieve higher scalability

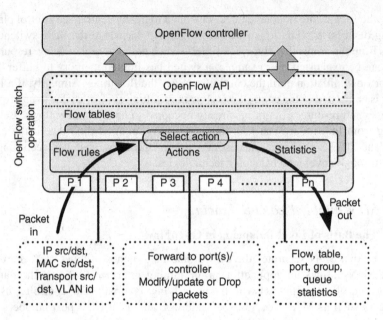

Figure 13.5   Architecture of OpenFlow switch.

besides maintaining a balanced load on the controller. These techniques use algorithms that compute concise wildcard rules that achieve target distribution of the traffic and automatically adjust to changes in load balancing policies without disturbing existing connections.

13.2.2.2   Server Load Balancing

Data centers host a huge variety of online services on their servers. These services can also be offered to other operators for CAPEX and OPEX cost savings. Since the magnitude of services in normal data centers is huge, these data centers use front-end load balancing technologies to direct each client request to a particular server replica. However, dedicated load balancers are expensive and can easily become a single point of failure and congestion. The currently used variant of SDN, that is, the OpenFlow [2] standard, provides an alternative solution where network switches divide traffic among the servers. An OpenFlow controller installs the packet handling rules in OpenFlow switches on run-time and can change these rules immediately if changes are required.

OpenFlow-based server load balancing is proposed in Ref. [5] for Content-Centric Networks (CCNs). One of the important functionalities in server load balancing is imposing policies that balance client requests in CCNs. Server load balancing in Ref. [5] proposes three load balancing policies to balance client request on servers. The first policy uses per client request scheme that maps every new request to a fixed content server. This client-based policy forwards the Address Resolution Protocol (ARP) reply of the least loaded server to a new client that initiates the ARP request. In case the client is not new, then the same server will reply to the user.

The second policy balances the load based on the OpenFlow switch statistics, which are checked periodically by the OpenFlow controller, and the load is estimated. The controller in this case finds the statistics of the amount of data sent through the existing flows and estimates the load of traffic handled from each server. Hence, the traffic is distributed among the available content servers using this load-based policy. Whenever an overloaded server is detected, the most demanding content requests are switched to another less congested server leading to efficient distribution of traffic among all the servers. The third policy is proximity based in which clients are assigned to the servers having the most quick response using first-come, first-served technique. However, this technique is useful in low network traffic having negligible traffic delays [5].

Use Case: Live VM Migration

Live VM migration provides an efficient way for data centers to perform load balancing by migrating VMs from overloaded servers to less heavily loaded servers. Administrators can dynamically reallocate VMs through live VM migration techniques without significant service interruptions. However, live VM migration in legacy networks is still limited due to two main reasons. First, live VM migration is limited to LAN since the IP does not support mobility without session breakups. Second, network state is unpredictable and hard to control in the current network architectures.

SDN enables live VM migration since the control plane is centralized having global visibility of the network and is independent of the layered IP stacks. Since the SDN controller has information about the underlying network topologies, SDN-based VM migration would diminish the chances of migration breakups due to topological complexities existing in legacy networks. For example, to migrate a VM, the new end-to-end forwarding paths can be easily established without interrupting the service by pushing new forwarding rules in the switch flow tables. Modifying the existing flow rules in OpenFlow switches would hardly require temporary storage of the existing flow packets compared to session breakups in the current networking environments.

SDN had made it possible to migrate a whole system comprising of VMs, the network, and the management system to a different set of physical resources. For example, the LIve Migration Ensemble (LIME) [6] design leverages from the control–data plane separation logic of SDN to migrate an ensemble of VMs, the network, and the network management. LIME clones the data plane state to a new set of OpenFlow switches and then incrementally migrates the traffic sources. OpenFlow-based interdomain VM migration is illustrated in Ref. [7] where it is shown that OpenFlow data center can be configured on the fly regardless of the complexity of its topology.

13.2.2.3 Load Balancing as SDN Applications

Online services, network function applications, and management plane functionalities are implemented in the application plane in SDNs. These application plane functionalities are implemented on high-end servers. To properly load balance among multiple servers, front-end load balancing mechanisms could be used, which typically directs various requests to the right servers and their replicas.

Most of the load balancing mechanisms in SDN reside in the SDN application plane working on top of the control plane. For example, Aster*x [1] is a NOX application that uses the OpenFlow architecture to measure the state of the network and directly control the paths taken by flows. As shown in Figure 13.6 [1], the Aster*x load balancer relies on three functional units:

- *Flow Manager*: This module manages the routes of flows based on the chosen load balancing algorithm.

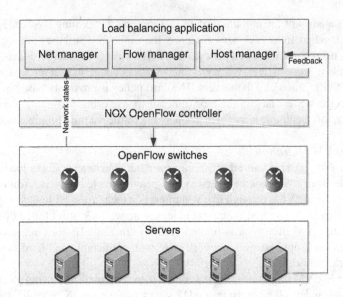

Figure 13.6   Aster*x load balancing architecture and its functional units.

- *Net Manager*: This module is responsible for keeping track of the network topology and its utilization level.
- *Host Manager*: This module keeps track of the servers and monitors their state and load.

Aster*x enables service providers to load balance their network based on different types of applications. The options that applications have include proactive versus reactive load balancing, load balancing on individual versus aggregated flow requests, and static versus dynamic load balancing. These choices make Aster*x a scalable distributable load balancing architecture.

13.2.2.4   Control Plane Load Balancing

In SDN, the controller implementing the control plane functionalities installs flow rules in the data path. Since a controller can set up a limited number of flows in the forwarding devices, it is suggested to use multiple controllers working in a logically centralized fashion. Hence, the latest versions of OpenFlow support multiple controllers in a single network domain where switches can have simultaneous connections to multiple controllers. Therefore, distributed OpenFlow controller architectures such as HyperFlow [8] and Onix [9] are proposed to implement multiple controllers to manage large networks. Load balancing among such distributed controllers plays a vital role to maintain a fair workload distribution of the control plane and ensure quick response. Load balancing in such scenarios will also enable maximum aggregate controller utilization and mitigate the risks of controller being a single point of failure or bottleneck.

Distributed Control Plane

BalanceFlow [10] is a controller load balancing architecture for wide-area OpenFlow networks. BalanceFlow works at the granularity of flows and partitions control traffic load among multiple controller instances in a large network. All the controllers in this architecture maintain

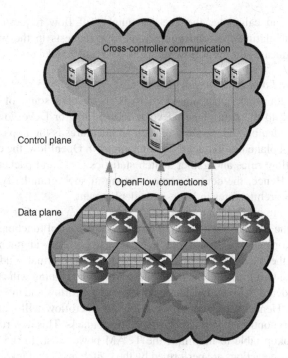

Figure 13.7 Architecture of the BalanceFlow controller.

their own load information, which is published periodically with other controllers. The controller architecture is hierarchical where one controller acts as a supercontroller to keep a balance of load on the rest of the controllers in the domain. When the traffic conditions change, the supercontroller partitions the traffic and allocates controllers to different flow setups to maintain a balance of workload among the working sets of controllers. This architecture also minimizes the flow setup delay since the nearest controller to the switch is allocated for the flow setup in the switch. Figure 13.7 shows the BalanceFlow load balancing architecture.

The BalanceFlow architecture has two requirements, that is, simultaneous multiple controller connections and controller X actions extension in the OpenFlow switches. The controller X extension in the switches allows sending flow requests to particular controllers. A controller, let's say controller k, maintains an $N \times N$ matrix M_k, where N is the number of switches in the network. Elements in the ith row and jth column denote the average number of flow requests from switch i to switch j. When a flow request packet is received, the controller first learns the switch from which the packet has arrived. After checking the destination address of the packet, the controller locates the corresponding egress switch for that flow, and the relevant element in the matrix is updated periodically. The average number of flow requests from switch i to switch j is calculated using the following formula:

$$R_{\text{avg}}(i,j) = (1-w) R_{\text{avg}}(i,j) + w T(i,j) \tag{13.1}$$

where w is the weighted coefficient and $T(i, j)$ is the number of flow requests from switch i to switch j in a certain period of time. The supercontroller collects the flow request matrixes from all the controllers and calculates the average number of flow requests handled by each controller. After calculating the total number of flow requests in the whole network, the supercontroller reallocates different flow setups to different controllers.

Control–Data Plane Load Distribution

Another approach for load balancing in SDN is to devolve some of the control plane responsibilities back to the data plane. Devolved OpenFlow or DevoFlow [11] is one such example. The main idea behind developing such architectures is the implementation costs of involving the control plane too frequently. For example, in OpenFlow, the controller might be required to install flow rules and gather switch statistics (byte and packet counters) in very quick successions. Hence, the control plane working at such granularity would hinder the deployment of SDN architectures in large-scale deployment.

Two mechanisms are introduced to devolve the control from controller to a switch. The first is rule cloning and the second one is localizing some of the control functions in the switch. For rule cloning, the action part of wildcard rules in OpenFlow packets is augmented with a Boolean CLOONE flag. If the flag is clear, the switch follows the normal wildcard mechanisms; otherwise, the switch will locally clone the wildcard rule. The cloning will create a new rule by replacing all the wildcard fields with values matching the microflow and inheriting other aspects of the original rule. Hence, subsequent packets for the microflow will match the microflow-specific rule and thus contribute to microflow-specific counters. This new rule will be stored in the exact-match lookup table to minimize the TCAM power cost. In the local action set of DevoFlow, local routing actions are performed by the switch instead of involving the controller. The set of local actions include multipath support in the switch and rapid rerouting in the switch. DevoFlow enables multipath routing by allowing the clonable wildcard rule to select an output port for a microflow according to some probability distribution. The rapid rerouting would enable a switch to use one or more fallback paths if the designated output port goes down.

Load Balancing in Case of Controller Failure

In SDN, it is highly probable that the network fails due to a controller being the single point of failure. To avoid single point of failures, the use of multiple controllers is suggested. However, proper load balancing is required to redistribute the traffic of the failed controller among other controllers. Otherwise, if the load is evenly distributed, a controller already loaded to its capacity will fail, and such process can lead to cascading failures of the controllers [4]. Therefore, the optimal strategies for handling the controller failure in multicontroller environment must satisfy the following requirements:

- The whole network must have enough capacity to tolerate the load of a failed controller.
- The initial load must be balanced with respect to the capacity of controllers.
- After failure of a controller, the load redistribution must not cause overload to another controller having being already working to its full capacity. Rather, proper load balancing algorithms must be used to deploy less extra load on heavily loaded controllers and vice versa.

13.2.2.5 Data Plane Load Balancing

Open Application Delivery Networking (OpenADN) [12] enables application-specific flow processing. It requires packets to be classified into application flow classes using cross-layer communication techniques. The cross-layer design allows application traffic flows' information

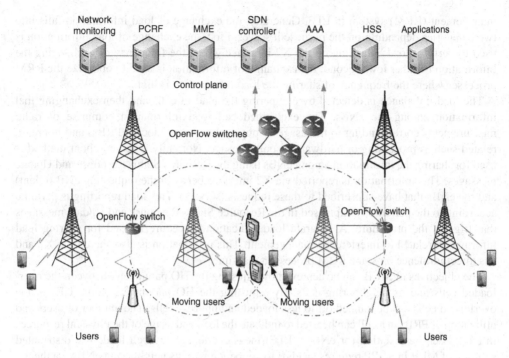

Figure 13.8 Software defined mobility load balancing among cells.

to be placed in the form of a label between network and transport layers. This Application Label Switching (APLS) layer forms layer 3.5 that is handled by the OpenADN switches enabling application traffic to be handled at the packet layer. Hence, it is possible to enable flow-based load balancing in the OpenFlow switches.

13.2.2.6 MLB

SDN provides a common control protocol such as OpenFlow that works across different wireless technologies with minimal changes. This capability of OpenFlow has made the integration of SDN into the current wireless networks straightforward where the data path remains the same, but the network control and logical elements such as MME, PCRF, and control part of SGW/PGW are abstracted in the control plane as shown in Figure 13.8. This makes it easy to use the available standardized mobility mechanisms in the radio part with the SDN-featured control plane having novel MLB algorithms implemented on top of the control plane. MLB in legacy networks followed by SDMN-enhanced features is described in the following text.

MLB in Cells

Load balancing in the cells of cellular is carried out with the help of eNBs. The aim of load balancing through eNBs is to keep a balance of load among the neighboring cells in order to improve the overall system capacity. Hence, load information is shared among the eNBs for maintaining a fair distribution of workload among the pool of eNBs. The information is shared via the X2 interface directly between eNBs since there is no central radio resource management (RRM) system in LTE. Generally, the exchange of load information falls into two categories depending on the purpose it serves. First, the exchange of load information is used to perform load balancing on the X2 interface where the frequency of exchanging the information is rather low. Second, the exchange of information is used to optimize the RRM processes where the frequency of sharing the load information is high.

The load imbalance is detected by comparing the load of cells and then exchanging that information among the eNBs. The exchanged cell load information comprise of radio measurements corresponding to the usage of physical resource blocks (PRBs) and nonradio related such as processing or hardware resources usage. Normally, a server–client method is used for sharing the information among eNBs using the Resource Status Response and Update messages. This information is reported via X2 interface between the requesting eNB (client) and the eNBs that have subscribed to these requests (servers). This load reporting is periodic according to the periodicity expressed in the Resource Status Response and Update messages that triggers the procedure. A separate load indication procedure is used for sharing load information related to interference management. This information is also shared via X2 and has direct influence on some RRM process in real time.

The objectives of MLB can be achieved by adjusting the HO parameters between the overloaded cells and its neighboring cells. By adjusting the HO parameters, some UEs in the overloaded cells can be handed off to less loaded neighboring cells. The number of users and utilization of PRBs in a cell can be used to indicate the load and usage of the physical resources in LTE. Each base station (or eNB in LTE) measures its serving cell load. The distributed solution of MLB in 3GPP requires an eNB to cooperate with its neighboring eNBs via the X2 interface. Overloaded eNBs obtain its neighboring cell loads and adjust the HO parameters via X2 to force some of UEs to hand off from the current cell to the neighboring cells. In LTE, the HO decision is generally triggered by the event A3 simplified as

$$M_n > M_s + \text{HO}_{\text{margine}}, \tag{13.2}$$

where M_n is reference signal received power (RSRP) in dBm or reference signal received quality (RSRQ) in dB for a neighboring cell, M_s is RSRP or RSRQ of the serving cell, and $\text{HO}_{\text{margine}}$ is a margin between M_n and M_s in DB. Each cell can have its own value of $\text{HO}_{\text{margine}}$. HO decision is based on formula (13.2) by measuring these parameters in UEs. When an eNB detects that its serving cell is overloaded, the $\text{HO}_{\text{margine}}$ to its neighboring cells will be adjusted to trigger handoff of UEs from the current cell to the neighboring cell. For efficient and precise MLB, the measurement reports from UEs can be used to predict the cell loads after adjusting the HOmargine.

Since these measurement reports from UEs can contain M_s and M_n in the formula (13.2), the eNB can collect information of M_s and M_n of each UE located at cell edge between the serving cell and its neighboring cells. Hence, the eNB can measure PRB utilization of UEs at the serving cell. Therefore, the PRB utilization at the neighboring cell can be estimated in accordance with the user throughput and modulation and coding scheme on the neighboring cell. This would enable to allocate users to the neighboring cells while not congesting the cell and effectively balance load among neighboring cells. However, in such mechanisms and techniques, the eNBs itself adjust the $\text{HO}_{\text{margine}}$ of the serving cell cooperating with its neighboring eNBs. The eNBs are required to cooperate and exchange information that makes the network rather complex and difficult to scale and maintain. SDN on the other hand centralizes all the control plane functionalities where centralized servers collect the network information and direct individual entities such as eNB to set HO parameters and perform HOs when required. An SDMN architecture having mobile users in neighboring cells is shown in Figure 13.8. Due to mobility, the PRB usage of resources will either increase or decrease. The centralized control plane in SDMN will collect information regarding usage of PRBs in neighboring cells and hence be able to easily compare the load in the two cells. Since implementing new functionalities in SDN requires writing software logic on top of the control plane, the mobility management algorithms can be implemented on top of the control plane, which will utilize global visibility of the network physical resources. For example, the MLB algorithm can be implemented as an SDN application that has real visibility of the physical resources of the neighboring cells and will be in better state to adjust the HO margins between neighboring cells. Another advantage of such centralized MLB would be to distribute the unbalanced cell load when several cells close to each other are overloaded.

MLB in MME

In cellular networks, a UE is associated with one particular MME for all its communications where the MME creates a context for that UE. The MME is selected by the Nonaccess Stratum (NAS) Node Selection Function (NNSF) in the first eNB from which the UE connected to the network. When a UE becomes active with an eNB, the MME provides the UE context to that eNB using the Initial Context Setup Request message. With the transition back to idle mode, a UE Context Release Command message is sent to the eNB from the MME to erase the UE context, which then remains only in the MME.

For mobility within the same LTE system or inter-eNB HO, the X2 HO procedure is normally used. However, when there is no X2 interface between the two eNBs or if the source eNB is configured to initiate HO toward a particular eNB, the S1 interface is used. In HO process through the X2 interface, the MME is notified only after completion of the HO process. The process of HO and control flow during HO on the S1 interface is shown in Figure 13.9. The HO process comprises of a preparation phase where resources at the core network are prepared for HO, followed by an execution phase and a completion phase. Since MME is actively involved in HOs and context maintaining, it is very important to perform load balancing among the MMEs in a cellular network.

The aim of MME load balancing is to distribute traffic among the MMEs according to their respective capacities. S1 interface is used to perform load balancing among MMEs in a pool of MMEs in cellular networks. MME carries out three types of load management procedures over the S1 interface. These include a normal load balancing procedure to distribute the traffic, an overload procedure to overcome a sudden rise in load, and a load rebalancing procedure to either partially or fully off-load an MME. The MME load balancing depends on the NNSF present in each eNB, which contains weight factors corresponding to the capacity of each MME node. A weighted NNSF carried out at each eNB in the network achieves statistically balanced distribution of load among MMEs. However, there are some specific scenarios that require specific load balancing actions. First, if a new MME is introduced, the weight factor corresponding to the capacity of this node may be increased until it reaches an adequate level of load. Similarly, if an MME is supposed to be removed, the weight factor of this MME should be gradually decreased so that it catches minimum traffic and its traffic must be

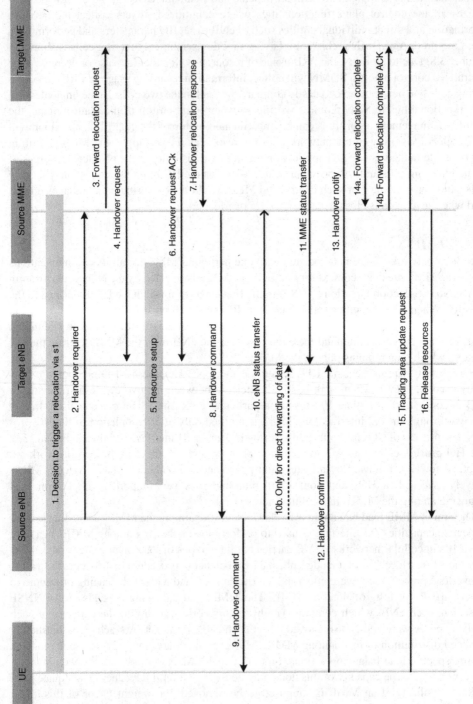

Figure 13.9 S1 handover control flow.

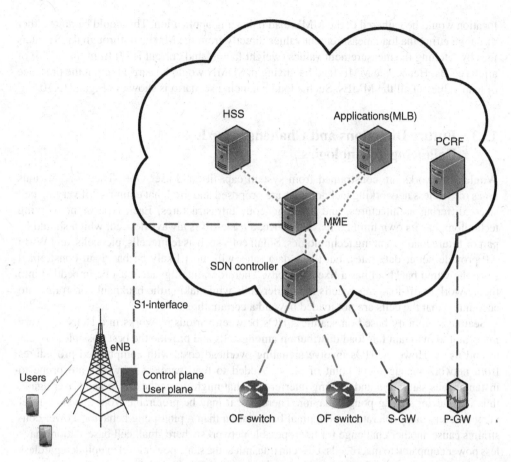

Figure 13.10 MME load balancing in SDMN.

distributed among the remaining MME nodes. Second, if there is an unexpected peak in the load, an overload message can be sent over the S1 interface to eNBs to temporarily restrict certain types of traffic to that particular MME. The MME can also adjust the number of eNBs and restrict the types of traffic it needs to avoid. Third, if an MME wants to rapidly remove the UEs, it will use the rebalance function to force UEs to reattach to other MMEs using a specific cause value in the UE Release Command S1 message [13].

In SDN, MME becomes part of the control plane and interacts with eNBs through the S1 interface. Along with MME, the control planes of the SGW and PGW are also abstracted to the control plane. Since MME is now a logical entity in the SDMN-centralized control plane, checking its load and maintaining a fair load on MME would be easy. There can be a separate application for load balancing among MMEs, or the MME load balancing algorithms can be part of the overall load balancing application. Since the current MME load balancing is dependent on the load measurement values of NNSF in the eNB, these values represent only the load of the MMEs, which are attached to that particular eNB. Thus, each eNB has limited visibility of those MMEs that are not listed in its NNSF. This makes the current MME load balancing rather inefficient. In SDMN, the load of all the MMEs in a certain geographic location would be gathered to the MME load balancing application. This could be done either through getting the load measurement values directly from the MMEs or through the S1 interface by fetching the measurement values (weight factors and current load) from the NNSF of all the eNBs. Hence, the MME load balancing in SDMN would be carried out in the presence of load values of all the MMEs. Such a load balancing scenario is shown in Figure 13.10.

13.3 Future Directions and Challenges for Load Balancing Technologies

Wireless networks are constrained from system capacity and user QoE. Therefore, various types of wireless networking technologies are proposed and used that comprise of varying cell sizes, differing architectures, and heterogeneous infrastructures. Each type of networking technology has its own limitations, and hence, a trade-off is always desired, which should be part of future load balancing technologies. Small cells such as femtocells, picocells, and Wi-Fi AP provide better data rates, but these data rates will most likely be backhaul constrained since the wired backhaul has a fixed capacity. Therefore, intelligence must be imbedded into the network to off-load macrocells to smaller cells while taking the backhaul constraint into account so that the cells are not loaded beyond a certain threshold.

Seamless mobility based on seamless HOs between various networks in a HetNet is very important to maintain fair load distribution among cells and provide the best possible services to end users. However, HOs involve signaling overhead costs with complicated procedures from network management point of views. Added to that, wireless networks are prone to instantaneous saturation and varying interference that might force the UE to hand over again, thus introducing a ping-pong behavior. Therefore, it may be preferable from a system-level view to temporarily tolerate a suboptimal base station than a ping-pong behavior. Power constraints cause another challenge for interoperable networks where small cell base stations have less power compared to macrocells. UEs can transmit at the same power level in uplink regardless of the base station type where the need of strong coordination among the femto-, pico-, micro-, and macrocell base station is very important for load balancing. Such asymmetries require a centralized control mechanism where load balancing mechanisms direct the base station and maintain a balanced load among the cells irrespective of their transmit power capabilities. 3GPP has initiated work items on device-to-device (D2D) communication allowing direct communication between cellular users. Such direct communication opens up new directions for load balancing technologies motivated by off-loading the load from cellular networks.

Although network security is an integral part of network management, it has been rarely researched in parallel to network load balancing. It is extremely important in SDMN to develop load balancing architectures that work according to network security policies. For example, security lapses of the controller can introduce delay in setting flow rules in the switches leading to congestion in switches with unsolicited traffic flows. Therefore, it is necessary to consider network security while designing and deploying traffic load balancing technologies in SDMN.

References

[1] Handigol, Nikhil, Mario Flajslik, Srini Seetharaman, Ramesh Johari, and Nick McKeown. "Aster*x: Load-balancing as a network primitive." In ninth GENI Engineering Conference (Plenary), Washington, DC, 2010.

[2] McKeown, Nick, Tom Anderson, Hari Balakrishnan, Guru Parulkar, Larry Peterson, Jennifer Rexford, Scott Shenker, and Jonathan Turner. "OpenFlow: Enabling innovation in campus networks." ACM SIGCOMM Computer Communication Review, vol. 38, no. 2 (2008): 69–74.

[3] Jarschel, Michael, Simon Oechsner, Daniel Schlosser, Rastin Pries, Sebastian Goll, and Phuoc Tran-Gia. "Modeling and performance evaluation of an OpenFlow architecture." In Proceedings of the 23rd International Teletraffic Congress, San Francisco, CA, USA, pp. 1–7, 2011.

[4] Yao, Guang, Jun Bi, and Luyi Guo. "On the cascading failures of multi-controllers in software defined networks." 21st IEEE International Conference on Network Protocols (ICNP), pp. 1–2, October 7–10, 2013. DOI:10.1109/ICNP.2013.6733624.

[5] Choumas, Kostas, Nikos Makris, Thanasis Korakis, Leandros Tassiulas, and Max Ott. "Exploiting OpenFlow resources towards a cContent-cCentric LAN." In Second European Workshop on Software Defined Networks (EWSDN), pp. 93–98. IEEE, October 10th–11th, 2013, Berlin, Germany.

[6] Keller, Eric, Soudeh Ghorbani, Matt Caesar, and Jennifer Rexford. "Live migration of an entire network (and its hosts)." In Proceedings of the 11th ACM Workshop on Hot Topics in Networks, pp. 109–114. ACM, Redmond, WA, 2012.

[7] Boughzala, Bochra, Racha Ben Ali, Mathieu Lemay, Yves Lemieux, and Omar Cherkaoui. "OpenFlow supporting inter-domain virtual machine migration." In Eighth International Conference on Wireless and Optical Communications Networks, pp. 1–7. IEEE, Paris, 2011.

[8] Tootoonchian, Amin and Yashar Ganjali. "HyperFlow: A distributed control plane for OpenFlow." In Proceedings of the 2010 Internet Network Management Conference on Research on Enterprise Networking, pp. 3–3. USENIX Association, San Jose, CA, 2010.

[9] Koponen, Teemu, Martin Casado, Natasha Gude, Jeremy Stribling, Leon Poutievski, Min Zhu, Rajiv Ramanathan, Yuichiro Iwata, Hiroaki Inoue, Takayuki Hama, Scott Shenker. "Onix: A distributed control platform for large-scale production networks." Ninth USENIX Conference on Operating Systems Design and Implementation, vol. 10, Vancouver, BC, Canada, pp. 1–6, 2010.

[10] Hu, Yannan, Wendong Wang, Xiangyang Gong, Xirong Que, and Shiduan Cheng. "BalanceFlow: Controller load balancing for OpenFlow networks." In IEEE 2nd International Conference on Cloud Computing and Intelligent Systems (CCIS), vol. 2, pp. 780–785. IEEE, Hangzhou, 2012.

[11] Andrew R. Curtis, Jeffrey C. Mogul, Jean Tourrilhes, Praveen Yalagandula, Puneet Sharma, and Sujata Banerjee. "DevoFlow: scaling flow management for high-performance networks." In Proceedings of the ACM SIGCOMM 2011 Conference (SIGCOMM '11), pp. 254–265. ACM, New York, 2011. DOI:10.1145/2018436.2018466.

[12] Paul, Subharthi and Raj Jain. "OpenADN: Mobile apps on global clouds using OpenFlow and software defined networking." In Globecom Workshops (GC Wkshps), 2012 IEEE, Palo Alto, CA, USA, pp. 719–723, 2012.

[13] Alcatel-Lucent, "The LTE network architecture: strategic white paper." Available at: http://www.cse.unt.edu/~rdantu/FALL_2013_WIRELESS_NETWORKS/LTE_Alcatel_White_Paper.pdf (accessed on February 19, 2015), 2013.

Part IV

Resource and Mobility Management

14

QoE Management Framework for Internet Services in SDN-Enabled Mobile Networks

Marcus Eckert and Thomas Martin Knoll
Chemnitz University of Technology, Chemnitz, Germany

14.1 Overview

In order to achieve acceptable service quality, the broad spectrum of Internet services requires differentiated handling and forwarding of the respective traffic flows in particular within increasingly overloaded mobile networks. The 3GPP procedures allow for such service differentiation by means of dedicated GPRS Tunneling Protocol (GTP) tunnels, which need to be specifically set up and potentially updated based on the client-initiated service traffic demand. The software defined networking (SDN)-enabled quality monitoring (QMON) and enforcement framework for Internet services presented in this chapter is named Internet Service quality Assessment and Automatic Reaction framework and will be abbreviated as ISAAR herein. It augments existing quality of service functions in mobile as well as software defined networks by flow-based network-centric quality of experience monitoring and enforcement functions. The framework is separated in three functional parts, which are QMON, quality rules (QRULE), and quality enforcement (QEN). Today's mobile networks carry a mixture of different services. Each traffic type has its own network transport requirements in order to live up to the user expectations. To observe the achieved transport quality and its resulting user service experience, network operators need to monitor the QoE of the respective services. Since the quality of service experienced by the user is not directly measurable within the network, a new method is required, which can calculate a QoE Key Performance Indicator (KPI) value out of measurable QoS parameters. The most challenging and at the same time most rewarding service QoE estimation method is the one for video streaming services. Therefore, the chapter will focus on

Software Defined Mobile Networks (SDMN): Beyond LTE Network Architecture, First Edition.
Edited by Madhusanka Liyanage, Andrei Gurtov, and Mika Ylianttila.

video QMON and estimation, not limiting the more general capabilities of ISAAR for all sorts of service KPI tracking. YouTube is the predominant video streaming service in mobile networks nowadays, and ISAAR is consequently delivering a YouTube-based QoE solution first. The KPI extraction and mapping to a measurable QoE value like the Mean Opinion Score (MOS) is done by QMON. The QRULE is supplied with the flow information and the estimated QoE of the corresponding stream by the QMON entity. The QRULE module also contains a service flow class index in which all measurable service flow types are registered. The enforcement actions for the required flow handling are determined based on subscription and policy information from the subscriber database and the general operator policy rule set. The third functional block in the ISAAR framework is QEN where the flow manipulation is performed. QRULE, that is, requests to change the per-flow behavior (PFB) of data streams with low QoE and QEN, reacts accordingly by applying suitable mechanisms to influence the transmission of the respective data frames or packets of those flows. One possibility to influence the data transmission is to use the PCRF/PCEF and trigger the setup of dedicated bearers via the Rx interface. A second option is to deploy layer 2 and layer 3 frame/packet markings. As a third option—in case that the predefined packet handling configuration of the routers should not be used—the ISAAR framework is also able to perform a fully automated router configuration. With the SDN approach, there is a fourth possibility to influence data flows by using OpenFlow capabilities.

The first two sections state the current situation followed by the explanation of the ISAAR architecture in Section 14.4 and its internal realization in Sections 14.5, 14.6, and 14.7. In Section 14.7, the SDN demonstrator is presented and the summary and outlook are given in Section 14.9.

14.2   Introduction

Internet-based services have become an essential part of private and business life, and the user experienced quality of such services is crucial for the users' decision to subscribe and stay with the service or not. However, the experienced service quality results from the whole end-to-end lineup from participating entities. It starts from the service generation, covers potentially several transport entities and finishes up in the application displaying or playing the result on the end device's screen or audio unit. However, the contributing performances of the individual service chain parties can often not be separately assessed from the end user perspective. Sluggish service behavior can thus stem from slow server reaction and transport delay or losses due to congestion along the forwarding path as well as from the end device capabilities and load situation during the information processing and output. More insight can be gained from the mobile network perspective, which potentially allows for a differentiated assessment of the packet flow transport together with a transparent and remote quality of experience (QoE) estimation for the resulting service quality on the end device. User satisfaction and user experienced service quality are strongly correlated and lead—from an Internet service provider point of view—either to an increase in subscription numbers or to customer churn toward competitors. Neither the capabilities and load situations on end devices nor the performance of content provider server farms nor the transport performance on transit links can be influenced by the operator of a mobile network. Therefore, this QoE framework will concentrate on the monitoring and enforcement capabilities of today's mobile networks in terms of differentiated packet flow processing and potentially software defined networking

(SDN)-enabled forwarding. Since all competing providers will face similar conditions on either end of the service chain, the emphasis on the provider own match between service flow requirements and attributed mobile network resources in a cost efficient manner will be key for the mobile operator business success. That applies especially for SDN-enabled networks, where a split between control and data path elements is made. This way, functions traditionally realized in specialized hardware can now be abstracted and virtualized on general-purpose servers. Due to this virtualization, network topologies as well as transport and processing capacities can be easily and quickly adopted to the service demand needs under energy and cost constraints. One of the SDN implementation variants is the freely available OpenFlow (OF) standard [1]. With OF, the path of packets through the network can be defined by software rules. OF is Ethernet based and implements a split architecture between so-called OF switches and OF controllers. A switch with OF control plane is referred to as "OF switch." The switch consists of the specialized hardware (flow tables), the secure channel for communication between switch and OF controller, and the OF protocol, which provides the interface between them [2]. The Internet Service quality Assessment and Automatic Reaction (ISAAR) QoE framework takes this situation into account and leverages the packet forwarding and traffic manipulation capabilities available in modern mobile networks. It focuses on LTE and LTE-Advanced networks but is applicable to the packet domains in 3G and even 2G mobile networks as well. Since different services out of the broad variety of Internet services will ideally require individual packet flow handling for all possible services, the ISAAR framework will focus only on the major service classes for cost and efficiency reasons. The set of tackled services is configurable and should sensibly be limited to only the major contributing sources in the overall traffic volume or the strong revenue-generating services of the operator network. The current Sandvine Internet statistic report [3], for instance, shows that only HTTP, Facebook, and YouTube services alone cover about 65% of the overall network traffic.

14.3   State of the Art

The standardization of mobile networks inherently addresses the topic of quality of service (QoS) and the respective service flow handling. The 3GPP-defined architecture is called Policy and Charging Control (PCC) architecture, which started in Release 7 and applies now to the Evolved Packet System (EPS) [4]. The Policy and Charging Rules Function (PCRF) is being informed about service-specific QoS demands by the application function (AF). Together with the Traffic Detection Function (TDF) or the optionally available PCRF intrinsic Application Detection and Control (ADC), traffic flow start and end events are detected and indicated to the PCRF. This in turn checks the Subscription Profile Repository (SPR) or the User Data Repository (UDR) for the permission of actions as well as the Bearer Binding and Event Reporting Function (BBERF) for the current state of already established dedicated bearers. As can be seen here, the 3GPP QoS control relies on the setup of QoS by reserving dedicated bearers. These bearers need to be set up, torn down for service flows, or modified in their resource reservation, if several flows are being bundled into the same bearer [5]. Nine QoS Class IDs (QCI) have been defined by 3GPP for LTE networks, which are associated with such dedicated bearers. Today, IP Multimedia Subsystem (IMS)-based external services and/ or provider own services make use of this well-defined PCC architecture and setup dedicated service flow-specific reservations by means of those bearers. Ordinary Internet services, however, are often carried in just one (default) bearer without any reservations and thus experience considerable quality degradations for streaming and real-time services. Therefore, network operators need to address and differentiate service flows besides the standardized QoS mechanisms of the 3GPP. HTTP-based adaptive streaming video applications currently amount the highest traffic share (see Ref. [3]). They need to be investigated for their application behavior, and appropriate actions should be incorporated in any QoS enhancing framework architecture. An overview of HTTP-based streaming services can be found in [6, 7]. There are many approaches found in the literature, which address specific services and potential enhancements. HTTP Adaptive Streaming Services (HAS) [8], for instance, is a new way to adapt the video streaming quality based on the observed transport quality. Other approaches target the increasing trend of fixed–mobile convergence (FMC) and network sharing concepts, which inherently require the interlinking of PCRF and QoS architecture structures and mechanisms (see, e.g., Ref. [9]). This architectural opening is particularly interesting for the interlinking of 3GPP and non-3GPP QoS concepts, but has not yet been standardized for close QoS interworking. The proposed interworking of WiMAX and LTE networks [10] and the Session Initiation Protocol (SIP)-based next-generation network (NGN) QoE controller concept [11] are just examples of the recent activities in the field. The ISAAR framework presented in this chapter follows a different approach. It aims for service flow differentiation either within single bearers without PCRF support or PCRF-based flow treatment triggering dedicated bearer setups using the Rx interface. This way, it is possible to use ISAAR as a stand-alone solution as well as aligned with the 3GPP PCRF support. The following sections document the ISAAR framework structure and work principle in detail.

14.4   QoE Framework Architecture

The logical architecture of the ISAAR framework is shown in Figure 14.1. The framework architecture is 3GPP independent but closely interworks with the 3GPP PCC. If available, it also can make use of flow steering in SDN networks using OF. This independent structure generally allows for its application in non-3GPP mobile networks as well as in fixed line networks. ISAAR provides modular service-specific quality assessment functionality for selected classes of services combined with a QoE rule and enforcement function. The assessment as well as the enforcement is done for service flows on packet and frame level. It incorporates PCC mechanisms as well as packet and frame prioritization in the IP, the Ethernet, and the MPLS layer. MPLS as well as OF can also be used to perform flow-based traffic engineering to direct flows in different paths. Its modular structure in the architecture elements allows for later augmentation toward new service classes as well as a broader range of enforcement means as they are defined and implemented. Service flow class index and enforcement database register the available detection, monitoring, and enforcement capabilities to be used and referenced in all remaining components of the architecture. ISAAR is divided into three functional parts, which are the QMON unit, the QoE rules (QRULE) unit, and the QEN unit. These three major parts are explained in detail in the following sections. The interworking with 3GPP is mainly realized by means of the Sd interface [10] (for traffic detection support), the Rx interface (for PCRF triggering as AF and thus triggering the setup of dedicated bearers), and the Gx/Gxx interface [11] (for reusing the standardized Policy and Charging Enforcement Function (PCEF) functionality as well as the service flow to bearer mapping in the BBERF).

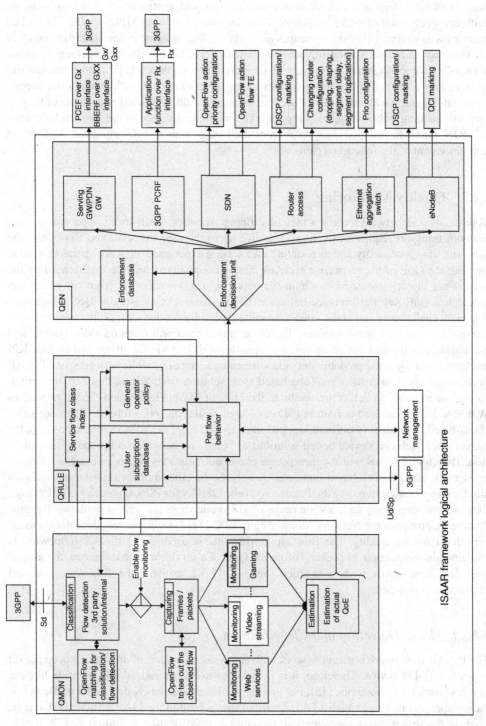

Figure 14.1 SDN-enabled ISAAR framework.

Since ISAAR is targeting default bearer service flow differentiation also, it makes use of DiffServ code point (DSCP) markings, Ethernet prio markings, MPLS traffic class (TC) markings, as well as OF priority changes as available. This is being enforced within the QEN by Gateway and Base Station (eNodeB)-initiated packet header priority marking on either forwarding direction inside or outside of the potentially deployed GTP tunnel mechanism. This in turn allows all forwarding entities along the packet flow path through the access, aggregation, and backbone network sections to treat the differentiated packets separately in terms of queuing, scheduling, and dropping. The modular structure of the three ISAAR units (QMON, QRULE, and QEN) allows for a centralized as well as a decentralized deployment and placement of the functional elements.

14.5 Quality Monitoring

Today's mobile networks carry a mix of different services. Each traffic type has its own network transport requirements in order to live up to the user expectations. To observe the achieved transport quality and its resulting user service experience, network operators need to monitor the QoE of the respective services. Since the quality of service experienced by the user is not directly measurable within the network, a new method is required, which can calculate a QoE Key Performance Indicator (KPI) value out of measurable QoS parameters. The most challenging and at the same time most rewarding service QoE estimation method is the one for video streaming services. Therefore, this chapter will focus on video QMON and estimation, not limiting the more general capabilities of ISAAR for all sorts of service KPI tracking. YouTube is the predominant video streaming service in mobile networks, and ISAAR is consequently delivering a YouTube-based QoE solution first. Within this YouTube monitoring, we are able to detect and evaluate the QoE of MP4, Flash Video (FLV), as well as WebM video in standard-definition (SD) and high-definition (HD) format. There are some client-based video quality estimation approaches around (e.g., the YoMo application [12]), but we consider such end device bound solutions as being cumbersome and prone to manipulation. Therefore, ISAAR will not incorporate client-side solutions but concentrates on simple, transparent, and network-based functionality only. Some other monitoring solutions follow a similar way of estimation, like the Passive YouTube QMON for ISPs approach [13]. However, they are not supporting such a wide range of video encodings as well as container formats. Another approach is the Network Monitoring in EPC [14] system, but this does not focus on flow level service quality. The flow monitoring that is used in the ISAAR framework is explained in this section. However, before the QoE of a service can be estimated, the associated data flow needs to be identified. Section 14.5.1 explains the flow detection and classification in detail.

14.5.1 Flow Detection and Classification

The ISAAR framework is meant to work with and without support of an external deep packet inspection (DPI) device. Therefore, it is possible to use a centralized DPI solution like the devices provided by Sandvine [15]. For unencrypted and more easily detectable traffic flows, the cheaper and more minimalist DPI algorithm that is built in the ISAAR framework can be used. In the first demo implementation, the build in classification is limited to TCP traffic, focusing on YouTube video stream detection within the operator's network. Extended with SDN support, there is a third possibility: given the proper configuration, the matching function from OF could be used to identify the supported service flows within the traffic mix. In the centralized architecture, the flow detection and classification is most suitably done by a commercial DPI solution. In this case, the QMON units have to be informed that a data stream was found and the classification unit has also to tell them the data stream-specific "five-tuple." Contained in the five-tuple are the source and destination IP address as well as the source and destination port and the used transport protocol. The QoE measurement starts as soon as the flow identification information (five-tuple) is available. Due to the new SDN features provided by OF, it is not only possible to identify specific data flows within the Internet. OF is also capable of teeing out a stream, which matches a specific pattern. Thereby, the QoE estimation could be distributed to different monitoring units, for example, depending on the specific Internet application. OF disposes the right flows to the right monitoring unit.

14.5.2 Video Quality Measurement

Traditionally, video QMON solutions were focusing on fine-grained pixel error and block structure errors. However, such KPIs are not suitable for progressive download video streams, since YouTube and other popular video portals are using the so-called pseudo streaming scheme that downloads the video file without losses into a playout buffer first and plays it out from there. Due to the data correctness ensured by TCP and the equalized transport delays by the buffering, pixel errors due to bad QoS transport parameters can no longer occur. The main cause for bad quality of progressive download videos are therefore stalling events due to delayed data reception and resulting buffer depletion times. Thus, QMON focuses on the occurrence and duration of playback stalls only. To determine these events, it is necessary to estimate the fill level of the playout buffer and to detect depletion events. Due to the fact that QMON does not have access to the user's end device, it relies on the data that can be observed at a measurement point within the network. The required information needs to be extracted out of TCP segments since YouTube and other progressive download streaming services are based on HTTP/TCP transport. Therefore, the TCP segment information and the TCP payloads of the video flow have to be analyzed. This analysis of the TCP-based video download derives the estimated buffer fill level based on the video timestamps encoded within the video payload of the respective TCP flow. For this extraction, it is necessary to decode the video data within the payload. After determining the playout timestamp, it is compared to the observation timestamp of the corresponding TCP segment [16].

The estimation process is shown in Figure 14.2. The result of this comparison is an estimate of the fill level of the playout buffer within the client's device. This estimation is done without access to the end device. The network-based QoE measurement setup is shown in Figure 14.3.

14.5.3 Video Quality Rating

A 5-point Mean Opinion Score (MOS) is used as a common scale for user experience. The MOS is calculated due to the occurrence of stalling events. Each stall decreases the MOS. The impairment decrease of a single stall event depends on the number of previously occurred

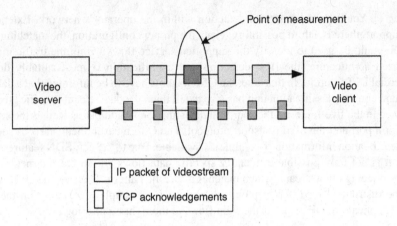

Figure 14.2   Video quality estimation scheme.

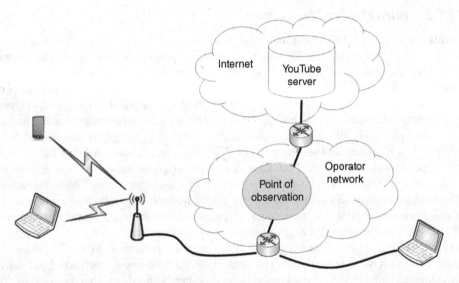

Figure 14.3   QoE measurement setup.

stalls and follows an e-function that best reflects the quality perception of human beings. For each video, an initial buffering stall is taken into account, which does not influence the perceived quality if it is below 10 s. The exact quality estimation function is shown below where x represents the number of stalling events:

$$\mathrm{MOS} = e^{-x/5 + 1,5} \tag{14.1}$$

The amount of buffered video time hits the zero line five times; therefore, five stalling events occurred during the video playback. The video stalling events take place at 18, 27, 45, 59, and 75 s, and each stalling event decreases the video MOS according to Equation 14.1. The resulting video quality is shown in Figure 14.4.

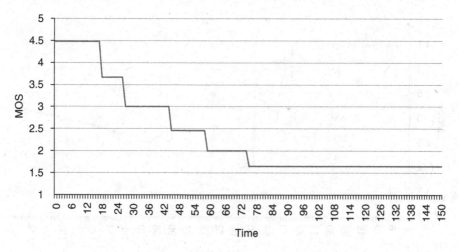

Figure 14.4   MOS video example according to Equation 14.1.

However, the user experience in reality is not as simple as shown in the figures before. One of the problems is the memory effect [17] of the QoE. That means the MOS is also improving over time if no further impairment has happened. Therefore, this effect has to be taken into account within the quality estimation formula. A time dependency of the influence of stalling events has been modeled again with a weighted e-function. That means the MOS estimate can recover if the video is running smoothly. To incorporate the memory effect portion, Equation 14.1 has been changed as shown in Equation 14.2, where x represents again the number of stalling events, t depicts the time since the last stall happened in seconds, and α is a dimensioning parameter, which adjusts the influence of the memory effect. α has been set to 0.14 for the shown figures:

$$\mathrm{MOS} = e^{x/5 + 1,5 - \alpha \sqrt[4]{t}} \tag{14.2}$$

where x = number of stalls; t = time since last stall; and α = memory parameter.

Figure 14.5 shows the calculated video score respecting the memory effect as shown in Equation 14.2.

14.5.4   Method of Validation

Two methods of validation have been used to compare the QMON estimates with real user experiences. First, a group of test persons had been involved in the evaluation of the estimation method and the demonstrator. A test consisting of 17 YouTube videos (in all available resolutions) was set up. The videos were watched on laptops with mobile network access by the test users, and the data traffic was recorded at the Gi interface as the measurement point within the mobile operator's network. During the assessment, the users had to note down the occurrence of stalling events as well as their duration. Later, the recorded packet capture (PCAP) traces

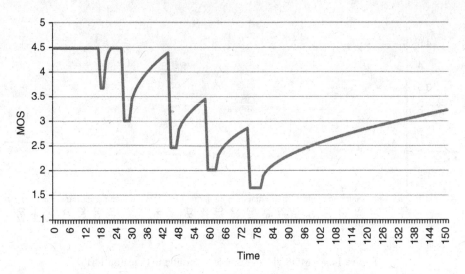

Figure 14.5   MOS video QoE according to Equation 14.2.

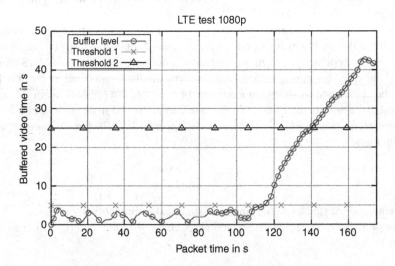

Figure 14.6   Example video buffer fill-level estimation.

were processed with QMON. The results of both, the user assessment and the outcomes of the QMON calculation, were compared to each other to validate the functionality of QMON. In a second step, online monitoring was deployed (mainly in Long-Term Evolution (LTE) networks), where the live watching of videos was augmented with the QMON graph of the estimated buffer fill level (see, e.g., Figure 14.6) as well as the respective quality score (Figure 14.5). The comparison of observed stalling events in the video player and the zero-level buffer estimates in the QMON graph was used for evaluation.

14.5.5 Location-Aware Monitoring

Due to the fact that it is not possible to measure all streams within an operator network, a subset of flows has to be chosen either randomly or in a policy-based fashion. For example, the sample flows could be selected based on the criteria of which tracking area the flow goes into. If it is possible to map the eNodeB cell IDs to a tracking area, the samples also can be selected in a regionally distributed fashion. With that, it could be decided whether a detected flow is monitored or not due to the respective destination region. Over the time, this sample selection procedure can shift the policy focus to regions with poor QoE estimation results in order to narrow down the affected regions and network elements.

14.6 Quality Rules

In this section, the QRULE entity of the ISAAR framework is presented. The QRULE gets the flow information and the estimated QoE of the corresponding stream form the QMON entity. It also contains a service flow class index in which all measurable service flow types are stored. The enforcement actions for the required flow handling are determined based on information from the subscriber database and the general operator policy. Also, the enforcement database within the QEN is taken into account. Combining all this information, the QRULE maps the KPIs to the per-flow behavior (PFB) for each data stream managed by ISAAR. PFBs are defined by appropriate marking of packets and frames. Each PFB has to be specified. For video streams, three possible PFBs (corresponding to three different markings) are provided. These PFBs depend on the buffer fill level. In the example (Figure 14.7), two

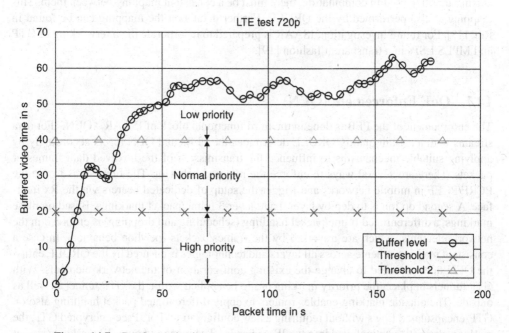

Figure 14.7 Per-flow behavior dependent on the buffer fill level (YouTube example).

buffer fill-level thresholds are defined: th1 = 20 s and th2 = 40 s. If the QoE is poor, that is, the video buffer fill level is below threshold 1 ($t<$ th1), the EF class (101 110) should be used. If the fill level is between thresholds 1 and 2 (th1 $<t<$ th2), a DSCP value like CS5 (101 000) should be chosen, because the video QoE is sufficient. Finally, if the fill level exceeds threshold 2 (th2 $<t$), a DSCP value with a lower priority like BE (000 000) or LE (001 000) is taken, so that other flows might get preferred access to the resources.

QRULE also decides which kind of marking is deployed depending on the networking technology. It is possible to apply IP DiffServ, Ethernet priority, MPLS TC marking, and QCI tunnel mapping for GTP. The rules unit has to ensure that there are no oscillating effects in the network. Oscillating could occur on flow level if one flow that is lifted up in priority causes quality impairments for the neighboring flows. Thus, the second flow will also require enforcement actions, which in turn causes the first one to deteriorate again. To overcome this effect, QRULE has to consider which flows were manipulated and in which location they are. Continuous action triggering is an early indication for such race conditions, which results in QRULE dampening of enforcement actions. That is, the transport impairment is such that ever-increasing priority is simply not solving the issue. Oscillating could also occur not only on flow level but on local area level within the network. Thus, regional impairment mitigation should not cause increased levels of impairments in neighboring regions. If this is being detected by location-aware QMON, QRULE should also dampen enforcement actions. Close interworking of ISAAR with network management systems fosters this detection of oscillation situations and provides vital information for root cause analysis. If the majority of the traffic would need to be precedented in priority, ISAAR has simply hit its limitation. If there are OF-enabled switches within the network, it is also possible to influence the priority of the frames belonging to a critical flow by changing the OF actions for that stream. As these mechanisms are often used in combination, there must be a consistent mapping between them. This mapping is also performed by the QRULE. Further details on the mapping can be found in Ref. [18]. For future investigation, ISAAR is prepared to incorporate the interworking of GTP and MPLS LSPs in a transparent fashion [19].

14.7   QoE Enforcement (QEN)

The enforcement of the PFB is done in the third functional block of ISAAR, "QEN." For data streams with a certain quality QRULE determines the PFBs and QEN reacts accordingly by applying suitable mechanisms to influence the transmission of the involved data frames or packets. There are several ways to enforce the required behavior. The first one is to use the PCRF/PCEF in mobile networks and trigger the setup of dedicated bearers via the Rx interface. A second option is to deploy layer 2 and layer 3 frame/packet markings. Based on these markings, a differentiated frame/packet handling (scheduling and dropping) is enforced in the network elements, which are traversed by the frames/packets (per-hop behavior). In case a consistent marking scheme across all layers and technologies is ensured by the QRULE entity, the QEN does not need to change the existing configuration of the network elements. With GTP tunnels in place, the priority marking has to be applied within the GTP tunnel as well as outside. The outside marking enables routers to apply differentiated packet handling also on GTP-encapsulated flows without requiring a new configuration. For IPsec-encrypted GTP, the marking also has to be included into the IPsec header. The inner and outer IP markings are set in downstream and in upstream direction based on the flow information (five-tuple) and the

PFB obtained from QMON. As a third option—in case that the predefined packet handling configuration of the routers should not be used—the ISAAR framework is also able to perform a fully automated router configuration. With that, the QEN may explicitly change the router packet handling behavior (e.g., packet scheduling and dropping rules) to influence the flows. With the SDN approach, a fourth possibility to influence data flows is realized by using OF features. For example, the priority of a flow can be changed in the forwarding configuration directly in an OF switch action list configuration. Furthermore, flow-specific traffic engineering could be realized. In order to use the OF features for flow enforcement, ISAAR is connected to the control interfaces of the SDN switches.

14.8 Demonstrator

To illustrate the QoE measurement, a demonstrator was used to process an example HD YouTube video for MOS calculation. The demo setup consisting of three laptops that are forming the SDN switch and SDN controller, another laptop where the QoE monitor was running, and two PCs that are generating background traffic is shown in Figure 14.8. The video is streamed from the video server to the video client through the SDN setup. The video traffic is copied out to the QMON device, which is evaluating the QoE of the video flow as described in Section 14.5. The video detection is done by matching rules within the SDN

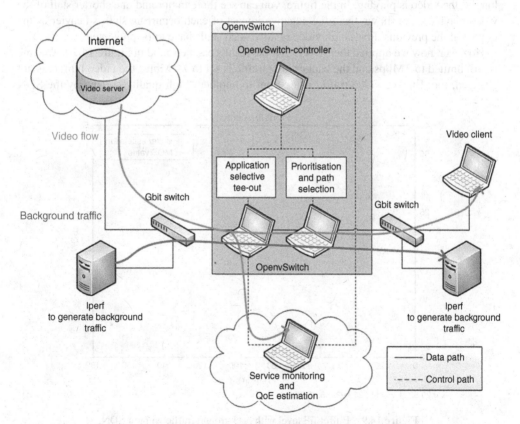

Figure 14.8 SDN demonstrator setup.

switch. The switch is also used to change the priority of the video flow in case of high traffic loads. Therefore, two queues have been created inside the switches: one for the video stream and one for the other traffic. The SDN switch sorts the data packets into the right queue due to matching and action rules, which are configured by the controller.

Within the test, the video buffer was set to 10 s. The outgoing line to the video client has a data rate of 2 Mbps, the used video has an average bit rate of 800 kbps, and the background traffic is set to 1.4 Mbps. Therefore, without any traffic engineering, the line has to get congested due to an overuse of 200 kbps. In this experiment, no background traffic is applied to the network; only the video was transmitted. The SDN matching as well as the SDN enforcement had been switched off in that test. In the figure, it can be seen that the video buffer is filled with a plenty amount of data during the whole video playback, due to the 2 Mbps line that is only used by 800 kbps. Hence, there have no stalling events occurred and the QoE was not decreased. The second test is driven out without the SDN functionality but with applied background traffic. The results are shown in Figure 14.9. It can be seen that after the initial buffer event, the video playout is consuming the buffered data until the buffer level hits the zero line. In this moment, the video gets stuck and the MOS value and with it the QoE is decreasing. The stalling event itself reduced the QoE and the negative impact gets even higher each second the video is not playing. Therefore, the MOS value is falling until the video playback is restarted. After the playback is resumed, the memory effect kicks in and the MOS value is increasing as long as the video is playing. In the figure, you can see three major and one shorter stall of the video playback. As shown there the negative impact of each occurring stall is heavier as the impact of the previous. For a high video quality, such stalls have to be prevented.

However, now we applied the SDN QEN; the results can be found in Figure 14.10. The line is still limited to 2 Mbps and the background traffic is set to 1.4 Mbps; the video bit rate is not changed, too. But the video traffic can be put to another "high-quality" queue by the SDN

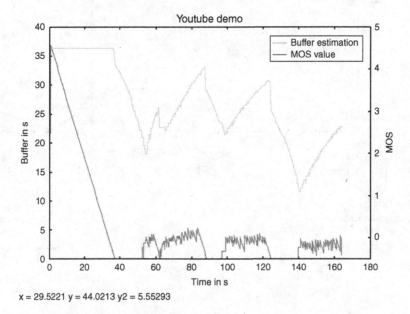

Figure 14.9   Buffer fill level with background traffic without SDN.

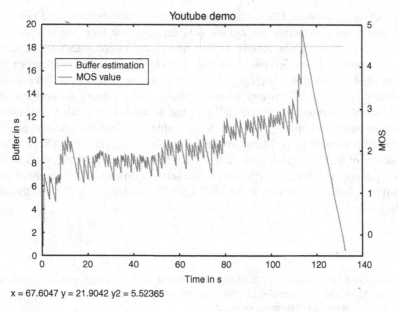

Figure 14.10   Buffer fill level with background traffic with SDN.

controller. Therefore, the video buffer is filled with sufficient data over the whole video play-back and no stalling events occurred. The demonstrator shows that it is possible to use SDN functions to detect specific traffic, copy it out, and enforce the needed QoE to it.

14.9 Summary

The ISAAR framework presented in this chapter addresses the increasingly important QoE management for Internet-based services in mobile networks. It takes the network operator's position to optimize the transport of packet flows belonging to most popular video streaming, voice, Facebook, and other Web services in order to satisfy the customer's service quality expectations. The framework is aware of the 3GPP standardized PCC functionality and tries to closely interwork with the PCRF and PCEF functional entities. However, 3GPP QoS control is mainly based on dedicated bearers and observations in today's networks reveal that most Internet services are carried undifferentiated within the default bearer only. ISAAR therefore sets up a three-component logical architecture, consisting of a classification and monitoring unit (QMON), a decision unit (QRULE), and an enforcement unit (QEN) in order to selec-tively monitor and manipulate single service-specific flows with or without the standardized 3GPP QoS support. This is mainly achieved by priority markings on (potentially encapsu-lated) service flow packets making use of the commonly available priority and DiffServ capabilities in layer two and three forwarding devices. In the case of LTE networks, this involves the eNodeBs and SGWs/PGWs for selectively bidirectional marking according to the QRULE-determined service flow behavior. More sophisticated mechanisms for location-aware service flow observation and steering as well as direct router respectively OF switch configuration access for traffic engineered flow routing are optionally available within the modular ISAAR framework. Due to the strong correlation between achieved video streaming QoE and customer satisfaction for mobile data services, the high traffic volume share of YouTube video streaming services is tackled first in the ongoing ISAAR implementation activity. An optimized network-based precise video QoE estimation mechanism is coupled with automated packet flow shaping and dropping means guided by a three-level playout buffer fill-level estimation. This way, a smooth playout with reduced network traffic demand can be achieved. To prove the functionality of the network-based video QoE estimation, a demonstrator has been implemented, which is capable of offline packet trace analyses from captured traffic as well as real-time online measurements. Since ISAAR is able to work independently of 3GPP's QoS functionality, it can be used with reduced functionality in any IP-based operator network. In such setups, the service flow QoS enforcement would rely on IP DiffServ, Ethernet priority, and MPLS LSP TC marking as well as SDN-based flow forwarding only.

References

[1] The OpenFlow Switch Specification. Available at http://OpenFlowSwitch.org. Accessed February 16, 2015.
[2] IBM; OpenFlow: The next generation in networking interoperability; 2011.
[3] Sandvine: Global Internet Phenomena Report; 2011.
[4] 3GPP: TS 23.203 Policy and Charging Control Architecture. 3GPP standard (2012).
[5] Ekström, H.: QoS Control in the 3GPP Evolved Packet System. In: IEEE Communications Magazine February 2009, pp. 76–83. IEEE Communications Society, New York (2009).
[6] Ma, K. J., Bartos, R., Bhatia, S., Naif, R.: Mobile Video Delivery with HTTP. In: IEEE Communications Magazine April 2011 pp. 166–175. IEEE Communications Society, New York (2011).
[7] Oyman, O., Singh, S.: Quality of Experience for HTTP Adaptive Streaming Services. In: IEEE Communications Magazine April 2012, pp. 20–27. IEEE Communications Society, New York (2012).
[8] Ouellette, S., Marchand, L., Pierre, S.: A Potential Evolution of the Policy and Charging Control/QoS Architecture for the 3GPP IETF-Based Evolved Packet Core. In: IEEE Communications Magazine May 2011, pp. 231–239. IEEE Communications Society, New York (2011).
[9] Alasti, M., Neekzad, B., Hui, L., Vannithamby, R.: Quality of Service in WiMAX and LTE Networks. In: IEEE Communications Magazine May 2010, pp. 104–111. IEEE Communications Society, New York (2010).
[10] Sterle, J., Volk, M., Sedlar, U., Bester, J., Kos, A.: Application-Based NGN QoE Controller. In: IEEE Communications Magazine January 2011, pp. 92–101. IEEE Communications Society, New York (2011).
[11] 3GPP: TS 29.212 Policy and Charging Control (PCC) over Gx/Sd reference point. 3GPP standard (2011).
[12] Wamser F., Pries R., Staehle D., Staehle B., Hirth M.: YoMo: A YouTube Application Comfort Monitoring Tool; March 2010.
[13] Schatz R., Hossfeld T., Casas P.: Passive YouTube QoE Monitoring for ISPs. 2nd International Workshop on Future Internet and Next Generation Networks (Palermo, Italy): June 2012.
[14] Wehbi B., Sankala J.: Mevico D5.1 "Network Monitoring in EPC," Mevico Project (2009–2012). http://www.mevico.org/Deliverables.html. Accessed February 16, 2015.
[15] Sandvine Incorporated ULC: Solutions Overview. (2012); http://www.sandvine.com/solutions/default.asp. Accessed February 16, 2015.
[16] Rugel S., Knoll T. M., Eckert M., Bauschert T.: A Network-based Method for Measurement of Internet Video Streaming Quality; European Teletraffic Seminar Poznan University of Technology, Poland 2011; http://ets2011.et.put.poznan.pl/index.php?id=home. Accessed February 16, 2015.
[17] Hoßfeld T., Biedermann S., Schatz R., Platzer A., Egger S., Fiedler M.: The Memory Effect and Its Implications on Web QoE Modeling; ITC '11 Proceedings of the 23rd International Teletraffic Congress 2011, pp. 103–110.
[18] Knoll T. M.: Cross-Domain and Cross-Layer Coarse Grained Quality of Service Support in IP-based Networks; http://archiv.tu-chemnitz.de/pub/2009/0165/. Accessed February 16, 2015.
[19] Windisch, G.: Vergleich von QoS- und Mobilitätsmechanismen in Backhaul-Netzen für 4G Mobilfunk. Technische Universität Chemnitz, Chemnitz; 2008.

15

Software Defined Mobility Management for Mobile Internet

Jun Bi and You Wang
Tsinghua University, Beijing, China

15.1   Overview

This chapter proposes to use software defined networking (SDN) to address mobility management in the Internet. This chapter first reviews existing mobility protocols in the Internet and points out their drawbacks. Then this chapter explains why SDN is a promising way to solve these problems, followed by a description of SDN-based mobility management architecture for mobile Internet. This chapter also presents an instantiation of this architecture that is designed and implemented using OpenFlow, as well as related evaluation and comparison with existing Internet mobility solutions to illustrate the advantages of the proposal.

15.1.1   Mobility Management in the Internet

Internet mobility has been an active research topic for over two decades. Along with the evolution of the Internet, especially the growing of mobile data due to more and more mobile devices and applications, many research efforts have been paid to address Internet mobility. However, so far, there is no consensus on how to provide mobility support in the Internet, making it remain an open issue.

15.1.1.1   Mobility Management in the Internet and Cellular Networks

Internet mobility research is different from that in cellular networks. Although cellular networking has been providing mobility support to global users, it may not replace the role of Internet mobility support because of their disparate bandwidths, costs, service models, etc. [1, 2]. Moreover,

Software Defined Mobile Networks (SDMN): Beyond LTE Network Architecture, First Edition.
Edited by Madhusanka Liyanage, Andrei Gurtov, and Mika Ylianttila.
© 2015 John Wiley & Sons, Ltd. Published 2015 by John Wiley & Sons, Ltd.

mobility in the Internet is showing new features comparing with that in cellular networks, that is, Internet mobility refers to not only movement from one point of attachment to another but also inter-ISP handover and even interdevice switching.

On the other hand, mobility management research in the Internet and cellular network is closely related, especially after IP is considered as the core part of future cellular networks [3]. Since the evolution trend of cellular networking is moving toward an all-IP-based infrastructure, which means all traffic leaving base stations becomes IP based and is delivered over packet-switched networks, IP mobility management becomes a key role to support future wireless systems.

Many IP mobility solutions serve as candidates to provide mobility management in cellular networks [4, 5], and some proposals have already been integrated into cellular networking. For example, Proxy Mobile IPv6 (PMIPv6), which is a typical IP mobility solution, has been adopted by the cellular core network Evolved Packet Core (EPC) in the 3rd Generation Partnership Project (3GPP) [6]. Besides, cellular backhaul technologies are also evolving toward IP-based designs, such as femtocells, which deploys home-located cellular base stations to provide connectivity to cellular users over their IP networks [7].

Current and future researches on Internet mobility management will continue to contribute to cellular networking. Recently, along with the development of 3GPP Long-Term Evolution (LTE), there is a growing trend to provide a more flexible and dynamic mobility management, which is also a concern in the Internet research area. The Internet Engineering Task Force (IETF) has already been standardizing related protocols, which may be introduced into 3GPP to replace its existing mobility management functions [8]. Due to the same reasons, we believe that this chapter also offers beneficial references for developing mobility management systems in current and future cellular networks.

15.1.1.2   Existing Internet Mobility Solutions

Generally, supporting Internet mobility means to offer uninterrupted Internet connectivity to mobile nodes (MNs) (mobile devices, users, or other entities), which change their attachment points while roaming in the network. Internet mobility is difficult to realize because it was an unforeseen feature when the Internet was built, and unfortunately, this feature contradicts with the current Internet architecture: due to the tight coupling of TCP and IP [9], changing IP addresses will cause interruption of TCP sessions on MNs, which may seriously impact experiences of mobile users.

Basically, Internet mobility solutions can be divided into two categories, that is, routing-based approach and mapping-based approach [10]. Routing-based approach makes an MN use the same IP address while roaming and thus requires dynamic routing to keep reachability of the MN. On the contrary, mapping-based approach allows an MN to change IP addresses but keep a piece of stable information, known as *identifier*, which does not change during movement. To reach the MN using its identifier, a mapping mechanism is introduced to resolve identifier to the MN's current *locator* (normally represented by its IP address). In these solutions, TCP sessions are always bound to identifiers instead of IP addresses; thus, they can keep survivability facing changing IP addresses. As discussed by Zhang et al. [1], routing-based approach is not suitable to provide mobility support in the global Internet, because the whole network requires to be informed of each MN's movement that may not scale well in large networks. Therefore, this chapter focuses on mapping-based approach.

Among all related proposals, Mobile IP (MIP) [11, 12] and its extensions [13, 14] are the earliest and most well-known protocols. MIP is an IETF-standardized protocol that allows MNs to keep session survivability while roaming around and changing IP addresses. Later, a large number of MIP derivatives have been proposed to improve its basic functionality [15–18]. Recently, there are also many individual IP mobility protocols [19–22] as well as future Internet architectures [23–28] that arise to address mobility problems in the Internet.

15.1.2 Integrating Internet Mobility Management and SDN

From one point of view, the key of providing mobility support in the Internet is to properly distribute the MN's identifier-to-locator mapping within the network so that its correspondents can reach the MN directly or indirectly. Although there exist various ways to realize such a function, they have drawbacks in different aspects, making it remain an unsolved issue. This chapter tries to address the problem using SDN and OpenFlow. SDN is an emerging network architectural approach, while OpenFlow is one of the most well-known instantiations of SDN [29]. In SDN, network structures, functions, and performance can be defined in a simpler way, which is usually achieved by providing programmable devices and a centralized control logic. As will be shown in this chapter, network functions or services required to support IP mobility can also be realized in a software defined way.

SDN helps to solve problems in IP mobility protocols for the following reasons: firstly, programmable SDN devices enable the *flexibility* of SDN-based mobility solution, which is a lack of existing IP mobility solutions. Specifically, the mapping of each MN can be flexibly placed on any SDN device instead of fixed Home Agents (HAs) or CNs, and this feature provides the basis to make SDN-based mobility solution become *adaptive* to diverse mobility scenarios. Secondly, centralized control let SDN-based mobility solution be aware of all kinds of mobility details, for example, how the MN moves, how the CN-to-MN traffic flows, etc. These details help to generate *optimal* strategies to handle different mobility scenarios via a lightweight algorithm without introducing complex distributed protocols. Thirdly, IP mobility in SDN architecture requires less host involvement. Most mobility functions can be realized on the network side as will be shown in Section 15.3, and this implies faster handoff without IP reconfiguration, less signaling overhead especially on wireless links, as well as higher security and privacy assurance.

15.1.3 Chapter Organization

The remainder of this chapter is organized as follows: Section 15.2 gives a classification and overview of related Internet mobility solutions. To further make clear of the problem this chapter addresses, Section 15.2 also presents a study on the mobility management functions of related solutions and discusses on their pros and cons. Section 15.3 proposes an SDN-based mobility management architecture for mobile Internet, together with an instantiation of this architecture that is designed and implemented using OpenFlow as well as performance evaluations and experiments.

15.2   Internet Mobility and Problem Statement

This section gives an overview of Internet mobility solutions and then discusses on the drawbacks of these solutions to further make clear of the problem this chapter focuses. Finally, a brief discussion on addressing Internet mobility in an SDN way is presented. Detailed protocol design, implementation, and evaluation of the solution this chapter proposes are given in the next section.

15.2.1   Internet Mobility Overview

One of the earliest Internet mobility solutions is MIP, which began its standardization in the IETF about two decades ago. Since then, various MIP derivatives have been proposed to improve the original protocol in order to adapt to the evolving Internet. Another category of solutions mainly relies on end hosts to realize mobility management. These protocols belong to "Identifier/Locator Split" (ILS) designs. ILS is an architectural model that points out that IP address has embedded both identifier and locator semantics and a split of the two is necessary. The concept of ILS had also been discussed many times during the past two decades and currently has got wide acceptance [30–32].

Besides MIP and ILS solutions, many future Internet architecture proposals also try to provide Internet mobility support. However, future Internet architecture proposals are usually clean slate and require substantial changes to the current Internet. They even do not rely on IP to work, which makes it difficult and inappropriate to compare them with current IP-based mobility solutions. Therefore, this chapter does not go further into mobility management proposals in future Internet architectures.

15.2.1.1   MIP and Its Derivatives

MIP derivatives are based on two origin protocols: MIP [11] and Mobile IPv6 (MIPv6) [12]. The core idea of MIP is illustrated in Figure 15.1a by taking MIPv6 as an example: the protocol uses a special type of IP address called Home Address (HoA) to identify an MN. When an MN moves to a new network, it obtains a Care-of Address (CoA), which can be used to reach the MN. Then it communicates with the HA located in its home network to update the binding cache that maps the MN's HoA to its current CoA. Since a CN does not know the MN's CoA, it sends packets to the MN using its HoA; thus, the packets are forwarded to the HA. With up-to-date binding cache, the HA can then encapsulate and redirect packets toward the MN's current CoA.

MIP centralizes both mobility signaling and data forwarding functions into a single HA, which increases signaling cost when MN is not within the home network. To address the problem, some extensions to MIP are proposed. Hierarchical Mobile IPv6 (HMIPv6) [13] deploys Mobility Anchor Points (MAP) in the network and uses them to localize mobility signaling when the MN is away from HA. Specifically, MN attaches to a nearby MAP, which is located using a Regional CoA (RCoA), and then the MAP is responsible for keeping the bindings between the MN's HoA and a Local CoA (LCoA), which is exactly the MN's current location, and tunneling packets to the MN. When attaching to a new MAP, MN informs HA of the new MAP's RCoA to keep reachability of the MN. PMIPv6 [14] is a similar solution, and

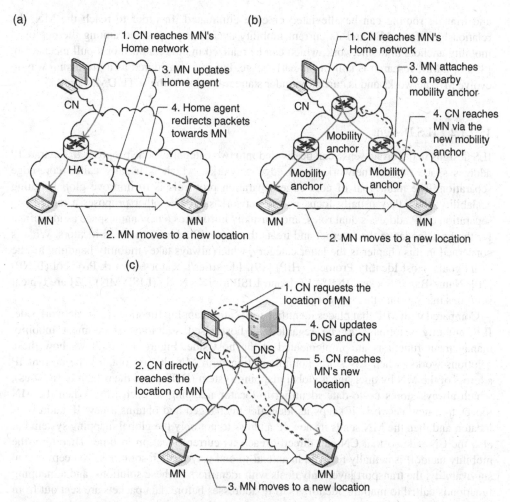

(a)

1. CN reaches MN's Home network

3. MN updates Home agent

CN

4. Home agent redirects packets towards MN

HA

MN   MN

2. MN moves to a new location

(b)

1. CN reaches MN's Home network

3. MN attaches to a nearby mobility anchor

CN

4. CN reaches MN via the new mobility anchor

Mobility anchor

Mobility anchor   Mobility anchor

MN   MN

2. MN moves to a new location

(c)

1. CN requests the location of MN

4. CN updates DNS and CN

CN   DNS

2. CN directly reaches the location of MN

5. CN reaches MN's new location

MN   MN

3. MN moves to a new location

Figure 15.1 Illustration of (a) Mobile IPv6, (b) distributed mobility management, and (c) identifier/locator split designs.

it frees MNs from mobility signaling and employs Mobile Access Gateways (MAG) to perform mobility management functions on behalf of MNs.

The major drawback of MIP and its extensions mentioned earlier is that all the packets from CN to MN have to take a detour to pass the HA, which is known as the triangle routing problem. Triangle routing can result in routing path stretch, which means the actual routing path is longer than the shortest one, as well as heavy load on HA. In recent years, a series of MIP derivatives [15–18], which follow Distributed Mobility Management (DMM) architectural paradigm [8, 33, 34], arise to address the problem. As shown in Figure 15.1b, DMM solutions distribute the functionality of HA to multiple mobility anchors deployed in the network so that the MN can always choose a nearby mobility anchor to maintain its binding cache and perform packet redirection. Thus, the MN's HoA never represents a fixed location, and triangle routing can be alleviated or even eliminated. In order to reach the MN, the relationship of the MN and its current mobility anchor is propagated among the deployed mobility anchors in the network, which can be realized in either a push or a pull mechanism [8, 33]. DMM research is still in the early stage, but it is considered as a promising way to evolve MIP networks and is currently under standardization in the IETF DMM group.

15.2.1.2 ILS Designs

ILS designs in the broad sense can be divided into two categories: one proposes to separate IP address space of core networks from edge networks, which is usually called core-edge separation. The main goal of core-edge separation protocols is to improve global routing scalability; thus, they usually focus on the network side. The other proposes a more clear separation of IP address's dual roles, and it usually introduces a new namespace as identifiers for the hosts/nodes in the Internet and treats the entire IP address space as locators. What is concerned in this chapter is the latter category, which always takes mobility handling as one of its goals. Host Identity Protocol (HIP) [19], Identifier/Locator Network Protocol (ILNP) [20], Name-Based Sockets (NBS) [21], and LISP Mobile Node (LISP-MN) [22] are typical solutions that fall into this category.

Compared with MIP that places identifier-to-locator mapping functions at the network side, ILS mobility solutions can be regarded as host-based solutions, since most mobility management functions are implemented at the host side. Figure 15.1c shows how these solutions works: when CN starts communication with MN, it first obtains the current IP address of the MN by querying a global mapping system (DNS plays the role in most cases), which always stores up-to-date identifier-to-locator mapping of each MN. When the MN moves to a new network, it keeps its identifier unchanged and obtains a new IP address as locator, and then the MN sends its new IP address to not only the global mapping system but also the CN side so that CN can directly reach its current location in time. Therefore, the mobility handoff is actually realized in an end-to-end way in such solutions. To keep session survivability, the transport layer only deals with identifiers in these solutions, and a mapping function is called to map the identifiers to IP addresses before data packets are sent out from the network layer. To realize such a function, these solutions either introduce a new layer or modify existing layers in the TCP/IP stack.

Though ILS protocols share the same core idea, they differ in ways to realize the idea including the formatting of identifiers, implementation of mapping functions on the host side and in the global mapping system, etc. HIP uses self-authenticating identifiers, called Host Identity (HI), to identify mobile hosts. HI is obtained by hashing the public key of a key pair that belongs to the host. With self-authenticating identifiers, each host is able to prove ownership to its HI through cryptographic methods. HIP use HI together with a port number to uniquely identify a transport layer session. To handle the mapping between HI and IP addresses on the host side, HIP inserts a new layer, called Host Identity Layer, between the network and transport layer in the protocol stack. HIP utilizes DNS together with additional rendezvous points to form a global mapping system, which stores the HI-to-IP address mappings of all mobile hosts. When communication begins, the initiator sends DNS request and fetches correspondent's HI and location of related rendezvous point. Then, the first data packet goes via the rendezvous point to reach the correspondent. After the initiator receives a data reply from the correspondent, following data stream travels directly between both communicating ends.

ILNP does not introduce new namespaces but utilizes IPv6 address space to identify mobile hosts, whose idea is derived from earlier research related to ILS [35]. ILNP splits IPv6 address space into an identifier part and a locator part: the first 64 bits of IPv6 address remains to be used for routing in the network, while the last 64 bits are used to uniquely identify mobile hosts. ILNP modifies the transport layer of mobile hosts to ensure that the session state only contains the identifier part of the entire IPv6 address (with port number). Different from HIP, ILNP completely relies on DNS to store the identifier-to-locator mappings.

NBS proposes to use domain names as identifiers of mobile hosts. NBS adopts a different approach to realize the mapping function by inserting a layer above the transport layer, which gives applications new socket interfaces and calls existing interfaces of TCP and UDP for data delivery. Using this method, mobility is hidden from the applications, and no change to TCP/IP stack is required. However, applications need to be redesigned to adapt the new socket interface, which implies that stale application cannot benefit from the mobility features offered by NBS. NBS also makes DNS as its global mapping system.

LISP-MN is developed based on research on Locator Identifier Separation Protocol (LISP) [36], which is a core-edge separation design. LISP proposes a separation of endpoint identifiers (EID) from routing locators (RLOC) and deploys ingress/egress tunnel routers (TR) to maintain EID-to-RLOC mappings. LISP-MN utilizes EID as identifier and RLOC as locators of mobile hosts and implements a lightweight TR on each mobile host to realize the mapping functions. LISP-MN does not rely on DNS, but makes use of several alternative map servers proposed by LISP as its global mapping system.

15.2.2 Problem Statement

Although there exist various methods to support Internet mobility, they have drawbacks in different aspects including triangle routing, large handoff latency, heavy signaling overhead, etc. This subsection gives a problem statement of existing Internet mobility solutions by analyzing mobility management functions of current solutions and pointing out a trade-off between routing path stretch and handoff efficiency.

15.2.2.1 Mobility Management Analysis

One of the main differences among existing Internet mobility solutions is how they implement handoff management functions. Handoff management is responsible for maintaining session survivability when the communicating nodes are moving. Different handoff management approaches can be classified into three categories. MIP comprises the first category, which we call local-scope handoff management, since the handoff signaling is always confined within a limited scope, that is, between the MN and the HA (or a local HA in HMIPv6 and PMIPv6). Thus, CNs only know one way to reach the MN, that is, via the HA. On the contrary, ILS designs adopt global-scope handoff management. The MN always sends mapping updates to the CN side, making all the CNs know the MN's exact location and send packets directly to the MN.

DMM solutions fall into the third category, which is a hybrid of the first two approaches: local-scope handoff signaling is always triggered to propagate the MN's mapping to a close HA, but global-scope handoff signaling is also required when the MN leaves one HA and attaches to another. Therefore, packets from the CNs can always be forwarded to an intermediary node that is close to the MN and knows how to reach the MN. Then the packets are routed to the MN based on its current IP address. Note that usually some agent near the CN is responsible for handling the signaling on behalf of the CN, making the handoff process transparent to CN.

There exist similarities in all three approaches. During handoff management, the MN must announce its up-to-date mappings into the network so that the CNs can reach the MN directly or indirectly. Specifically, some identifier-aware nodes in the network, we call rendezvous, must receive the mapping announcement and store the mappings, and then the CNs are able to reach the MN via the rendezvous (note that CNs themselves can also be rendezvous). The difference among these approaches is the scope of the mapping announcements: local scope, global scope, or a mixture of the two.

15.2.2.2   Routing Path Stretch and Handoff Efficiency

Existing efforts to handle Internet mobility expose a trade-off between routing path stretch and handoff efficiency: if the scope of mapping announcement is limited, as the MIP approach does, CNs may have to take a detour to reach the MN, which then causes routing path stretch; while if the mappings are announced to the CN side, as the ILS approach does, it may bring heavy overhead and large latency during the handoff, because the CNs can be distance from the MN and the number of CNs can be large.

A simple explanation of the trade-off is given here by drawing an analogy to the Internet routing. One common understanding in the routing research area is a fundamental trade-off between the routing table size and routing path stretch in a static network [37–39]. This trade-off implies a node in the network must store one routing table entry for each of the other nodes in the worst case to achieve the shortest path routing. Otherwise, one can only trade off an increase of the routing path stretch for a drop of the routing table size, because once a node loses the routing table entry for some remote node, it may not be able to forward packets to that node via the optimal path. The situation in mobility management is analogous: to ensure optimal routing path, mapping announcements must reach the CN side in the worst case; while if the scope of mapping announcement is limited to reduce the signaling overhead and latency, CNs will lose the exact location of the MN and have to reach the MN via indirection, which may lead to potential routing path stretch.

Internet mobility solutions take different ways to make the trade-off and get their own pros and cons: MIP only announces mappings to the HA; thus, it gains potential large routing path stretch but low handoff latency and overhead especially when MIPv6 extension (e.g., HMIPv6) is applied; ILS approach announces mappings to all the CNs; thus, it always gets none routing path stretch but may suffer from large handoff latency and overhead; DMM solutions are seeking a balance between the two.

It is still an open question on how to make the trade-off in DMM solutions. If an MN moves slowly and seldom changes HA, deploying HAs close to MNs can indeed reduce handoff latency and overhead without bringing routing path stretch. However, it may not be the case in real mobility scenarios. Considering the scenario that an MN simultaneously connects to multiple ISPs (e.g., the MN have both Wi-Fi and 3G/4G accesses), switch between different ISPs may become more common. Furthermore, a mobile user in future Internet may be able to switch ongoing communications from one device to another, which may also lead to frequent switch among ISPs. In both scenarios described earlier, there exists a large possibility that an MN changes HA frequently, which may significantly decrease the handoff efficiency of the hybrid handoff management. Therefore, it still needs investigation on how the third approach behaves when applied to more complex mobility patterns.

15.2.3 Mobility Management Based on SDN

To address the problems in current Internet mobility solutions, this chapter proposes to use SDN and OpenFlow. Before introducing the detailed protocol design, this subsection reviews existing research on SDN-based mobility and discusses benefits of handling Internet mobility using SDN.

15.2.3.1 Existing Research on SDN-Based Mobility

Researchers have already begun studying on how to offer better mobility support under SDN architecture. Yap et al. [40, 41] proposed OpenRoads to improve robustness during mobility handoff using multicast in OpenFlow networks. They showed how this is achieved by demonstration and also described their testbed deployment. In the following paper [42], they further abstract their idea as separating wireless services from infrastructures and rename OpenRoads to OpenFlow Wireless, which serves as a blueprint for an open wireless network. The focus in this chapter differs from the research earlier: this chapter focuses on improving basic IP mobility functions commonly adopted by existing protocols, while they paid more attention to adding new features, such as multicast, to basic mobility functions.

Pupatwibul et al. [43] proposed to enhance MIP networks using OpenFlow, which share similar goals to the proposal in this chapter. However, as will be shown, they only proposed one possible way to solve the problem, which may not be optimal in many scenarios, while this chapter abstracts the problem and gives a general discussion to seek for the best solution.

15.2.3.2 Benefits of SDN-Based Mobility

A summary of benefits of SDN-based mobility solution is already given in Section 15.1.2, while this subsection presents a further analysis based on the problem statement earlier. Recall the conclusion in the problem statement in Section 15.2.2.2, existing Internet mobility solutions adopt different ways to realize mapping announcement in handoff management and thus make different trade-offs between routing path stretch and handoff efficiency. SDN-based mobility solution can serve as a promising way to seek an ideal balance for the performance trade-off. It is because programmable devices and centralized control enable the flexibility to perform mapping announcement according to the MN's movement details. Specifically, since each device is programmable in SDN, they are all potential rendezvous for MNs, which makes SDN-based mobility management no longer restricted to local-scope or global-scope handoff management, but can perform mapping announcement in an arbitrary scope. Moreover, centralized control can decide in which scope the mappings should be announced according to the movement of MNs: if an MN moves within a limited area, only local-scope mapping announcement is sufficient, and along with the increasing of the MN's movement distance, the controller may find it necessary to announce the mappings to a larger scope.

Therefore, to seek an algorithm that optimizes the scope of mapping announcement in different mobility scenarios is one of the most important goals of this chapter. Section 15.3.3 describes how to seek such an algorithm, and it also proves optimality of the algorithm in terms of both optimal routing path and minimum handoff latency as well as signaling overhead.

15.3   Software Defined Internet Mobility Management

This section describes an SDN-based mobility management architecture for mobile Internet. Firstly, an overview of the architecture is given. Secondly, an instantiation of the architecture is presented that is designed using OpenFlow. Thirdly, an algorithm problem is addressed that serves as a key component of the architecture. At last, an implementation of the architecture is presented together with experiments to compare the proposal with existing Internet mobility solutions.

15.3.1   Architecture Overview

Internet mobility management functions based on SDN can be separated into control plane functions and data plane functions as demonstrated in Figure 15.2. In the control plane, two subfunctions are required to realize mobility management. One subfunction requires SDN controllers to collect the current location of each MN and maintain a mapping for each MN. The mapping dynamically binds identifier of an MN to locator of the MN. The definition of identifier is the same as that in ILS-related researches, that is, some stable information that does not need to change when the MN changes its location in the network. Identifiers do not have a restricted format but can be any field in the packet that can be recognized by SDN controllers and devices. MN's locators should be represented by some packet field that can be used to reach the MN's current location. Normally, IP addresses serve as locators of MNs.

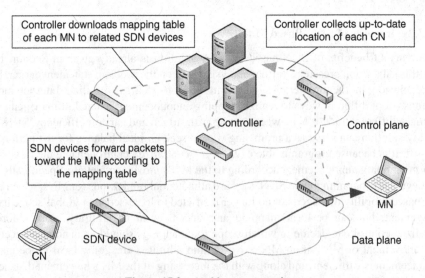

Figure 15.2   Architecture overview.

To realize this subfunction, SDN devices are required to inform controllers of the attachment and detachment of each MN. When an MN leaves one SDN domain and enters another, inter-SDN domain mechanism is required to synchronize mappings of the MN between controllers in different domains.

The other control plane subfunction requires the controller to download each MN's mapping to related SDN devices. It can be subdivided into two cases: in the first case, the controller downloads an MN's mapping to the SDN devices that are requesting the mapping; in the second case, after the controller updates an MN's mapping, it downloads the mapping to some SDN devices that has already stored the mapping. The purpose of the downloading in the second case is to replace stale mappings on SDN devices with up-to-date ones so that packets toward the MN can be forwarded to its current location in time. Note that both CN and MN are hardly involved in the control plane functions and most mobility management functions in the control plane are realized on the network side.

In the data plane, SDN devices directly receive packets destined to MNs from CNs and forward the packets according to the mappings downloaded from controllers. When an SDN device lacks required mapping for packet delivery, it triggers a control plane function to request the mapping from the controller.

The control and data plane functions described above comprise the basic mobility management functions based on SDN. Protocol details such as how the mappings are collected and downloaded are explained using an OpenFlow-based example in the following subsections.

15.3.2 An OpenFlow-Based Instantiation

This subsection describes an OpenFlow-based instantiation of the proposed architecture. Note that though the detailed protocol design described in this chapter is based on OpenFlow, the proposed architecture can also be designed and implemented in a similar way using other techniques that realize the idea of SDN.

15.3.2.1 Protocol Description

The OpenFlow-based design employs IP addresses to identify and locate MNs. Like all the other mobility protocols, a stable identifier is assigned to each MN. The identifier is also called HoA, which is nonroutable and should belong to a specific address block. An MN's location is represented by CoA, which is not owned by MN, but its first-hop OpenFlow switch. It means MNs never require to reconfigure IP addresses when attaching to new networks, but the network side helps to accomplish the work, which is similar to PMIPv6. CoAs are routable addresses and thus are used to reach MNs when they are moving around.

OpenFlow controller is responsible for maintaining binding cache that maps an MN's HoA to CoA. For each MN, a subset of OpenFlow switches in the network serve as indirection point for the MN. They store replica of the MN's binding cache in the form of flow table, which is downloaded from the controller, and redirect packets toward the MN according to the flow table.

Figure 15.3 illustrates how CN reaches MN in both communication initiation and handoff procedures. HoA of MN and CN are IP_M and IP_C, respectively. First, when switch S3 detects the attachment of MN, it learns the MN's HoA, assigns a CoA IP_S3 to the MN, and then sends a Binding Update message, which contains a (IP_M, IP_S3) tuple to its controller. The controller stores the binding locally and immediately downloads a flow table entry to S3 that indicates "for all packets with destination address IP_S3, rewrite their destination addresses to IP_M."

The communication initiation process is described as follows, assuming that CN is communicating with MN: since CN only knows HoA of the MN, the destination address in the packets it sends to MN is IP_M. When CN's first-hop switch S1 receives such a packet, it learns IP_M is nonroutable, and there are no local flow tables that match the address. Thus, it forwards the packet to its controller via packet-in. The controller (for simplicity, here, the controllers of S1 and S3 are assumed to be the same one, and the case with multiple controllers will be discussed in the following subsection) looks IP_M up in local binding cache table and gets the corresponding CoA. Then the controller forwards the packet to S3 via packet-out and at the same time places the binding cache to S1 by downloading a flow table entry indicating "for all packets with destination address IP_M, rewrite their destination addresses to IP_S3." Then the following packets can flow directly from CN to MN: first from S1 to S3 and then from S3 to the MN.

The handoff process is described as follows, assuming that the MN leaves S3 and attaches to S4: similarly, detecting the attachment, S4 assigns IP_S4 to the MN and sends Binding Update to the controller. The controller receives the update and learns that the MN has just moved; thus, it is responsible for accomplishing the handoff by modifying existing CN-to-MN flow path toward MN's new location. Take the scenario in Figure 15.3 as an example: since the controller knows how the flow goes from CN to MN, it places the new binding cache to S2

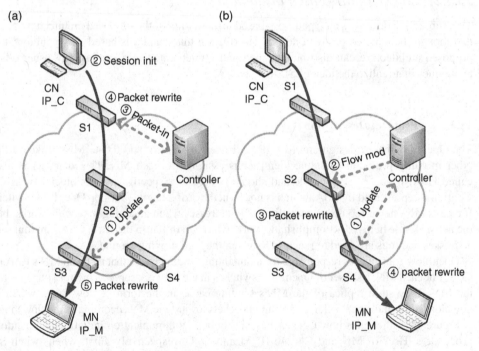

Figure 15.3   (a) Communication initiation and (b) handoff processes.

by downloading a flow table entry to S2 that indicates "for all packets with destination address IP_S3, rewrite their destination addresses to IP_S4." Then the new CN-to-MN flow will go through three redirections: S1 to S2, S2 to S4, and S4 to MN. In practice, there may exist various ways to place the binding cache, and Figure 15.3 only shows one possibility. The binding cache placement algorithm is further discussed in Section 15.3.3.

15.3.2.2 Discussions

If CN and MN are located far apart from each other or located in different domains, it is possible that the controllers of their first-hop switches are different. If the two controllers belong to the same administrative domain, the problem may become simpler since intradomain communication between controllers is more common. If the two controllers belong to different administrative domains, interdomain interactions between controllers are required, which may bring larger cost comparing with the intradomain case. Specifically, interdomain communication initiation between MN and CN is less costly than interdomain handoff of MN, because the former only requires a query and response of binding cache and is easier to handle, but the latter requires the controller to know the CN-to-MN flow path, which is not a preknowledge of the controller in interdomain case.

However, interdomain handoff is not common in practice. There are two common ways to trigger an interdomain handoff: in one case, the MN moves for a long distance and then leaves one domain and enters another, which can be quite infrequent; in the other case, the MN switches between different providers (e.g., different Wi-Fi or 3G/4G networks) without long-distance movement. As for the second case, its occurrence probability can be further reduced by making multiple heterogeneous local networks be controlled by one logical controller. Using this method, MN's switching among different access networks is analogous to intradomain handoff.

However, though infrequent, interdomain handover is unavoidable. To improve interdomain handover efficiency, the protocol can temporarily fall back to triangle routing that only requires one flow table downloading to the MN's previously attached switch. After MN–CN communication is restored, further operations can be performed to optimize the path between MN and CN.

15.3.3 Binding Cache Placement Algorithm

This subsection further researches into the binding cache placement during MN's handoff procedure. Theoretically, any switch on the CN-to-MN flow path before MN's movement (e.g., S1, S2, and S3 in Fig. 15.3) can serve as a candidate switch, which is called Target Switch (TS). However, choosing some TS may lead to serious performance drawbacks. For example, it is a straightforward idea to choose MN's first-hop switch before movement (e.g., S3 in Fig. 15.3) as TS, but this method will result in triangle routing in most cases. Another idea is to choose CN's first-hop switch (e.g., S1 in Fig. 15.3) as TS, but this method may result in a large number of flow table downloading and high handoff latency, which is analogous to the end-to-end Binding Update manner adopted by HIP-like protocols. Therefore, the binding cache placement problem (BCPP) requires further study. The following of this section formalizes and solves the problem.

15.3.3.1 BCPP

First, the goals of the binding cache placement algorithm are given as follows:

Goal 1: Keep optimal forwarding path. This goal ensures the shortest forwarding data path between MN and CN and avoids triangle routing.

Goal 2: Minimize the distance between MN and TS. The purpose of proposing this goal is to localize the signaling caused by MN's mobility events.

Goal 3: Minimize flow entry downloading per movement. This goal can help to both limit the mobility-related flow table maintained on switches and reduce the signaling overhead introduced by flow table downloading.

Then a general *BCPP* is defined as an optimization problem: given a set of TS to place the binding cache for an MN, the BCPP is to find a subset of the switches that optimizes some goals.

However, the proposed goals conflict with each other in many cases, for example, selecting MN's first-hop switch before movement as TS will always satisfy Goal 3 but has a large possibility to conflict with Goal 1. Thus, BCPP is further specified into the following two problems:

BCPP-1: BCPP that takes Goal 2 as optimization objective and Goal 1 as constraint.

BCPP-2: BCPP that takes Goal 3 as optimization objective and Goal 1 as constraint.

15.3.3.2 Problem Formalization and Solution

BCPP-1

Assume that during a handoff procedure, MN moves from switch s_n to $s_{n'}$ and CN stays attaching to switch s_1 as shown in Figure 15.4. A set of definitions are given before formalizing BCPP-1:

Definition 15.1

$path_{prev}$ is defined as a set of switches on the MN–CN path before movement of MN, for example, $\{s_1, s_2, ..., s_i, ..., s_n\}$ in Figure 15.4.

$path_{current}$ is defined as a set of switches on the MN–CN path after movement of MN, for example, $\{s_1, s_2, ..., s_i, ..., s_{n'}\}$ in Figure 15.4.

Path pair is a ($path_{prev}$, $path_{current}$) tuple.

Switch s satisfies path pair p means after placing the binding cache (of the MN) on s, the new MN–CN path $path_{new}$ equals to $path_{current}$. This ensures Goal 1, that is, optimality of the forwarding path (no triangle routing).

Then BCPP-1 is formalized as:

Problem 15.1

Given each path pair p, find a switch s that satisfies p and at the same time minimizes its distance to the MN.

The solution to Problem 15.1 is relatively simple. First, another group of definitions is given as follows:

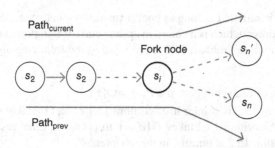

Figure 15.4 This figure helps to demonstrate Definitions 15.1, 15.2, and 15.3: path$_{prev}$ consists of nodes from s_1 to s_n, while path$_{current}$ consists of nodes from s_1 to $s_{n'}$, and s_i is the fork node.

Definition 15.2

Satisfactory switch set C_p for path pair p is defined as $\forall s \in C_p$, s satisfies p, for example, $\{s_1, s_2, \ldots, s_i\}$ is a satisfactory switch set for path pair (path$_{prev}$, path$_{current}$) in Figure 15.4.

Fork node of path pair p is defined as the node where two paths of the path pair "fork," for example, s_i is the fork node of path pair (path$_{prev}$, path$_{current}$) in Figure 15.4.

Then the solution to Problem 15.1 is given:

Algorithm 15.1

Given a path pair p, find its fork node s.

The complexity of the algorithm is $O(d \cdot n)$ where n represents number of path pairs and d represents length of the path. Related proof is omitted here.

BCPP-2

More definitions are given to formalize BCPP-2:

Definition 15.3

Switch s satisfies a set of path pairs P means $\forall p \in P$, s satisfies p.

A set of switches S satisfies a set of path pairs P means $\forall p \in P$, $\exists s \in S$ s.t. s satisfies p.

Problem 15.2

Given a set of n path pairs P, find the smallest set of switches S that satisfies P.

Problem 15.2 can be solved in two steps:

Step 1: For each path pair p, find the largest satisfactory switch set C_p.
Step 2: Find the smallest set S s.t. for each C_p, $S \cap C_p \neq \emptyset$.

The complexity of Step 1 is $O(d \cdot n)$. Step 2 can be reduced by *Set Covering Problem* that is NP hard; thus, Problem 15.2 is an NP-hard problem.

Obviously, Problem 15.2 can be solved using exhaustive search, but its complexity is $O(d^n)$ and is unacceptable. We find that under certain circumstance, Problem 15.2 can be solved using a simple algorithm. The circumstance is described as:

Assumption 15.1

Two paths to the same destination share identical "suffix" after they "meet."

Assumption 15.1 is satisfied as long as packet forwarding between MN and CN only relies on destination IP address, which is a common case in current intradomain scenarios. With this assumption, the solution to Problem 15.2 is proposed as the following algorithm:

Algorithm 15.2

Find the fork node s_i of each path pair $p \in P$, $S = \cup \{s_i\}$.

Actually, Algorithm 15.2 is similar to Algorithm 15.1 except that Algorithm 15.2 works on a set of path pairs. Algorithm 15.2 takes $O(d \cdot n)$ to get the optimal result. The proof of the optimality of Algorithm 15.2 is omitted in this chapter.

Obviously, Algorithm 15.2 also optimizes Problem 15.1. Thus, if Assumption 15.1 is satisfied, Algorithm 15.2 can generate optimal results for both Problems 15.1 and 15.2, which means all three goals can be simultaneously achieved.

When Assumption 15.1 cannot be satisfied, another algorithm is given to solve Problem 15.2:

Algorithm 15.3

Greedy Set Covering.

Step 1: Let $X = P$, $S = \varnothing$.
Step 2: Repeat the following process until $X = \varnothing$; find i s.t. S_i contains the largest number of elements in X, and then let $S = S \cup \{s_i\}$, $X = X \setminus S_i$.

According to existing research [44], Greedy Set Covering algorithm takes $O(d \cdot n^2)$ to get a result with approximation ratio $\ln n + 1$.

Note that Algorithm 15.3 also generates optimal result when Assumption 15.1 is satisfied. The proof is omitted in this chapter.

15.3.3.3 Evaluation

This subsection evaluates the previously proposed algorithms to see how they perform in real network topologies. Since it is difficult to get real intradomain routing data that conflicts with Assumption 15.1, only Algorithm 15.2 is evaluated under Assumption 15.1 using real intradomain topology with the shortest path routing. Two additional algorithms are introduced as comparatives:

Algorithm-random: For each path pair p, this algorithm randomly selects a switch s that satisfies p as TS.

Algorithm-CN: For each path pair p, this algorithm selects the first-hop switch of CN as TS.

All three algorithms satisfy Goal 1; thus, the comparison uses metrics from the other two goals: one metric is MN–TS distance, and the other metric is the number of binding cache downloaded per MN per movement.

The evaluation topology and routing data are calculated using intradomain topologies from RocketFuel [45] including AS1221, AS1755, AS6461, AS3257, AS3967, and AS1239. As evaluations based on different topologies show similar results, three of them are chosen to demonstrate the results, and they are AS1221 with 208 nodes, AS3257 with 322 nodes, and AS6461 with 276 nodes. To study the performance of the algorithm based on various topologies, another two topologies are generated to add differentiation: one is a 200-node hierarchical topology with a densely interconnected core network and several treelike edge networks, and the other is a 200-node flat topology in which nodes randomly connect to each other with an average degree.

For each one of the topologies above, the evaluation runs for 100 turns. In each turn, a different node in the topology is selected as the MN, and 10 randomly located nodes are chosen as CNs. The MN performs 10 movements per turn using a modified Markov chain-based random walk model: during each movement, the MN randomly attaches to a new node that is one hop away from its previous location.

Evaluation results are demonstrated in Figure 15.5. As shown in the figures, Algorithm 15.2 has the lowest value. Figure 15.5a shows that MN–TS distance of Algorithm 15.2 only takes 10–20% of the network diameter, which means TS is located about two hops away from MN on average, and this offers a good guarantee on the handoff efficiency. Algorithm-CN has the largest MN–TS value since it always pushes binding cache to the CN side. The value of Algorithm-random stays between Algorithm-CN and Algorithm 15.2. Also, when evaluation topology becomes more flat, MN–TS values of three algorithms approximate to each other. It is because with the "flatten" of topology, average distance between nodes also drops.

Figure 15.5b shows that, on average value, Algorithm 15.2 only generates about 0.3–0.5 flow table downloading per each CN in three intradomain topologies, while the other two algorithms always require downloading one flow table for each CN. Again, Algorithm 15.2 requires more flow table downloading in flat topologies. Just as the fact that hierarchical topology helps to reduce routing table size, it also helps to reduce binding cache maintenance.

15.3.4   System Design

This subsection describes system design of the proposal including an implementation based on Mininet and experiments that compares this proposal with another two representative Internet mobility protocols.

15.3.4.1   Implementation

The protocol is implemented based on Mininet 2.1.0 [46]. Pox [47] is chosen as the controller in the implementation. Figure 15.6a demonstrates the protocol flow. When switch S3 detects MN's attachment, it assigns a CoA to the MN and registers (HoA, CoA) tuple on the controller using port-status message. Receiving registration from S3, controller stores the binding cache locally and downloads a "rewrite" flow entry (the same as that in Section 15.3.2.1) to S3 using flow-mod message.

When switch S1 receives a packet toward an unknown host, it sends the packet to controller using packet-in message. Upon receiving packet-in, controller rewrites its destination IP address and resends it out using packet-out message. At the same time, controller downloads MN's binding cache to S1 using flow-mod message. In order to generate the path pairs used in the binding cache placement algorithm, the controller needs to keep a record of all the CN-to-MN paths, which we call *Path Record (PR)*. For example, in this case, when the controller downloads MN's binding cache to S1, it adds one entry into PR indicating that "There exists a flow from S1 to the MN." When the binding cache on S1 expires, S1 will acknowledge the controller, and then the controller will delete the related entry in PR.

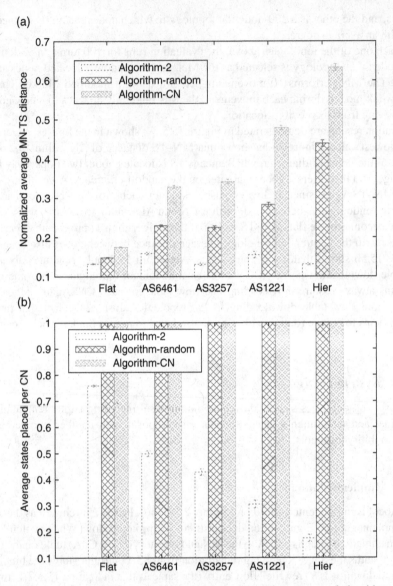

Figure 15.5 The two figures show results of comparison among three proposed algorithms based on five different topologies. (a) Shows normalized average MN–TS distance, and (b) shows average number of binding cache placed per CN. Algorithm 15.2 outperforms the other two in all scenarios.

After MN moves to S2, another similar procedure handles related registration and flow table downloading procedures. To deal with the handoff, the controller runs the binding cache placement algorithm discussed previously. It uses PR to get all path pairs related to the MN (only one path pair in this case). To obtain the path pairs, the controller looks up in the

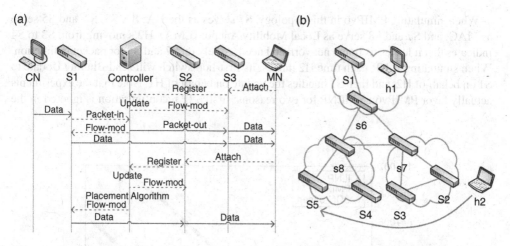

Figure 15.6   (a) Protocol flow and (b) experiment topology of the implementation.

PR and calculates the previous path from S1 to MN via S3, as well as the current path from S1 to MN via S2. After running the binding cache placement algorithm, the controller gets a set of switches that require updating. Then the controller downloads the MN's up-to-date binding cache to these switches using flow-mod messages. Note that the controller also needs to update PR after this turn of flow table downloading.

15.3.4.2   Experiment

Methodology
Several experiments are made based on the implementation to compare the proposed protocol with another two IP mobility protocols: PMIPv6 and ILNP. The two protocols are chosen because they serve as good representatives of the solutions reviewed in Section 15.2.1: a network-based protocol and a host-based protocol. To make comparisons, another two controllers are implemented to realize the basic mobility functions of PMIPv6 and ILNP, respectively. PMIPv6 is easier to implement based on Mininet since it is a network-based protocol. But ILNP is more difficult; thus, the protocol is simulated in an approximate way: the mobility functions are moved from hosts to their first-hop switches.

The experiment topology is shown in Figure 15.6b, which consists of one controller, two hosts, and eight switches. The topology is divided into three interconnected subdomains: (S7, S2, S3), (S8, S4, S5), and (S6, S1). Delays of intersubdomain links, intrasubdomain links, and "wireless links" (between H2 and attached switches) are 20, 2, and 10 ms, respectively. Bandwidths of the above three types of links are 100, 100, and 10 Mbps, respectively. Since in the current version of Mininet in-band control between switches and controller is not supported, thus, control traffic is out of band in the experiments. H2 serves as the MN and moves back and forth between switches S2 and S5. H1 serves as the CN and keeps immobile. The experiments run Iperf, which is a commonly used network testing tool, between the two hosts and collect end-to-end performance including round-trip time (RTT), packet loss rate, as well as throughput.

When simulating PMIPv6 in this topology, S7 serves as the HA; S2, S3, S4, and S5 serve as MAG; and S7 and S8 serve as Local Mobility Anchor (LMA). H2's moving from S3 to S4 indicates that it leaves its home network and needs to rely on S7 and S8 for packet indirection. When simulating ILNP, each time H2 moves, its first-hop switch will send Binding Update to S1 on behalf of H2, and then S1 handles the update on behalf of H1. Note that the experiments actually favor PMIPv6 and ILNP for two reasons: firstly, IP reconfiguration is ignored in the

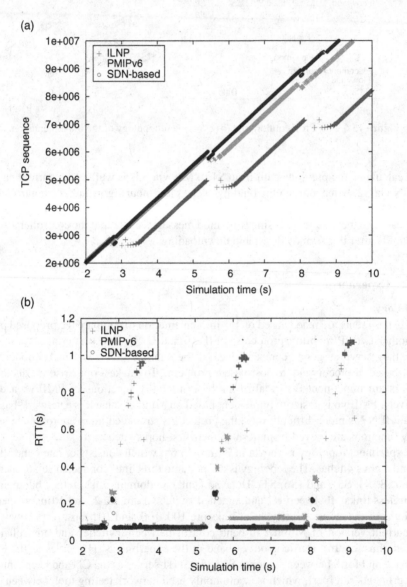

Figure 15.7 (a) TCP sequence and (b) round-trip time (RTT) of three simulated protocols during three handoff events in 10 s.

handoff process of the two protocols. Secondly, Binding Update process is simplified and only takes one-way delay: MN-to-HA delay in PMIPv6 case and MN-to-CN delay in ILNP case. Both simplifications help to improve handoff efficiency of the two protocols.

Experiment Results

The first experiment runs Iperf between H1 and H2 for 10 s during which period H2 moves from S2 to S5 and performs three handoffs. Figure 15.7a shows collected TCP sequence of PMIPv6, ILNP, and the SDN-based solution within the simulation time, from which we can infer that SDN-based solution generates smoother handoff than the other two: TCP based on ILNP experiences time-out and slow start during each handoff, which makes ILNP performs the worst in the experiment. It is because ILNP needs to send Binding Update from the MN side toward the CN side, and this may seriously degrade handoff efficiency especially when both sides are located away from each other. TCP based on PMIPv6 experiences only one time-out during the second handoff, as the other two handoffs can be handled locally by LMA, while the second handoff is an intersubdomain handoff and requires interactions with HA. In contrast, TCP on the SDN-based solution can always recover from packet loss during handoff using fast retransmit.

Figure 15.7b shows RTT of three protocols collected in the same experiment scenario, where we observe that RTT value of all three protocols temporarily raises to a higher value during handoff process. Besides, RTT of PMIPv6 stays at a higher value after the second handoff. It is because when H2 leaves the home network (after moving from S3 to S4), all packets to H2 are relayed by the HA S7, which results in triangle routing. The SDN-based solution avoids triangle routing as the binding cache placement algorithm ensures optimal forwarding path, and in this scenario, it is achieved by downloading binding cache to S6.

15.4   Conclusion

This chapter addresses mobility in IP network under SDN architecture. SDN has advantages in handling problems in current mobility protocols because of its programmable devices, centralized control, as well as other features. By proposing an SDN-based mobility management architecture together with an OpenFlow-based protocol design, implementation, and experiments, this chapter demonstrated that SDN enables the flexibility of mobility management, making it adaptive to various mobility scenarios in future mobile Internet.

References

[1] L. Zhang, R. Wakikawa, and Z. Zhu. Support Mobility in the Global Internet. In Proceedings of the 1st ACM Workshop on Mobile Internet through Cellular Networks, ACM, 2009. Beijing, China.

[2] P. Zhang, A. Durresi, and L. Barolli. A Survey of Internet Mobility. In International Conference on Network-Based Information Systems 2009, NBIS'09, 147–154. IEEE, 2009. Indianapolis, USA.

[3] F. M. Chiussi, D. A. Khotimsky, and S. Krishnan. Mobility Management in Third-Generation All-IP Networks. Communications Magazine, IEEE, 2002, 40(9): 124–135.

[4] D. Saha, A. Mukherjee, I. S. Misra, and M. Chakraborty. Mobility Support in IP: A Survey of Related Protocols. Network, IEEE, 2004, 18(6): 34–40.

[5] I. F. Akyildiz, J. Xie, and S. Mohanty. A Survey of Mobility Management in Next-Generation All-IP-Based Wireless Systems. Wireless Communications, IEEE, 2004, 11(4): 16–28.

[6] A. Lucent. Introduction to Evolved Packet Core. Strategic White Paper, 2009. Available from www.alcatel-lucent. com (accessed February 19, 2015).

[7] O. Tipmongkolsilp, S. Zaghloul, and A. Jukan. The Evolution of Cellular Backhaul Technologies: Current Issues and Future Trends. Communications Surveys & Tutorials, IEEE, 2011, 13(1): 97–113.

[8] J. C. Zuniga, C. J. Bernardos, A. de la Oliva, T. Melia, R. Costa, and A. Reznik. Distributed Mobility Management: A Standards Landscape. Communications Magazine, IEEE, 2013, 51(3): 80–87.

[9] J. N. Chiappa. Endpoints and Endpoint Names: A Proposed Enhancement to the Internet Architecture. 1999, Available from http://mercury.lcs.mit.edu/~jnc/tech/endpoints.txt (accessed January 24, 2015).

[10] Z. Zhu, R. Wakikawa, and L. Zhang. A Survey of Mobility Support in the Internet. RFC 6301, IETF, 2011.

[11] C. Perkins. IP Mobility Support for IPv4, Revised. RFC 5944, IETF, 2010.

[12] C. Perkins, D. Johnson, and J. Arkko. Mobility Support in IPv6. RFC 6275, IETF, 2011.

[13] H. Soliman, C. Castelluccia, K. ElMalki, and C. Castelluccia. Hierarchical Mobile IPv6 (HMIPv6) Mobility Management. RFC 5380, 2008.

[14] S. Gundavelli, K. Leung, V. Devarapalli, K. Chowdhury, and B. Patil. Proxy Mobile IPv6. RFC 5213, IETF, 2008.

[15] R. Wakikawa, G. Valadon, and J. Murai. *Migrating Home Agents towards Internet-scale Mobility Deployment.* In Proceedings of the 2006 ACM CoNEXT Conference. ACM, 2006. Lisboa, Portugal.

[16] M. Fisher, F.U. Anderson, A. Kopsel, G. Schafer, and M. Schlager. A Distributed IP Mobility Approach for 3G SAE. In 19th International Symposium on Personal, Indoor and Mobile Radio Communications (PIMRC 2008), IEEE, 2008. Cannes, French Riviera, France.

[17] R. Cuevas, C. Guerrero, A. Cuevas, M. Calderón, and C.J. Bernardos. P2P Based Architecture for Global Home Agent Dynamic Discovery in IP Mobility. In 65th IEEE Vehicular Technology Conference, IEEE, 2007. Dublin, Ireland.

[18] Y. Mao, B. Knutsson, H. Lu, and J. Smith. DHARMA: Distributed Home Agent for Robust Mobile Access. In Proceedings of the IEEE Infocom 2005 Conference, IEEE 2005. Miami, USA.

[19] R. Moskowitz and P. Nikander. Host Identity Protocol (HIP) Architecture. RFC 4423, IETF, 2006.

[20] R. Atkinson and S. Bhatti. Identifier-Locator Network Protocol (ILNP) Architectural Description. RFC 6740, IETF, 2012.

[21] J. Ubillos, M. Xu, Z. Ming, and C. Vogt. Name-Based Sockets Architecture. IETF Draft, 2011. Available from https://tools.ietf.org/html/draft-ubillos-name-based-sockets-03 (accessed February 19, 2015).

[22] D. Farinacci, D. Lewis, D. Meyer, and C. White. LISP Mobile Node. IETF Draft, 2013. Available from http://tools.ietf.org/html/draft-meyer-lisp-mn-09-12 (accessed February 19, 2015).

[23] National science foundation future internet architecture project. Available from www.nets-fia.net (accessed February 19, 2015).

[24] I. Seskar, K. Nagaraja, S. Nelson, D. Raychaudhuri. MobilityFirst Future Internet Architecture Project. In Proceedings of the 7th Asian Internet Engineering Conference. pp. 1–3. ACM. Bangkok, Thailand.

[25] A. Venkataramani, A. Sharma, X. Tie, H. Uppal, D. Westbrook, J. Kurose, and D. Raychaudhuri. Design Requirements of a Global Name Service for a Mobility-centric, Trustworthy Internetwork. In Fifth International Conference on Communication Systems and Networks (COMSNETS). pp. 1–9. IEEE, 2013. Bangalore, India.

[26] D. Han, A. Anand, F.R. Dogar, B. Li, H. Lim, M. Machado, A. Mukundan, W. Wu, A. Akella, D.G. Andersen, J.W. Byers, S. Seshan, and P. Steenkiste. XIA: Efficient Support for Evolvable Internetworking. In Proceedings of the 9th USEnIX NSDI. ACM, 2012. San Jose, USA.

[27] L. Zhang, A. Afanasyev, and J. Burke. Named Data Networking. Technical Report, 2014. Available from http://named-data.net/publications/techreports (accessed February 19, 2015).

[28] Z. Zhu, A. Afanasyev, and L. Zhang. A New Perspective on Mobility Support. Technical report, 2013. Available from http://named-data.net/publications/techreports (accessed February 19, 2015).

[29] N. McKeown, T. Anderson, H. Balakrishnan, G. Parulkar, L. Peterson, J. Rexford, S. Shenker, and J. Turner. OpenFlow: Enabling Innovation in Campus Networks. ACM SIGCOMM Computer Communication Review, vol. 38, pp. 69–74, 2008.

[30] J. Saltzer. On the Naming and Binding of Network Destinations. RFC 1498, IETF, 1993.

[31] E. Lear and R. Droms. What's in a Name: Thoughts from the NSRG. IETF Draft, 2003. Available from http://tools.ietf.org/html/draft-irtf-nsrg-report-10 (accessed February 19, 2015).

[32] D. Meyer, L. Zhang, and K. Fall. Report from the IAB Workshop on Routing and Addressing. RFC 4984, IETF, 2007.

[33] H.A. Chan, H. Yokota, P.S.J. Xie, and D. Liu. Distributed and Dynamic Mobility Management in Mobile Internet: Current Approaches and Issues. Journal of Communications, 2011, 6(1): 4–15.

[34] IETF. Distributed Mobility Management (DMM). IETF Working Group. Available from http://tools.ietf.org/wg/dmm/ (accessed January 24, 2015).

[35] M. O'Dell. GSE—An Alternate Addressing Architecture for IPv6. IETF Draft, 1997. Available from http://tools.ietf.org/html/draft-ietf-ipngwg-gseaddr-00 (accessed February 19, 2015).

[36] D. Farinacci, D. Lewis, D. Meyer, and V. Fuller. The Locator/ID Separation Protocol (LISP). RFC 6830, IETF, 2013.

[37] D. Peleg and E. Upfal. A Trade-off between Space and Efficiency for Routing Tables. In 20th Annual ACM Symposium on Theory of Computing (STOC), pp. 43–52. ACM, 1988. Chicago, IL, USA.

[38] C. Gavoillz and S. Pérennès. Memory Requirement for Routing in Distributed Networks. In Proceedings of the 15th PODC. ACM, 1996. Philadelphia, PA, USA.

[39] D. Krioukov, K.C. Claffy, K. Fall, and A. Brady. On Compact Routing for the Internet. ACM Computer Communications Review, 2007, 37(3): 41–52.

[40] K. Yap, T.Y. Huang, M. Kobayashi, M. Chan, R. Sherwood, G. Parulkar, and N. McKwown Lossless Handover with n-Casting between WiFi-WiMAX on OpenRoads. In ACM Mobicom. ACM, 2009. Beijing, China.

[41] K. Yap, M. Kobayashi, D. Underhill, S. Seetharaman, P. Kazemian, and N. Mckwown. The Stanford Openroads Deployment. In ACM Workshop on Wireless Network Testbeds, Experimental Evaluation and Characterization (WiNTECH), ACM, 2009. Beijing, China.

[42] K. Yap, R. Sherwood, M. Kobayashi, T.Y. Huang, M. Chan, N. Handigol, N. McKeown, and G. Parulkar. Blueprint for Introducing Innovation into Wireless Mobile Networks. In Workshop on Virtualized Infrastructure Systems and Architectures, pp. 25–32. ACM 2010. New Delhi, India.

[43] P. Pupatwibul, A. Banjar, A.A.L. Sabbagh, and R. Braun. Developing an Application Based on OpenFlow to Enhance Mobile IP Networks. Local Computer Networks (LCN) 2013 Workshop on Wireless Local Networks, IEEE 2013. Sydney, Australia.

[44] V. Chvatal. A Greedy Heuristic for the Set-Covering Problem. Mathematics of Operations Research, 1979, 4(3): 233–235.

[45] Rocketfuel: An ISP Topology Mapping Engine. Available from www.cs.washington.edu/research/networking/rocketfuel/ (accessed January 24, 2015).

[46] Mininet: An Instant Virtual Network on Your Laptop (or other PC). Available from http://mininet.org/ (accessed January 24, 2015).

[47] POX Controller. Available from www.noxrepo.org/pox/about-pox/ (accessed January 24, 2015).

16

Mobile Virtual Network Operators
A Software Defined Mobile Network Perspective

M. Bala Krishna

University School of Information and Communication Technology,
GGS Indraprastha University, New Delhi, India

16.1 Introduction

Wireless networks demand dynamic, innovative, and user-centric business models for emerging markets of mobile and wireless communication system. Current trends demand flexible and wide range of services for mobile end users. Mobile network operators (MNOs) with licensed and unlicensed spectrum function as the potential stakeholders in wholesale and retail business markets for mobile networking and communication services. MNOs are the primary licensed vendors and depend on virtual network operators (VNOs) to meet the demands of end users. Mobile virtual network operators (MVNOs) [1] lease the radio spectrum from MNOs and extend the infrastructure from MNOs and third-party service providers to potential mobile end user nodes (MEUNs). MVNOs are not assigned the licensed spectrum by regulating bodies and hence operate using the unlicensed radio spectrum for short-distance communication. MVNO's leased radio spectrum is based on service agreements (wholesale and retail) with MNOs and the density of MEUNs. MVNOs support value-added services like high-speed data, multimedia streaming, video conference, E-commerce, M-commerce, etc. Virtual private network systems (VPNS) enable the MEUNs to install and utilize the user services within the private domain by working in collaboration with third-party network operators [2].

MVNOs create a virtual interface framework to manage and monitor the services and requests from subscribed MEUNs, MNOs, and access points. MVNOs like active network operators use packet processing, filtering, and forwarding mechanism in the network. Virtual interfaces do not define the service packages, but maintain the complete business details of service providers and function as access point black box in the mobile network. Virtual interfaces address resource partitioning, multiplexing, and demultiplexing in the network. The

Software Defined Mobile Networks (SDMN): Beyond LTE Network Architecture, First Edition.
Edited by Madhusanka Liyanage, Andrei Gurtov, and Mika Ylianttila.
© 2015 John Wiley & Sons, Ltd. Published 2015 by John Wiley & Sons, Ltd.

number of MEUNs subscribed to MVNO depends on the existing business model such as one to one and one to many. The interaction between MNO and MVNO is based on Business Service Subsystem (BBS), Network Service Subsystem (NSS), Application Service Subsystem (ASS), and User-Supportive Subsystems (USS) and configured through the attributes of software defined network (SDN). Figure 16.1 illustrates various types of services between MNO and MVNO controller by SDN. SDN controller configures and maps the services between MNO and MVNO as per the service requirements of MEUNs. Table 16.1 highlights the SDN configuration metrics and their respective attributes used in MNO and MVNO business model.

The salient features of each subsystem as illustrated in Figure 16.1 are elucidated as follows:

BSS: This subsystem defines the business framework for existing market demands, license agreements with MNOs, types of services, and pricing tariffs to MEUNs. BSS includes spectral sharing factor and quality of bandwidth during peak traffic periods. MVNOs define several branded services with respect to multiple MNOs.

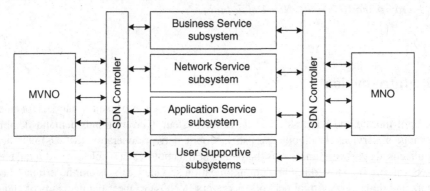

Figure 16.1   Services between MNO and MVNO controllers by software defined network.

Table 16.1   SDN Configuration Metrics and Attributes used in MNO-MVNO Business Model

Configuration Metrics	Attributes
Configure Control Plane	• Set logical paths • Set alternate paths • Set fault tolerant paths
Configure Data Plane	• Flow of voice and data packets (size, path, etc.) • Prioritize data packets
Load Balancing	Balance the load across MNOs and MVNOs
Spectrum Allocation Schemes • Cooperative • Non-Cooperative Spectrum Access Technique • Overlay • Underlay	• MNO regulates the allocated spectrum range with respect to the number of MVNOs • Resource allocation is prioritized for commercial and value added MEUNs • Wholesale and retail pricing schemes are based on dynamic spectrum allocation methods • Efficient switching methods for guaranteed MVNO-MNO networks
Security	Firewalls, VPN, etc.
Application Service	Value added services, Voice mail, etc.

NSS: This subsystem defines the networking model of MVNO with mobile bearer services such as PSTN, GPRS, GSM, UMTS, LTE, and LTE-A. The channel specifications (such as bit transfer rate, SNR, transmitting power, receiving power, multiplexing, multiple accessing techniques, etc.) and user-specific parameters (such as bandwidth, time delays, packet transfer rate, power specifications, etc.) are the primary attributes of this subsystem.

ASS: This subsystem defines the service specifications for MEUNs with pricing and tariff schemes. MVNO physical location is estimated as per the business strategy and density of MEUNs. MVNO offers the application services such as voice, high-speed data, video, SMS, multimedia, Internet applications, etc. to MEUNs. The business and networking specifications with QoS parameters are manifested in this subsystem. The services offered in this subsystem are classified as follows:

Class I Excel: Applications that demand high-speed and efficient services with maximum bandwidth (longer or shorter duration) are used in this class. Such premium services are used in commercial applications for multination organizations, financial and banking sector, weather forecast, etc.

Class II On Demand: Applications that require consistent services during catastrophic events such as fire accidents, earthquake, tsunami, etc. are used in this class. These applications demand maximum bandwidth for shorter duration with high-speed SMS and data transfer rates.

Class III Economical: Applications with high-speed data services and adaptive bandwidth allocation used in this class. Such flexible services are used in medium-scale organizations such industries, hospitals, corporate offices, etc.

Class IV Regular: Application service requirements of MEUNs based on PSTN, IPTV, Access Point services, etc. are used in this class.

USS: This subsystem addresses the potential demands of MEUNs and facilitates additional services and packages offered by the neighboring MNOs. Retail tariff and pricing schemes are defined in this subsystem.

16.1.1   Features of MVNO

MVNOs are a collection of distinct unlicensed operators that outsource their radio operations to reliable networks and support value-added services to MEUNs. Unique identity is assigned to the MVNO subscriber such as the subscriber identity module (SIM). The MVNO transactional cost is reduced by subdividing the task and distributing it to neighboring MVNOs. Time-frame negotiations limit the potential hazards in transaction and support the ACID features [1]. Software defined network (SDN) [3] adds the reconfigurable mobile network parameters to the existing MVNO and enhance the services to MEUN. Reconfigurable parameters used by SDN are given as follows:

- Network capacity specifications defined by traffic controller and maximum transmission unit (MTU)
- Cognitive and dynamic switching techniques for UMTS, GPRS, and TCP/UDP-based services

- Gateway connectivity and route discovery protocols
- Update of MEUN database and lookup tables
- Update of MEUNs mobility management services
- Secure key management for encrypted data packets.

Software defined mobile network (SDMN) [4] architecture for future carrier networks decouples the data and control planes in mobile network. SDMN supports VNO interface and application program interfaces (API) to enhance the services of carrier network, network coverage, and proxy gateway nodes.

16.1.2 Functional Aspects of MVNO

The functional aspects of MVNOs are elucidated as follows:

- **Bearer Services and Gateway Interfaces**: Earlier business models of MVNO were designed for UMTS [5] that support medium-scale data services. The service aspect of MVNOs is a function of bearer services and gateway interfaces.
- **Centralized and Distributed MVNOs**: The services of centralized and distributed MVNOs are based on the density of MEUNs, the number of active MNOs, and the interface coordination units in the network. Various functions of MVNO servers are given as follows:

  (i) MVNO server maintains the user profile and service set.
  (ii) MNO server maintains the specifications of service set controlled by the network.
  (iii) Network server manages and controls the tasks of MNO and gateway nodes internetworking protocols.
  (iv) Proxy server addresses the license agreements and connectivity between heterogeneous API and network service providers.

16.1.2.1 Software Defined Network Perspective

The functional aspects of MVNO based on SDN perspective are illustrated in Figure 16.2. The functionality of each component is explained as follows:

- **Scalable and Secured Interfaces**: MVNOs define scalable network interface for virtual schemes [6] in varying traffic conditions. MNOs maintain the list of local and global operators to facilitate connectivity in wide coverage area and high-speed data access. For fair negotiations between MVNO and MEUNs, the parameters such as trust, privacy, compatibility, and sustainability are defined in the service set. Negotiation-based schemes [7] effectively satisfy machine-to-machine requirements, trust, and cooperation with existing VNOs.
- **Customer-Driven Server**: MVNOs operate based on customer-driven services (such as buy service, share service, value-added service, Internet-enabled service, etc.) that are flexible and adaptable in competitive business market. MVNO's user-centric approach [8] concentrates on the economic relations between MEUNs and MNOs. Horizontal marketing schemes by MNO (wireless access networks, WLAN, WWANs, etc.) use a different set of license agreements and offer their services to MVNO.

- **Regulation and Management Schemes**: Regulation schemes define access rights, QoS parameters, extended service sets, policy issues, and price tariffs for the interconnection network. Regulation schemes [9] in mobile communication market are a function of time–size–growth of business and variations in market strategy. Management schemes control the number of potential VNOs and compatibility with existing business models based on technologies like 3G, 4G, WiMAX, LTE-A, etc.
- **Spectral Utilization Services**: Efficiency in bandwidth utilization is implemented [10] by identifying the unused spectrum regions and allocating this region to VNOs with shared services.
- **Emergency-Supportive Services**: MVNO supports emergency services such as disaster management, natural calamities, etc. by establishing a flash network of cooperative MVNOs. Network operators at affected areas and critical service points are prioritized to access the MVNOs and base station (BS).
- **Extending Services of Virtual Enablers:** MVNOs support virtual enablers to provide the access rights for mobile virtual network enabler (MVNE) subscribers in the network.

16.1.3   Challenges of MVNO

MVNO comprises of infrastructureless network and integrated services of network service providers that support advanced mobile technologies such as LTE-A, 4G, etc. Various challenges of MVNO are given as follows:

Flexible Services—MVNOs offer flexible services to MEUNs and support scalable services for multidimensional MVNOs and MNOs in a distributed environment.

Ceaseless Connectivity and Mobility Support—With homogeneous and heterogeneous MNOs, MVNOs support seamless connectivity for internetwork and intranetwork systems. MVNOs support mobility across the various VNOs in the network.

Cost Effectiveness—MVNO business model specifies the service set parameters for MNOs, uses the unlicensed spectrum, and minimizes the operational cost of MEUNs. This potentially increases the marketing scope of VNOs.

Service Agreement—Sequence of MVNO service agreement plan is as follows: (i) sign the business agreement with MNOs, (ii) establish a business relationship with neighboring MVNOs, and (iii) support the MEUNs across various countries.

Security—Customer profile and service specifications are kept confidential in the virtual network. Based on MVNE requirements, MVNO configures the secure VPN to establish a secure communication channel.

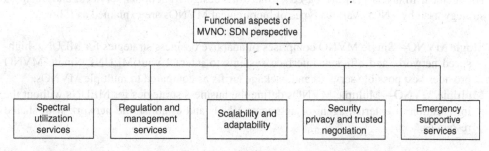

Figure 16.2   Functional aspects of MVNO.

16.2   Architecture of MVNO: An SDMN Perspective

SDMN comprises of SDN controller, SDN configuration interface, access points, HLR–VLR updates, mobile switching center, gateway servers, and BS. Figure 16.3 illustrates the architecture of SDMN-based MVNO that comprises of a set of MEUNs, access points, SDN interface, SDN controller, HLR–VLR components, MNO service point, gateway servers, and MVNO servers. User-defined unit comprising of authentic MEUNs and access points is connected to SDN configuration unit. SDN controller unit comprises of subcontroller units that configure the SDN parameters such as radio spectrum range, bandwidth, data rate, resource allocated, and service set. MEUN profiles are maintained in the subcontroller units that control and coordinate the services (Class I Excel, Class II On Demand, Class III Economical, and Class IV Regular) offered by MNO. MVNOs are connected to MNO service point and interact with MVNO server units through the gateway network.

16.2.1   Types of MVNOs

Various types of MVNOs [11] and their respective operations are highlighted as follows:

Full MVNO—Full MVNOs established by governing regulatory bodies and telecom operators are equipped with core network, access network infrastructure, and functional aspects such as routing, interconnection, and executable service list. The business model of full MVNO comprises of customer care service, billing, handset management, marketing and sales, etc.

Intermediate MVNO—Intermediate MVNOs access the radio spectrum and services of MNO. The business model of intermediate MVNO comprises of application services, customer care, billing, handset management, marketing and sales, etc.

Thin MVNO—Thin MVNOs support enhanced application services and increase the level of adaptability for MEUNs. Thin MVNOs support the services of MNO. The business model offers the best services for potential MEUNs.

Special Purpose MVNO—Special purpose MVNOs use specific MNOs with partial infrastructure for private and confidential applications. Special purpose MVNOs are used in corporate offices, business establishments, medium-scale industries, etc.

16.2.2   Hierarchical MVNOs

Hierarchical MVNOs [12] are based on network design, functional services, and business strategy used by VNOs. Various types of hierarchical MVNOs are explained as follows:

Single MVNO—Single MVNO comprises of adaptive business strategies for MEUNs, high-speed network, and efficient interface systems to interact with MEUNs. Single MVNO provides best possible services and package tariffs as compared to multiple MVNOs.

Multiple MVNO—Multiple MVNOs define the business strategies for MEUNs without the intervention of external agent or aggregate VNOs and establish the network with shared resources.

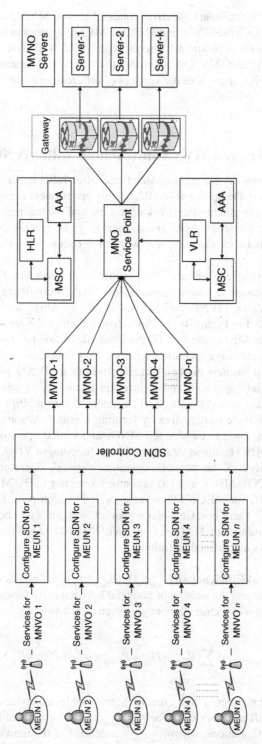

Figure 16.3 Architecture of SDMN-based MVNO.

Aggregator MVNO—To facilitate collective requests between MVNOs, aggregator MVNO acts like a bridge between the MNOs and MEUNs. MVNO aggregator comprises of distributed MVNO components, sophisticated interfaces, and MEUN's service request set. Mobile virtual network aggregator (MVNA) negotiates the license agreement, network capacity, and services with MNO, aggregates the services of different host MNOs, and specifies the price tariffs to hosting MVNOs.

16.3 MNO, MVNE, and MVNA Interactions with MVNO

MNOs and MVNOs sign the business contract and offer reliable services to retain their respective brand names in the open market. Regulating government organizations assign the licensed radio spectrum to potential network operators, and the business establishments are carried out at (i) network operator level and (ii) service point level. MNOs share the radio spectrum, control the network capacity, and work in coordination with MVNOs, MVNE, and MVNA.

Cognitive and SDN strategies enhance the performance of MVNO, MVNE, and MVNA by reducing the signal-to-noise ratio and packet loss. SDMN configures the radio and service parameters of MVNO, MVNE, and MVNA at distinguishing levels (level 1, level 2, and level 3) as illustrated in Figure 16.4. This scheme enables the network components to select the best available MNO based on (i) the available radio spectrum, (ii) connectivity with distinct MVNOs, and (iii) application services. This approach competes with existing service providers [10] in business market. Virtual resources at MVNO, MVNE, and MVNAs use "allocate-on-demand" approach and reduce the ambiguity in connecting with high-speed MEUNs. Cognitive access points (CAP) connected with multiple network operators extend their services in the coverage area by forming a grid. CAPs are context aware and define distinctive business services between MNOs and virtual operators such as MVNO, MVNE, and MVNA. MNO business strategy [13] is a function of VNO, authentic MEUNs, and value-added services of the network operator. Network operators are classified as (i) primary operator (PO), MNO, and (ii) secondary operator (SO), MVNO, MVNE, and MVNA. The price tariffs for MEUN SO are categorized as high-price (U_{hp}), medium-price (U_{mp}), and low-price (U_{lp}) services. Consider a network with an N number of active MEUNs distributed over an R number of MNO regions. With MVNO as the potential operator per region, SDN parameters are defined as follows:

Regional-Level Spectral Efficiency ($RL_{se\text{-}MVNO}$): $RL_{se\text{-}MVNO}$ is defined as the sum of products of transactional power and the number of potential licensed users for each service provided by MNO. Regional-level spectral efficiency is given as follows:

$$RL_{se\text{-}MVNO} = \sum (P_{T-hp} \times U_{hp} + P_{T-mp} \times U_{mp} + P_{T-lp} \times U_{lp}) \qquad (16.1)$$

U_{hp}, U_{mp}, and U_{lp} are the number of high-, medium-, and low-level services offered to MEUNs.
Average Business Tariff per Service ($BTariff_{Avg, Service}$): The business tariffs for each service vary with MNO configurations. $BTariff_{Avg, Service}$ is defined as the ratio of product of power

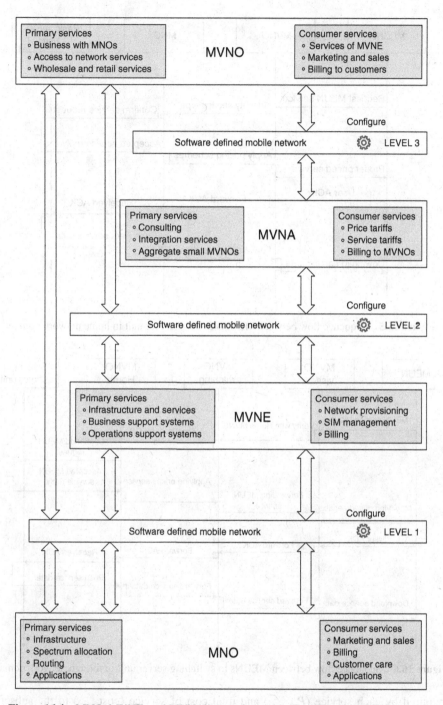

Figure 16.4 MNO, MVNE, and MVNA interactions with SDMN-based MVNO [11, 12].

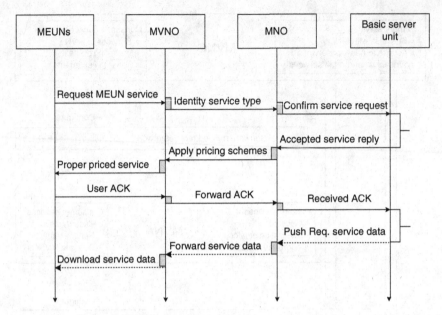

Figure 16.5 Sequence flow between MEUNs to basic server unit in home network domain.

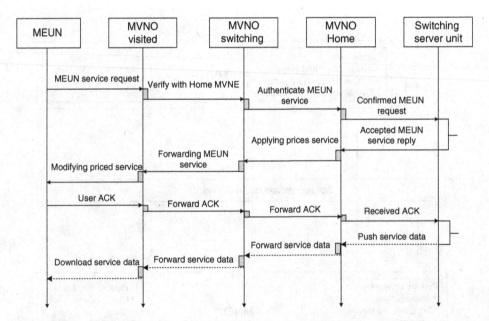

Figure 16.6 Sequence flow between MEUNs to switching server unit in foreign network domain.

required by each service ($P_{\text{T-service}}$) and total cost of service ($\text{cost}_{\text{service}}$) to the subscribed MEUN services (U_{service}). The average business tariff is given as follows:

$$BTariff_{\text{Avg, Service}} = \frac{P_{\text{T-service}} \times \text{cost}_{\text{service}}}{U_{\text{service}}} \qquad (16.2)$$

The sequence of events in home network domain from MEUNs to MVNO, MNO, and basic server unit is illustrated in Figure 16.5. In the foreign network domain, MVNO comprises of three components: visited MVNO, switching MVNO, and home MVNO. The sequence of events in foreign network is illustrated in Figure 16.6. Net business revenue ($BRNET$) for each service at MNO is defined as the ratio of product of average business revenue for each service in region R and tariff adjustment factor ($Tarif_{Adjustment\_Factor}$) to the total number of MEUNs (N). BR_{NET} is defined as follows:

$$BR_{NET} = \left(\frac{BR_{Avg,service} \times R}{N} \right) \times Tariff_{Adjustment\_Factor} \qquad (16.3)$$

16.3.1   Potential Business Strategies between MNOs, MVNEs, and MVNOs

Business model depends on the mutual contract between MNO, MVNE, and MVNO. Network enablers act as mediators between the network operator and VNO in home and foreign network. The primary factors affecting the quality of business are given as follows:

(i)   Number of MEUNs requesting for services in licensed and unlicensed spectrum
(ii) Business services offered at higher prices due to increase in spectrum range
(iii) Business services offered at lower prices due to collaboration between MVNOs and MNOs
(iv) Density of originated and terminated calls at MVNO.

The compactness of MNO and MVNO can be effectively reduced by using network enablers. MVNEs [14] implement self-defined business schemes to attract the business interests of VNO and service points. MVNOs are identified by their respective access network numbers (ANs) and allocated radio spectrum. New business models with additional services are incorporated in existing framework to increase the number of potential MEUNs. MVNEs function as potential service access points to multiple MVNOs and enhance the service requirements of MVNEs. MVNEs enable user-centric services with MVNOs and backup services with intra-MNO domain or inter-MNO domain. MNOs assign spectral bandwidths to MVNOs [15], and further MVNOs can request the MNOs for additional radio spectrum and reconfigure the bandwidth (based on the density of MEUNs). Efficient channel management techniques improve the quality of voice and data rates with enhanced radio spectrum, nature of unguided media, and functional aspects of MVNO. Separate bandwidth is assigned for retail and commercial services. The number of retail services offered by MNO depends on the number of MVNOs and the cooperation factor with MNOs. For reliable MEUN services, an optimum number of VNOs maintain the required QoS and network performance. MVNE [16] uses schemas to advertise new business strategies for VNO and service providers. Consumer model [17] facilitates the MVNOs to acquire licensed radio spectrum from multiple MNOs and adapt a cooperative approach to share the licensed radio spectrum in business market. MVNO revenue ($MVNO_{Revenue}$) is a function of retail services, commercial services, net profit, and net loss. $MVNO_{Revenue}$ is defined as follows:

$$MVNO_{Revenue} = f\left(MVNO_{Retail\_Services}, MVNO_{Net\_Profit}, MVNO_{Net\_Loss} \right) +$$
$$g\left(MVNO_{Commercial\_Services}, MVNO_{Net\_Profit}, MVNO_{Net\_Loss} \right) \qquad (16.4)$$

MVNO net profit ($MVNO_{Net\_Profit}$) is a function of active users ($MVNO_{Active\_Users}$), data rate ($DataRate_{pkts/sec}$), and subscribed services ($MVNO_{Subscribed\_Services}$). $MVNO_{Net\_Profit}$ is defined as follows:

$$MVNO_{Net\_Profit} = \sum_{i=1}^{n} MVNE^{i}_{Active\_Users} \times DataRate_{pkts/sec} + \sum_{j=1}^{k} MVNO^{j}_{Subscribed\_Services} \quad (16.5)$$

MVNO net loss ($MVNO_{Net\_Loss}$) is a function of passive users ($MVNO_{Passive\_Users}$), allocated spectrum, and unsubscribed services ($MVNO_{Unsubscribed\_Services}$). $MVNO_{Net\_Loss}$ is defined as follows:

$$MVNO_{Net\_Loss} = \sum_{i=1}^{n} MVNE^{i}_{Passive\_Users} \times Allocated\ Spectrum + \sum_{j=1}^{k} MVNO^{j}_{Unsubscribed\_Services} \quad (16.6)$$

16.3.2 Performance Gain with SDN Approach

SDN approach enhances the functions of network virtualization by classifying the MVNO business model into the network plane, the filtering plane, and the pivotal plane. Network plane comprises of switches, routers, and gateways to control the network traffic. It monitors and controls the packet arrival rate to reduce congestion in the network. Authentication and validity of MEUNs are verified at filtering plane that comprises of MVNO firewalls. Pivotal plane comprises of access control, load balancing, and network virtualization. In SDN approach, VNOs protect unused license spectrum and divert the network traffic to underutilized spectrum resulting in energy-efficient network virtualization. SDN configures and improves the resource utilization in WiMAX and EDGE bearer services of 3G networks. With extended infrastructure and resources available, MVNOs offer voice, data, and multimedia services on behalf of MNOs. MVNO business markets in Europe and North America are much larger as compared to traditional mobile business services [18]. In retail business, MVNOs initiate the services or collaborate with MNOs that offer similar type of services.

16.3.3 Cooperation between MNOs and MVNOs

Business agreement involves sharing of licensed radio spectrum and network resources between MNOs, SPs, and MVNOs. Resource sharing can be symmetric (uniform resource distribution across virtual operators) or asymmetric (nonuniform resource distribution across virtual operators). MVNO pricing schemes vary with service type (high-speed data, multimedia live streaming, video conferencing, etc.) and business contract with multiple MNOs. In asymmetric resource allocation, the following business approach is considered: (i) MVNOs with high bargaining and investing capabilities bid for additional spectrum bandwidth and priority services, and (ii) MVNOs with low bargaining and investing capabilities bid for allocated channel bandwidth during nonpeak traffic conditions. MNO architecture emphasizes on leased infrastructure and licensed spectrum allocation from Internet service provider (ISP). In cooperative-based MVNO models [19], the bargaining strategies are prioritized with the trade-off services of VNOs. Network traffic is a function of resource bargain factor and density of MEUNs for a given MNO domain. The min–max transaction cost is specified by MVNO controller. MVNO channel utility function ($MVNO_{Channel\_Utility\_Function}$) is low when the cooperation across MVNOs ($MVNO_{Co\text{-}operation}$) is low and the number of MEUNs is more than the threshold limit. $MVNO_{Channel\_Utility\_Function}$ is defined as follows:

$$MVNO_{Channel\_Utility\_Function} = low$$
$$\text{when } (MVNO_{Co\text{-}operation} = low) \text{ and } (MEUNs > threshold) \tag{16.7}$$

MNVO performance gain is estimated using (i) quality of VNO voice and data services, (ii) MVNO utility functions, and (iii) MNO response time [20]. Based on MVNO service requirements and number of MEUNs, SDN approach enables the MNO to select the available spectrum range for wireless channels. SDN isolates the services of data and control plane [21] to (i) reduce ambiguity in decision making (such as selecting a channel with nonoverlapping bandwidth) and (ii) reduce collisions in the communication channel. This approach reroutes the network traffic through different channels and reduces the communication overhead. MVNO price tariff [22] is a function of structural and operational expenditure of MNOs. Greedy approach extends the support of flexible network capacity and services as required by MEUNs and current traffic conditions. MVNO capacity ($MVNO_{Capacity}$) is a function of aggregate bit rate ($BR_{Aggregate}$), location category ($LOC_{Category}$), and the number of active MEUNs ($N_{Active\_Members}$) per area A. $MVNO_{Capacity}$ [22] is defined as follows:

$$MVNO_{Capacity} = f\left(BR_{Aggregate}, LOC_{Category}\right) \times g\left(\frac{N_{Active\_Members}}{A}\right) \tag{16.8}$$

Greedy approach validates the current MVNO capacity to be at par with the previous capacity levels. The cooperating features of MNO and MVNO are given as follows:

(i) Integrating capabilities to synchronize with multiple BS
(ii) Controlling and coordinating with heterogeneous network to meet the predefined QoS parameters
(iii) Maintaining consistency and integrity services with MEUNs.

16.3.4 Flexible Business Models for Heterogeneous Environments

Virtual ad hoc network operators use network virtualization [23] that shares the licensed spectrum and create an ad hoc network for emergency services such as military monitoring, disaster management, vehicular monitoring, etc. Network interface is further extended to support the services of UMTS, GPRS, Wi-Fi, WiMAX, LTE-A, etc. Flexible business model [24] evaluates the mobile price tariffs, MEUN services, and device technology developments of the smartphone. Generic business model comprises of multiple marketing schemes such as E-commerce, M-commerce, and payment services for end-to-end vendors. The functional aspect of payment tariffs is based on the business relationship between MNOs, MVNOs, and MEUNs. Table 16.2 describes various entities and their respective attributes used in existing business models. Distinctive features of actor and business model are given as follows:

(i) Interdependence between human, device specific, and business attributes
(ii) Cohesiveness between E-commerce and M-commerce applications
(iii) Revenue source and transactions performed by the active actors.

Business model [25] enhances the services of MNO based on the following features:

(i) MNOs define the business services and requirements for infrastructure network.
(ii) Service providers manage the resources between multiple MNOs.
(iii) List of MEUNs subscribed to corresponding MVNO.

Table 16.2 MVNO module type, access service, entity type and operational contract service [24] [27]

Module Type	Access Service	Entity Type	Operational Contract Service
		Primary	Guaranteed bandwidth
	Licensed Spectrum	Secondary	Guaranteed bandwidth with refunds
		Classified	Opportunistic access with pre-defined prices
Structural Module	Un-Licensed Spectrum	Regular	Opportunistic access with dynamic pricing schemes
		Wired or Cellular networks	Guaranteed or Opportunistic bulk access
	Networking Features	Wireless or Mobile networks	Guaranteed or Opportunistic bulk access
		Value Configured Services	Guaranteed bandwidth
	Wholesale Services	Value Proposed Services	Guaranteed or Opportunistic bulk access
Financial Module		Operator defined or branded	Guaranteed or Opportunistic bulk access, Region or Service based
	Retail Services	User specified	Opportunistic and Service oriented
	Threat Handling	Firewalls and Proxy Servers	Included with operational charges
Security Module	Authorized Access Control	Multiple purpose user identities	Included with operational charges

SDN [26] approach to backhaul MNO systems extends the support of LTE and LTE-A networks. Backhaul network pool and spectrum resources support uninterrupted QoS to potential, authentic, and licensed MEUNs. High-speed networks deploy micro and macro BS at various locations in mobile network. The spectral ensemble of current market trends [27] represents the existing business models for efficient bandwidth and fair connectivity with MEUNs. The spectrum access for mobile business model is given as follows:

(i) MNOs access the primary spectrum.
(ii) VNOs access the secondary spectrum through network enablers.
(iii) MEUNs function as ternary spectrum access providers.

MVNO reservation is based on license registration and regulation schemes subscribed to MNOs. A well-defined pricing scheme provides guaranteed and consistent services to MEUNs. The business approach in hierarchical spectrum market supports risk-return trade-offs and flexible pricing schemes to sustain the market growth from local and global competitors.

16.4   MVNO Developments in 3G, 4G, and LTE

Increasing competition in MNOs facilitates different pricing schemes for individual and group members based on content-oriented services. The limitations of MVNO business models are given as follows:

(i) Inaccessibility of existing cellular and Wi-Fi networks
(ii) Authentication process for active sessions
(iii) Nonflexible Mobile IP services to MEUNs
(iv) Sustainable services for high-speed mobile Internet.

This section elucidates various techniques to improve the MVNO' business model based on mobility support, multiple interfaces, and SDN approach to 3G, 4G, and LTE mobile networks.

16.4.1   MVNO User-Centric Strategies for Mobility Support

In MNO deployment phase, the infrastructure is configured with predefined set of rules between the vendor, bidder, and auctioneer. SDN metrics are configured to incorporate the updates received from MVNO. Negotiation schemes for object migration access [7] support resource flexibility and mobility services. Negotiated network resources meet the current requirements of MEUNs and estimate the future requirements. A prototype of middleware architecture is designed to support dynamic MEUNs with loosely coupled domains that function as the federation of grid resources.

Virtualizing personal and home networks [28] reveal the issues associated with Third Generation Partnership Project (3GPP), International Telecommunication Union (ITU), and Universal Mobile Telecommunications Systems (UMTS). Mobility in heterogeneous nodes establishes connectivity using pico, micro, and macro resources with increasing order of coverage and coordination levels. For a given application scenario, MEUNs can be loosely or tightly coupled between the home and foreign network. Technoeconomic evaluation [29] measures the net value of cash flows into the system over a period of time, and the cost evaluation model measures the current trends in mobile business market. Periodically, the discount rates offered to customers (for audio, video, Internet, etc.) and the rate of return are evaluated in this method. User-centric services [30] are the best connected services in the MVNO business model.

Virtual private mobile network operator (VPMNO) [31] is a three-phase functional model that ensures virtualization and subdivisive aspects of network management. VPMNO enhances the addressing mechanism and reduces the complexity in mobile network infrastructure. VPMNO further extends the functionality of MVNO business model by service replication and partitioning of MEUNs across MNOs. The unused bandwidth known as backbone bandwidth increases the business opportunities in MNO.

16.4.2   Management Schemes for Multiple Interfaces

Network resource manager (NRM) [32] allocates the bandwidth and manages services to meet the specified QoS configured by MNO. The business models are primarily categorized as follows:

(i) Radio resource management scheme that enables coverage services and network bandwidth
(ii) Service management scheme that meets the requirements of subscribed MEUNs with a wide range of service packages

Business management framework [33] uses incremental method from the physical layer (radio management) to the application layer (trusted services) to alter the decision-making process as per the demands of MEUNs. 3G MVNO [34] supports mobile bearer services such as 2G, UMTS, LTE, LTE-A, etc. in European markets. Network operators are categorized into rural and urban area networks based on the number of licensed MVNOs and the corresponding MEUNs associated with MVNOs. The business model with utility functions [35] estimates the gains of individual operations and determines the best utility package for MVNOs. MVNOs are grouped into clusters [36] with common aspects such as business development strategy, global scope, and application services. SDN-based spectrum allocation techniques dynamically distribute [37] the resources in MVNO business model. The collaboration between different stakeholders is primarily based on allocated channel frequency and rate of packet transmission. Due to increasing potential services of the Internet and LTE [38], mobile network virtualization explores multiplexing and multiuser diversity gain for a set of MVNOs. Wireless infrastructure [39] is configured with innovative business market models by using the MIMO-enabled MEUNs.

16.4.3   Enhancing Business Strategies Using SDN Approach

SDN configures the distributed virtual network and achieves price tariff trade-off for the services subscribed by MEUNs. Wide-area virtual service migration [40] supports adversaries in dynamic traffic conditions. VNO [41] categorizes the virtual components as follows:

(i) **Link Virtualizers**—Support virtual links that share the physical link. Virtual links define the tags (explicit and implicit) with allocated time slots and bandwidth.

(ii) **Node Virtualizers**—Differentiate the network protocols other than IP. Node virtualizers configure, manage, monitor, and resolve the network complexities in an active session.

MVNOs support multiple operating systems with common interface services and mapping procedures. Optimal prioritization of time schedulers and generic bandwidth framework [42] enhances the performance of MVNOs. SILUMOD [43] implements node mobility and ignores virtualization. The interface components support concurrent heterogeneous operators and subscribers. MEUNs elect a domain-specified language that exactly maps the mobility services in an active session. After initialization, the MEUNs invoke virtual operations that are controlled by VIRMANEL engine. This helps to locate the position of MEUNs in home or foreign network. The price tariffs defined by MNO services and their respective economic consequence on MVNO [44] are evaluated to estimate the performance of MEUNs in global market. This business model further optimizes the MNO service and increases the profit margins in local and global markets. VNO as cloud with data-centric services strengthens the resource pool spread across the network for voice and data services.

Operator migration for mobility-driven processing [45] extends the infrastructure of MVNO to support cloud computing and fog networking resources for end-to-end delivery services. This business model enhances the performance by supporting the network operators to use partial bandwidth during (a) peak traffic duration and (b) migration period. Spectrum sharing schemes and performance of MVNO are based on the pricing schemes [46] of the femtocell market.

16.5   Cognitive MVNO

16.5.1   Cognitive Radio Management in MVNOs

Cognitive radio management (CRM) resolves the functional differences in heterogeneous multiradio systems, coordinates with MVNOs, and efficiently utilizes the resources in multiuser subsystem. CRM comprises of tight coupling and coordination between the physical layers and software defined radio attributes of MVNO. MNOs use dynamic prioritizations to support multioperational schemes. Hence, multiagent MNOs are used for data sharing and resource allocation. Enhancements in MVNO design [47] support UMTS and WCDMA network with roaming in MEUNs and network consistency in heterogeneous systems. Based on the number of offered services, VNOs act as single or multiple carriers. High pricing schemes are invoked by dedicated carrier signals that serve multiple VNOs in rural and remote areas. Low pricing schemes are invoked by fixed capacity carrier signals that offer affordable pricing schemes to a large number of MEUNs. Tolerant levels of MVNO servicing scheme are a function of the number of active channels and service load. MVNO service rate ($MVNO_{Service\_Rate}$) for VNO is the ratio of load offered ($MVNO_{Load\_Offered}$) and channel capacity ($MVNO_{Channel\_Capacity}$). $MVNO_{Service\_Rate}$ is defined as follows:

$$MVNO_{Service\_Rate} = \frac{MVNO_{Load\_Offered}}{MVNO_{Channel\_Capacity}} \qquad (16.9)$$

MVNOs extend their consistent services and cooperate with other operators to form a grid of available resources. MNO accumulates the statistics of communication link and, if possible, reduces the network bandwidth. MVNO lookup table consists of the following attributes:

(i) Number of MEUNs associated with MVNO
(ii) Available resources
(iii) Network load
(iv) Available bandwidth
(v) Average energy required for maximum number of transactions.

New service requests from MEUNs can be accepted or rejected based on the service priority and network traffic conditions [48]. Cognitive MVNO (C-MVNO) [49] with OFDMA technique for downlink criteria supports the increasing demand of high-speed data transmission. Instant decisions based on available bandwidth, existing pricing schemes, and network traffic conditions are used in this model. The user requirements are broadly classified into service set, pricing schemes, and validity period. This approach becomes more realistic by monitoring the behavior of primary spectral management schemes (such as the spectral usage and time delays in packet transmissions) and decides the secondary spectral management schemes with proper pricing schemes. MEUN data packets that arrive at frame windows with overlapping time durations result in collision. This triggers a false alarm ($MVNO_{False\_Alarm\_Capacity}$) since the actual collision might have occurred in previous frame window that is much prior to the current transaction. The number of MVNO packets that were not detected ($MVNO_{Missed\_Detection}$) due to false triggering, increases the operational cost of MVNO. MVNO operational cost is defined as follows:

$$MVNO_{Operational\_Cost} = \begin{cases} high & when \left(MVNO_{False\_Alarm\_Capacity} \ \& \ MVNO_{Missed\_Detection} \right) \Rightarrow low \\ low & when \left(MVNO_{False\_Alarm\_Capacity} \ \& \ MVNO_{Missed\_Detection} \right) \Rightarrow high \end{cases}$$

(16.10)

This method certainly reduces the complexity as compared to the existing online pricing schemes in MVNO business market.

16.5.2   Cognitive and SDN-Based Spectral Allocation Strategies in MVNO

The price tariffs are assigned to MVNO with variations in radio spectrum allocation. Low-price tariff depends on wholesale prices offered by MNO and the density of MEUNs. High-price tariff is limited to bandwidth and used in high-speed commercial applications. MVNOs are adaptive to unattended market services and extend their services to widespread market. MVNO performance can be improved by using cognitive and SDN methodologies in interference mitigation and MNO switching. Based on the number of MEUNs and respective services, MNO partitions the allocated spectrum into standard and optimized services. The subscriber services are highlighted as follows:

(i) **Standard Subscriber Services (SSS)**: This service by MVNO includes allocation of premium channel bandwidth for short duration in low network traffic conditions. This service supports multimedia applications for MEUNs.

(ii) **Optimal Subscriber Services (OSS)**: This service by MVNO includes allocation of premium channel bandwidths for maximum time duration. Licensed MEUNs with allocated spectral resources are given high priority and achieve maximum throughput rate. The pricing tariffs for optimal services will be higher in licensed services as compared to unlicensed services.

Discrete MVNO service optimization (MSOP) is a function of:

(i) Authentic MEUNs associated to MNO
(ii) Set of services (SoS) enabled to MEUNs
(iii) Data rates open to MEUNs
(iv) Set of business relations to estimate the cost of service.

16.6   MVNO Business Strategies

MVNO business model comprises of generic services, services available at home domain, service requests by heterogeneous MEUNs, and additional service requests by MVNEs. MVNEs purchase the licensed spectrum from MNOs and coordinate the services with MVNOs.

Depending on the availability of MVNE services, appropriate tariff plans are facilitated to MVNOs and group MVNOs as illustrated in Figure 16.7. MVNOs implement business strategies based on market analysis, service requests by MEUNs, and degree of coordination between multiple MNOs. Table 16.3 highlights the MVNO service model and their respective features.

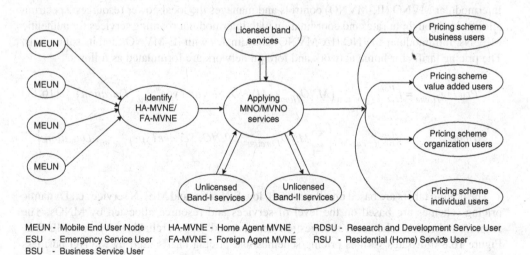

MEUN	- Mobile End User Node	HA-MVNE	- Home Agent MVNE	RDSU	- Research and Development Service User
ESU	- Emergency Service User	FA-MVNE	- Foreign Agent MVNE	RSU	- Residential (Home) Service User
BSU	- Business Service User				

Figure 16.7 Business model of mobile virtual network operator.

Table 16.3 MVNO service model and features

Service Model	Feature
Business Service Model	Business MVNO
	M2M MVNO
	Advertising MVNO
	Ethnic MVNO
Consumer Service Model	Value added services
	customer support
	Billing Process
	Flexible Packages
	Tariff Bundles and Packages
	Audio, video and text
Enhanced Service Model	Intelligent Network
	Next Generation Intelligent Network
	(voicemail, call forward, roam forward, VPN)
Application Service Model	Voice
	Data
	SMS
	Multimedia
User Supportive Service Model	Adaptable with GSM, CDMA, WiFi and WiMax technologies
	Time-based Tariff packages
	Group Tariff packages

16.6.1 Services and Pricing of MVNO

The emerging technology in WiMAX, Wi-Fi, WPAN, etc. significantly increases the role of MVNOs as trusted mediators. MVNOs offer intermodular and intramodular services. Intermodular MVNO (IE-MVNO) controls and manages the local power resources, spectrum allocation, throughput rate, and coordination with intermodular roaming services for authentic MEUNs. Intramodular MVNO (IA-MVNO) coordinates with IE-MVNOs and its subregions. The pricing tariffs for home network and foreign network are formulated as follows:

$$Pricing_{Home} = f_{Tariff\_Function}^{Home} \left(MNO_{Home}, MVNO, ServiceType_{Home}, Duration \right) \quad (16.11)$$

$$Pricing_{Foreign} = g_{Tariff\_Function}^{Foreign} \left(\sum_{i=1}^{n} MNO_{Foreign}^{i}, MVNO, ServiceType_{Foreign}, Duration \right)$$
$$(16.12)$$

The pricing tariffs are based on offered load, low S/N ratio, and MEUN service set. Dynamic pricing schemes are based on the level of services and resource allocated by MNOs. The resource utilization schemes are congestion sensitive and support reliable services to MEUNs. Figure 16.8 illustrates various pricing schemes and pricing tariffs used by MVNO business model. The pricing schemes are categorized as long term and short term, and the pricing tariffs vary with service set package across the MVNO business market.

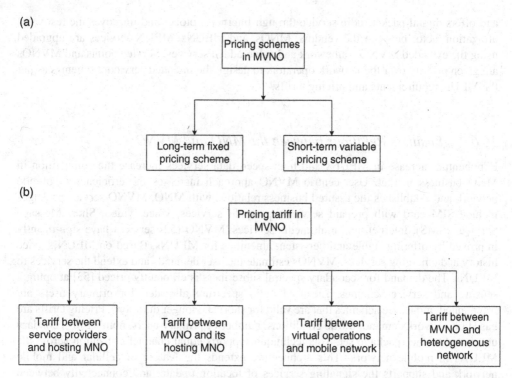

Figure 16.8 Pricing schemes and tariffs in MVNO business market. (a) Pricing schemes and (b) Pricing tariffs.

16.6.2 Resource Negotiation and Pricing

Resource negotiation is a function of number of hops, QoS within each hop, and degree of congestion. In Resource Negotiation and Pricing (RNAP) scheme [50], the MEUNs negotiate with MNOs and select (i) the resource set and (ii) the corresponding service set. Long-term fixed pricing scheme does not support dynamic resource allocation, and it leads to resource underutilization during abrupt and random traffic conditions. The attributes of RNAP pricing tariffs are: (i) MEUN set of services, (ii) time duration, and (iii) allocated spectrum bandwidth. The resources are distributed across MEUNs for high-speed audio, video, and Internet applications. In resource-constraint conditions, the video applications are given low priority because of inconsistent and intermittent data streaming conditions.

16.6.3 Pushover Cellular and Service Adoption Strategy

MENO supports voice services that can be extended to video conferencing using 3G, LTE-A, and 4G cellular networks. Military, marine, railway, and airline services use push-to-talk (PTT) services that function in full-duplex mode. This identifies the intruders in restricted areas like forest, valley, underground, underwater, etc. PTT uses very-high-frequency (VHF) signals to trace the recipient radio. Pushover cellular (PoC) [51] provides a standard platform and offers digital packet radio service through Internet protocol and improves the resources utilization factor between the vendors, MNOs, and MEUNs. MEUN services are upgraded using the extended MVNO framework and value-added services. Service points and MVNOs are given priority over the network operators to design the intended servicing schemes as per the MEUN requirements and pricing tariffs.

16.6.4 Business Relations between the MNO and MVNO

Exponential increase in MEUNs and high-speed data services increase the competition in MNO business market. User-centric MVNO approach increases the efficiency in mobile network and establishes the assured business relations with MNO. MVNO service packages include SIM card with prepaid services, postpaid services, voice, video, Short Message Services (SMS), Internet, and multimedia services. MVNO [52] services have significantly improved by offering stable and persistent amenities for MEUNs. Based on MEUN service history and emerging services, MVNOs estimate the user demands and extend the services to MEUNs. The demand for secondary spectral subscribers is efficiently priced [53] at optimal sensing and service schemes. Licensed radio spectrum allocated for primary users are fragmented into leased schemes that are valid for short and longer durations. Pricing tariffs are based on network demand, traffic conditions, number of primary users, number of secondary users, and service package set. This technique supports equal channel utilization across the MEUNs in mobile network. This architecture extends the feature of cellular and mobile network and supports the signaling services of location update, and connectivity between MEUNs and MNOs. Return on investment (ROI) [54] utilizes the unused shared spectrum region and enhances the resource utilization factor for existing MEUNs.

MVNOs reduce the risks in business support system [55] (flat and horizontal architectures) and accommodate high-speed network services in LTE-A and 4G systems. The emerging service agreements are classified and addressed by the cloud system. MVNO-connected [56] services are classified into full mode, multimode, and best connected mode based on packet switching, voice over LTE, and Evolved Packet Services (EPS). Business relation between the primary and secondary license holders in shared radio spectrum [57] enhances the services of existing economic policies. Marketing model is based on the service mitigation schemes between the bidders, the auctioneers, and the buyers. In the first phase, the bidding process is allowed by MVNOs, and in the second phase, the bidder fixes the unit price and assigns it to MVNOs. The business model emphasizes the following: (i) rational behavior between MVNOs, (ii) data distribution between MVNOs, and (iii) data integrity. The business model analyzes the revenue gain between the initial investment and final returns for primary and secondary radio spectrum services.

16.7 Conclusions

MVNOs are the primary contenders in the existing industry of mobile communication network. Due to available network resources and advanced infrastructure, MVNOs facilitate high-speed mobile services with reliable user-centric services at low-price tariffs as compared to MNOs and service providers. MVNOs adapt business strategies to extend the basic bearer services from GSM and UMTS to the latest LTE-A and 4G networks. Cognitive and SDN techniques in MVNO business models enhance the services of MNO, MVNE, and MVNA. This approach further facilitates extended services to MEUNs. This chapter elucidates mobile market policies and business relations between MNOs and VNOs. For enhanced services, MVNO business models use localized resource relocations such as the femto, pico, and micro BS for small-scale and large-scale mobile networks.

16.8   Future Directions

Future directions include enhance MVNO services that comprise of research development toward: (i) fault-tolerant MVNO (failure at one MVNO can disrupt the transactions of other MVNOs), (ii) time slag (the processing conditions and operational differences between heterogeneous MVNOs), (iii) cochannel interference with other broadband services, (iv) resource allocation and management in dynamic traffic conditions, (v) violations of pre-defined business contracts by MVNOs, and (vi) conflicts ascended across MVNOs.

References

[1] Svein U, Mobile Virtual Network Operators: A strategic transaction cost analysis of preliminary experiences, Elsevier Journal of Telecommunications Policy, October–November 2002, 26(9–10), pp. 537–549.

[2] Marcus B, Bernhard P and Rolf S, Establishing a framework allowing customers to run their own customized services over a provider's network, Communications of the ACM, April 2001, 44(4), pp. 55–61.

[3] Jyh-Cheng C, Jui-Hung Y, Yi-Wen L, Li-Wei L, Fu-Cheng C and Shao-Hsiu H, RAMP: Reconfigurable Architecture and Mobility Platform, In Proceedings of IEEE International Global Telecommunications Conference (Globecom), St. Louis, MO, USA, November 28–December 2, 2005, pp. 3564–3569.

[4] Kostas P, Yan W and Weihua H, MobileFlow: Toward software-defined mobile networks, IEEE Communications Magazine, July 2013, 51(7), pp. 44–53.

[5] Bartlett A and Jackson N N, Network planning considerations for network sharing in UMTS, In Proceedings of IET Third International Conference on 3G Mobile Communication Technologies, London, UK, May 8–10, 2002, pp. 17–21.

[6] Francisco B, Josep P, Fofy S and Monique G, Design and modelling of internode: A mobile provider provisioned VPN, Springer Journal of Mobile Networks & Applications, February 2003, 8(1), pp. 51–60.

[7] Peter K, Kris B and Kyle C, Enabling Virtual Organization in Mobile Worlds, In Proceedings of IEE Fifth International Conference on 3G Mobile Communication Technologies, Savoy Place, London, UK, October 18–20, 2004, pp. 123–127.

[8] Johan H, Klas J and Jan M, Business models and resource management for shared wireless networks, In Proceedings of IEEE Sixth Vehicular Technology Conference (VTC-Fall), Los Angeles, CA, USA, September 26–29, 2004, pp. 3393–3397.

[9] Park J S and Rye K S, Developing MVNO Market Scenarios and Strategies through a Scenario Planning Approach, In Proceedings of IEEE Seventh International Conference on Advanced Communication Technology (ICACT), Phoenix Park, Gangwon-Do, South Korea, February 21–23, 2005, pp. 137–142.

[10] Akyildiz Ian F, Won-Yeol L, Mehmet C. V and Shantidev M, NeXt generation/dynamic spectrum access/cognitive radio wireless networks: A survey, Elsevier Journal of Computer Networks, September 15, 2006, 50(13), pp. 2127–2159.

[11] Investelecom Inc., Technical Document, Innovation Services: Mobile Virtual Network Operator, Mobile Virtual Network Enabler—Strategy and Marketing, United Arab Emirates, 2009, pp. 1–8. Available at http://www.investele.com/documents/Mobile Virtual Network Operator.pdf (accessed February 17, 2015).

[12] Krzysztof K, White paper: How to Become an MVNO/MVNE www.mvnodynamics.com/wp-content/uploads/2011/05/whitepaper-howtobecomeanmvnoormvne-091119074425-phpapp02.pdf, Comarch: White paper Telecommunications, Comarch Headquarters, Poland, 2009, pp. 1–15.

[13] Jarmo H, Renjish Kumar K R, Thor Gunnar E, Rima V, Dimitris K and Dimitris V, Techno-economic evaluation of 3G and beyond mobile business alternatives, Springer Journal of Netnomics: Economic Research and Electronic Networking, October 2007, 8(1–2), pp. 5–23.

[14] Marc C, Interactions between a Mobile Virtual Network Operator and External Networks with regard to Service Triggering, In Proceedings of IEEE Sixth International Conference on Networking (ICN), Martinique, April 22–28, 2007, pp. 1–7.

[15] Philip K and Lars W, On the competitive effects of mobile virtual network operators, Elsevier Journal of Telecommunications Policy, June–July 2010, 34(5–6), pp. 262–269.

[16] Timo S, Annukka K and Heikki H, Virtual operators in the mobile industry: a techno-economic analysis, Springer Journal of Netnomics: Economic Research and Electronic Networking, October 2007, 8(1–2), pp. 25–48.

[17] Helene Le C and Mustapha B, Modelling MNO and MVNO's dynamic interconnection relations: Is cooperative content investment profitable for both providers?, Springer Journal of Telecommunication Systems, November 2012, 51(2–3), pp. 193–217.

[18] Aniruddha B and Christian M. D, Voluntary relationships among mobile network operators and mobile virtual network operators: An economic explanation, Elsevier Journal of Information Economics and Policy, February 2009, 21(1), pp. 72–84.

[19] Siew-Lee H and Langford B W, Cooperative resource allocation games in shared networks: symmetric and asymmetric fair bargaining models, IEEE Transactions on Wireless Communications, November 2008, 7(11), pp. 4166–4175.

[20] Imen Limam B, Omar C and Guy P, Performance Characterization of Signaling Traffic in UMTS Virtualized Network, In Proceedings of IEEE Global Information Infrastructure Symposium (GIIS), Hammemet, Tunisia, June 23–26, 2009, pp. 1–8.

[21] Nick F, Jennifer R and Ellen Z, The road to SDN: An intellectual history of programmable networks, ACM SIGCOMM Computer Communication Review, April 2014, 44(2), pp. 87–98.

[22] Gautam B, Ivan S and Dipankar R, A virtualization architecture for mobile WiMAX networks, ACM SIGMOBILE Mobile Computing and Communications Review, March 2012, 15(4), pp. 26–37.

[23] Peter D, Jeroen H, Ingrid M, Joris M and Piet D, Network virtualization as an integrated solution for emergency communication, Springer Journal of Telecommunication Systems, April 2013, 52(4), pp. 1859–1876.

[24] Key P and Yvonne H, Mobile payment in the smartphone age—extending the Mobile Payment Reference Model with non-traditional revenue streams, In Proceedings of ACM Tenth International Conference on Advances in Mobile Computing & Multimedia (MoMM), Bali, Indonesia, December 3–5, 2012, pp. 31–38.

[25] Joshua H, Lance H and Suman B, Policy-Based Network Management for Generalized Vehicle-To-Internet Connectivity, In Proceedings of ACM SIGCOMM Workshop on Cellular Networks: operations, challenges, and future design (CellNet), Helsinki, Finland, August 13, 2012, pp. 37–42.

[26] Dejan B, Eisaku S, Neda C, Ting W, Junichiro K, Johannes L, Stefan S, Hiroyasu I and Shinya N, Advanced Wireless and Optical Technologies for Small-Cell Mobile Backhaul with Dynamic Software-Defined Management, IEEE Communications Magazine, September 2013, 51(9), pp. 86–93.

[27] Pablo J C D C, Aparna G and Koushik K, Hierarchical Spectrum Market and the Design of Contracts for Mobile Providers, ACM SIGMOBILE Mobile Computing and Communications Review, October 2013, 17(4), pp. 60–71.

[28] Fawzi D and Seshadri M, Strategies for provisioning and operating VHE services in multi-access networks, IEEE Communications Magazine, January 2002, 40(1), pp. 78–88.

[29] Olsen B T, Katsianis D, Varoutas D, Stordahl K, Harno J, Elnegaard N K, Welling I, Loizillon F, Monath T, Cadro P. Technoeconomic Evaluation of the Major Telecommunication Investment Options for European Players, IEEE Network Magazine, July–August 2006 20(4), pp. 6–15.

[30] De Leon M P and Adhikari A, A user centric always best connected service business model for MVNOs, In Proceedings of IEEE Fourteenth International Conference on Intelligence in Next Generation Networks (ICIN), Berlin, Germany, October 11–14, 2010 pp. 1–8.

[31] Arati B, Xu C, Baris C, Gustavo de los R, Seungjoon L, Suhas M, Jacobus E, and Van der M, VPMN: virtual private mobile network towards mobility-as-a-service, In Proceedings of ACM Second International Workshop on Mobile Cloud Computing and Services, Washington, DC, USA, June 28, 2011,

[32] Jenq-Shiou L and Chuan-Ken L, On utilization efficiency of backbone bandwidth for a heterogeneous wireless network operator, Springer Journal of Wireless Networks, October 2011, 17(7), pp. 1595–1604.

[33] Hiram G-Z, Javier R-L, Pablo S-L, Ramon A, Joan S and Steven D, A business-oriented management framework for mobile communication systems, Springer Journal of Mobile Networks & Applications, August 2012, 17(4), pp. 479–491.

[34] Varoutas D, Katsianis D, Sphicopoulos Th, Stordahl K and Welling I, On the Economics of 3G Mobile Virtual Network Operators (MVNOs), Springer Journal of Wireless Personal Communications, January 2006, 36(2), pp. 129–142.

[35] Imen Limam B, Omar C and Guy P, Third-generation virtualized architecture for the MVNO context, Springer Journal of Annals of Telecommunications, June 2009, 64(5–6), pp. 339–347.

[36] Dong Hee S, Overlay networks in the West and the East: a techno-economic analysis of mobile virtual network operators, Springer Journal of Telecommunication Systems, April 2008, 37(4), pp. 157–168.

[37] Manzoor A K, Hamidou T, Fikret S, Sahin A and Barbara U K, User QoE influenced spectrum trade, resource allocation, and network selection, Springer Journal of International Journal of Wireless Information Networks, December 2011, 18(4), pp. 193–209.

[38] Yasir Z, Liang Z, Carmelita G and Andreas TG, LTE mobile network virtualization Exploiting multiplexing and multi-user diversity gain, Springer Journal of Mobile Networks & Applications, August 2011, 16(4), pp. 424–432.

[39] Kok-KiongY, Rob S, Masayoshi K, Te-Yuan H, Michael C, Nikhil H, Nick M and Guru P, Blueprint for Introducing Yuan H, Michael C, Nikhil H, Nick M and Guru P, Blueprint for Introducing infrastructure systems and architectures (VISA), New Delhi, India, September 3, 2010, pp. 28–32

[40] Bienkowski M, Feldmann A, Grassler J, Schaffrath G and Schmid S, The wide-area virtual service migration problem: A competitive analysis approach, IEEE/ACM Transactions on Networking, February 2014, 22(1), pp. 165–178.

[41] Jorge C and Javier J, Network Virtualization-A View from the Bottom, In Proceedings of ACM First International Workshop on Virtualized infrastructure systems and architectures (VISA), Barcelona, Spain, August 17, 2009, pp. 73–80.

[42] Ravi K, Rajesh M, Honghai Z and Sampath R, NVS: A Virtualization Substrate for WiMAX Networks, In Proceedings of ACM Sixteenth Annual International Conference on Mobile Computing and Networking (MobiCom), Chicago, IL, USA, September 20–24, 2010, pp. 233–244.

[43] Yacine B and Claude C, VIRMANEL: A Mobile Multihop Network Virtualization Tool, In Proceedings of ACM Seventh International workshop on Wireless network testbeds, experimental evaluation and characterization (WiNTECH), Istanbul, Turkey, August 22, 2012, pp. 67–74.

[44] Jeremy B, Rade S, Vijay E, Adriana I and Konstantina P, Last Call for the Buffet: Economics of Cellular Networks, In Proceedings of ACM Nineteenth Annual International Conference on Mobile Computing & Networking (MobiCom), Miami, FL, USA, September 30–October 4, 2013, pp. 111–121.

[45] Beate O, Boris K, Kurt R and Umakishore R, MigCEP: Operator Migration for Mobility Driven Distributed Complex Event Processing, In Proceedings of ACM Seventh International Conference on Distributed Event-Based Systems (DEBS), Arlington, TX, USA, June 29–July 3, 2013, pp. 183–194.

[46] Shaolei R, Jaeok P and Mihaela van der S, Entry and spectrum sharing scheme selection in femtocell communications markets, IEEE/ACM Transactions on Networking, February 2013, 21(1), pp. 218–232.

[47] Johansson K, Kristensson M and Schwarz U, Radio Resource Management in Roaming Based Multi-Operator WCDMA Networks, In Proceedings of IEEE Fifty Ninth Vehicular Technology Conference (VTC-Spring), vol. 4, Milan, Italy, May 17–19, 2004, pp. 2062–2066.

[48] Jiang X, Ivan H and Anita R, Cognitive Radio Resource Management Using Multi-Agent Systems, In Proceedings of IEEE Fourth International Consumer Communications and Networking Conference (CCNC), Las Vegas, NV, USA, January 11–13, 2007, pp. 1123–1127.

[49] Shuqin L, Jianwei H and Shuo-Yen Robert L, Dynamic profit maximization of cognitive mobile virtual network operator, IEEE Transactions on Mobile Computing, March 2014, 13(3), pp. 526–540.

[50] Xin W and Henning S, Pricing network resources for adaptive applications, IEEE/ACM Transactions on Networking, June 2006, 14(3), pp. 506–519.

[51] Timo A V and Sakari L, Service adoption strategies of push over cellular, Springer Journal of Personal and Ubiquitous Computing, January 2008, 12(1), pp. 35–44.

[52] Dong-Hee S, MVNO services: Policy implications for promoting MVNO diffusion, Elsevier Journal of Telecommunications Policy, November 2010, 34(10), pp. 616–632.

[53] Lingjie D, Jianwei H and Biying S, Investment and pricing with spectrum uncertainty: A cognitive operator's perspective, IEEE Transactions on Mobile Computing, November 2011, 10(11), pp. 1590–1604.

[54] Ashiq K, Wolfgang K, Kazuyuki K and Masami Y, Network sharing in the next mobile network: TCO reduction, management flexibility, and operational independence, IEEE Communications Magazine, October 2011, 49(10), pp. 134–142.

[55] Raivio Y and Dave R, Cloud Computing in Mobile Networks—Case MVNO, In Proceedings of IEEE Fifteenth International Conference on Intelligence in Next Generation Networks (ICIN), October 4–7, 2011, pp. 253–258.

[56] Rebecca C and Noel C, Modelling Multi-MNO Business for MVNOs in their Evolution to LTE, VoLTE & Advanced Policy, In Proceedings of IEEE Fifteenth International Conference on Intelligence in Next Generation Networks (ICIN), October 4–7, 2011, pp. 295–300.

[57] Shun-Cheng Z, Shi-Chung C, Peter B. L and Hao-Huai L, Truthful auction mechanism design for short-interval secondary spectrum access market, IEEE Transactions on Wireless Communications, March 2014, 13(3), pp. 1471–1481.

Part V
Security and Economic Aspects

17

Software Defined Mobile Network Security

Ahmed Bux Abro
VMware, Palo Alto, CA, USA

17.1 Introduction

Telecommunication has transformed rapidly into Infocommunication [1] in a short period and has forced us to digitalize our life and further derived changes for how we communicate, entertain, work, and socialize with other human beings.

This transformation to digital life has brought many opportunities as well as offers great challenges for how to protect our valued data, keep our privacy, and secure our Info-communication networks that are used to provide access to billions of users, devices, and things. Cyberattackers have frequently challenged the telecommunication industry and put company's reputation at risk. Increased use of mobile data has also introduced new security challenges and threat vectors.

Enhanced visibility, intelligence, and network wide control are required to protect from such threats while making sure mobile services are not compromised and always on.

Available vulnerabilities in an all-Internet-Protocol (IP)-based mobile network have made it easy for the attackers to use it as an attack Launchpad. Traditional security models were used to apply security on selected domains and place in the network (PIN) while keeping the rest of the network wide open.

Software defined mobile network (SDMN) has the potential to leverage the network as a tool to provide enhanced visibility, converged intelligence, centralized policy control, and real-time threat mitigation. Traditional security models may not address the needs for SDMN and the next generation of the mobile network. We need to think out of the box by leaving bolt-on security model and develop an inclusive and intrinsic security model across the mobile network.

Software Defined Mobile Networks (SDMN): Beyond LTE Network Architecture, First Edition.
Edited by Madhusanka Liyanage, Andrei Gurtov, and Mika Ylianttila.
© 2015 John Wiley & Sons, Ltd. Published 2015 by John Wiley & Sons, Ltd.

17.2   Evolving Threat Landscape for Mobile Networks

Phone hacking (also called as phreaking) was first spotted somewhere between the 1960s and 1970s when phreakers demonstrated their skills to manipulate the functions of a telephone network. Methods to attack telecommunication systems had evolved since then and have changed its shapes from war dialers to viruses, to worms, and to modern-day advance persistent threats (APTs). Tools to protect our telecommunication systems have also evolved from physical access control to antivirus to modern application and context-aware firewalls.

Increased use of smartphones for data services and applications has exposed these devices to the same security threats that were once known and dedicated to personal computers (PCs). Mobile devices have replaced legacy system and have changed our ways to learn, work, entertain, shop, and travel. Bring your own device (BYOD) and cloud technologies have further diminished the enterprise boundaries and often challenged security experts to work out-of-the-box strategies.

Motivations for attacking networks have also changed from fun-loving immature script kiddies to organized cybercrime rings and hacktivists with clear political and financial objectives. In this age of digitalization, after connecting humans using the Internet and mobile, we are talking about connecting things and machines. Mobile has not yet completely replaced the PC but has become an ideal place where personal information can be found for nefarious use.

Security needs to be architected to not only protect from the current threats but to address the increased and evolving threat landscape. Adequate security should include threat intelligence, visibility, and real-time protection.

17.3   Traditional Ways to Cope with Security
Threats in Mobile Networks

Mobile network has always been the target for security attacks. Legacy techniques to protect mobile networks from ongoing security threats are nothing but introducing new and unplanned security control for selected network segments, focusing perimeter security while leaving inside networks wide open, and finally building complex security systems that later cause impact to overall network operations and performance.

These techniques on one side were able to address selected threats but proved to be less efficient for new and advance threats where hackers follow a systematic and collaborative approach. The following figure demonstrate an ongoing attack or an attack vector that can consist of multiple attacks and what traditional security tools are used by a security admin to protect the mobile network. We can see how a collaborative attack can cause a network wide impact, while the security tools placed for selected PINs can be limited and less efficient to protect the network.

Let us review some of these legacy security tools and techniques that are applied in mobile network today (Fig. 17.1).

17.3.1   Introducing New Controls

Security is often deployed in a reactive manner. Separate security controls are deployed on various PINs to protect the mobile network and its services. Mobile network security includes various security controls such as firewalls, packet filters, network address translation (NAT),

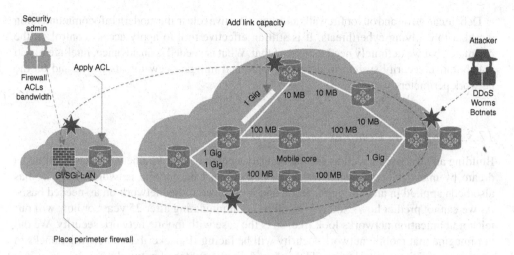

Figure 17.1 Mobile IP Core and Gi/SGi Network attack simulation and response.

packet inspections, antivirus, etc. Managing such controls remains a challenge and requires expensive technical resource involvement.

Most of the security controls are placed in a distributed fashion with a limited territory to protect. One of the major drawbacks of these security controls is that they all work in silos and offer close to none collaboration. A holistic and centralized control system is hard to imagine in today's mobile networks.

Software defined network (SDN) technology centralizes the network control plane and signaling using "controller" software. The controller sits in a central place and all network nodes communicate with and share network state and flow information with that controller. The SDN model also offers centralized policy and visibility to the entire network.

17.3.2 Securing Perimeter

Perimeter is an edge or a boundary where a mobile network can interface with an external network that can be another mobile or data network. Security for a mobile network has been most of the time heavily focused around securing the perimeter as it has been considered the most vulnerable point of the network. Access control is the major security tool to protect the mobile network perimeter. Perimeter security has traditionally been applied by placing network firewalls or packet filters that use IP address and protocol information to filter access to network. Perimeter firewalls or packet filters are usually placed on SGi and S8 (partner facing) interfaces.

Firewalls have served decades as the preferred technology to protect against network attacks on the perimeter and inside networks, but we have seen this technology as limited against the new era of coordinated and advance persistent threats.

Firewalls have proved to be successful in controlling traffic and protecting from certain network attacks, but it brings its own limitations, such as limited visibility and point protection.

Debate can go on and on for firewalls and their effective role in the modern Infocommunication world with evolving cyberthreats. It is still an effective tool to apply access control on the perimeter, but we definitely need more than that. What is needed is an advance, intelligent, and collaborative security system to detect, protect, and mitigate new threats inside and on the network perimeters.

17.3.3 Building Complex Security Systems

Building an effective as well as simple security system in a mobile network has remained a dream by many telecommunication security experts. Like mobile networks, security has also been applied in an evolving method and introduced to the network on as-needed basis. As we cannot predict how exactly we will be communicating after 25 years or how will our telecommunication networks look like, so is the case with mobile network security. We did not imagine that mobile network security will be facing IP packet-based network attacks in once time-division multiplexing (TDM)-based circuit-switched networks.

Some of the major factors driving complex security systems in telecommunication networks are:

- Interworking of various legacy (2G, 2.5G, and 3G) and new Long-Term Evolution (LTE) systems
- Convergence of voice, video, data, and other services
- Evolution of IP end-to-end network.

17.3.4 Throwing More Bandwidth

IP Core of the mobile network is used to provide faster backbone access between Evolved Packet Core (EPC) and other mobile network segments. One of the many schools of thoughts for security experts follows the strategy to keep the IP Core network clear from any security technology as this may impact the speed to delivery packets. We will not go into a debate on which school is better, but we will discuss the possible impact of one approach over the other.

The traditional method used by the operators to protect the core of the mobile network (also called mobile core) was to oversize the network capacity and overpopulate the resources to avoid services disruption.

That technique had proved successful in some situations, but it had its own drawbacks from a cost perspective and proved to be a short-term resolution.

17.4 Principles of Adequate Security for Mobile Network

Mobile networks need to adopt a security model that is not just based out of the common confidentiality, integrity, and availability (CIA) triad but also extended to address new security principles for centralized policy and enhanced visibility to offer better security and protect customer data (Fig. 17.2).

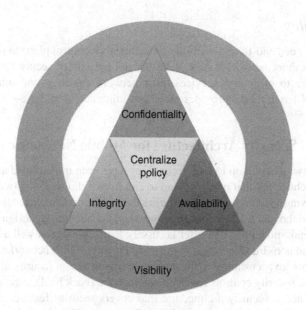

Figure 17.2 Mobile network security model.

17.4.1 Confidentiality

Confidentiality is to make sure that the least privileges are assigned and access is controlled for authorized applications and clients while denying any unauthorized access request. Principle needs to be equally applied on northbound and southbound traffic.

17.4.2 Integrity

It is critical to ensure that the information is not tempered and removed and remains integrated while transferred between different points inside the mobile network. Network can be compromised and data integrity can be lost if required security controls are not implemented on each layer.

17.4.3 Availability

Availability principle assures that the network components, services, and information are available as and when needed. Most of the mobile network offers service-level agreement (SLA) with 99.999% availability. Attacks such as distributed denial of service (DDoS) can cause damage to services and limited services for legitimate mobile users.

17.4.4 Centralized Policy

Centralized policy management and enforcement make it easy to control access to the network resources, services, and applications. It also helps to organize, manage, and associate security policies in a central location.

17.4.5   Visibility

Mobile networks need end-to-end visibility to the network control plane to monitor, optimize, and better troubleshoot network issues. Visibility not only helps secure the environment but also profile traffic to offer new services, plan network capacity, and introduce analytical capabilities. SDN-based model provides global visibility across all base stations.

17.5   Typical Security Architecture for Mobile Networks

Mobile networks with evolution to LTE technology have been transformed and built on all-IP-based network architecture that means IP end to end from radio access network (RAN) to core to data center. Having a flat architecture simplifies the mobile network but at the risk of increased vulnerabilities and threats. A flat architecture introduces challenges to design an effective security defense in depth model that can offer necessary fault isolation as well as protect the interwork of legacy and non-3rd Generation Partnership Project (3GPP) networks.

Mobile operators are accustomed to the traditional way to design security in a reactive manner by having separate security controls for RAN, aggregation (back haul), core, and data center.

3GPP has defined a security architecture that covers security features, mechanisms, and procedures for Evolved Packet System (EPS), EPC, and E-UTRAN. A detailed document for that security architecture can be found in the Refs. [2, 3]. Mobile network operators use this architecture to build a security layer around the access network, EPS/EPC, user authentication, application security, and finally security configuration.

The 3GPP security architecture (Fig. 17.3) is divided into five domains as below.

- Network Access Security (I): Defines security for users using USIM to securely access EPC resources and further protect RAN against various attacks
- Network Domain Security (II): Provides security features to offer secure communication over a wired network between different EPC nodes to protect user and signaling data

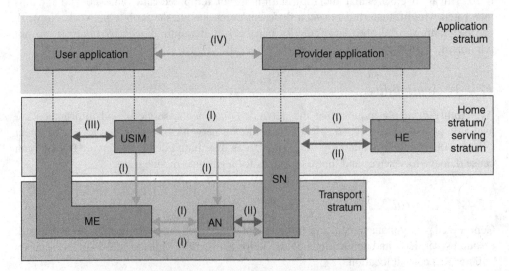

Figure 17.3   3GPP security architecture.

- User Domain Security (III): Covers the mutual authentication between a user equipment (UE) using USIM and mobile equipment (ME)
- Application Domain Security (IV): Includes necessary security features to protect user and provider application communication with the rest of the network
- Visibility and Configurability of Security (V): Offers users the visibility to their current security posture.

The 3GPP architecture covers security for different network types such as RAN and EPC network, secure communication for inter- and intranetwork nodes, and security for various interfaces (S1, S8, SGi, etc.) Inside mobile network is usually categorized into three domains:

- S1 Interface Security: To protect RAN-to-EPC communication
- SGi Interface Security: To protect the Internet facing links and interfaces
- S8 Security: To protect partner facing interface for secure roaming.

Various security controls are applied in this architecture to secure these interfaces (Fig. 17.4). The following figure maps the security controls with different PINs.

As seen in the above figure, security controls to protect these interface types are usually based on:

- Firewalls: Separate interface-specific firewall is installed for each domain, that is, S1 firewall installed between RAN and EPC to protect GPRS Tunneling Protocol (GTP) communication, a Gi firewall deployed near the Internet border to protect from Internet threats, and S8 firewall installed to protect communication with the roaming partners.
- Authentication and Authorization: Strong authentication and authorization are applied for the S1 interface to ensure only authorized evolved nodeB (eNB) is allowed access to the packet core network. Authentication and key management area has been well addressed by 3GPP by incorporating Authentication and Key Agreement (AKA). AKA offers mutual authentication along with integrity protection. Its variations are used to authenticate trusted non-3GPP clients. For trusted non-3GPP access, Extensible Authentication Protocol AKA (EAP-AKA) can be used for authentication.
- Encryption/IPSec: Encryption is applied to protect RAN-to-EPC communication and to ensure the network is protected from rogue and insecure eNBs.
- NAT: This is used to hide network core addresses while interfacing with the outside and partner world.
- Malware and Antivirus Protection: This is used to protect the network from viruses, worms, and other threats.

The abovementioned typical security architecture and security controls offers its own security benefits but has some limitations as well:

17.5.1   Pros

- Domain-specific security
- Strong authentication for UE
- Protects vulnerable points in the network.

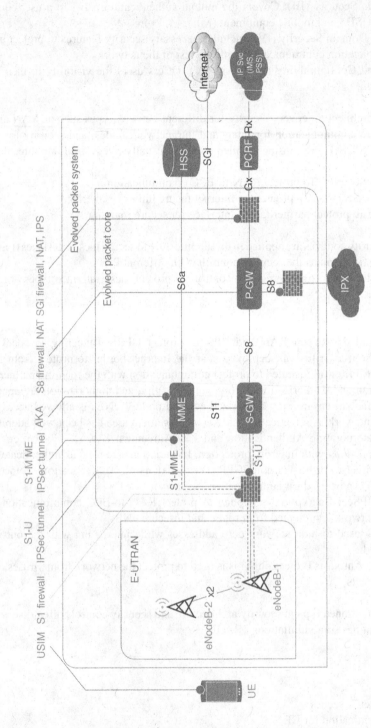

Figure 17.4 Typical security architecture for mobile network.

17.5.2   Cons

* Complex and fragmented security model
* Limited visibility
* Decentralize control
* Distributed security
* Lack of collaborative security model.

17.6   Enhanced Security for SDMN

An all-IP-based mobile network architecture has exposed us to new IP-centric threats on all layers of a mobile network. Threats such as IP spoofing, man-in-the-middle attack, and DoS have increased probability of attack success in this new environment.

At the time of the writing of this book, SDN is not fully embraced by the major telecommunication operators, but we are seeing a trend toward SDN and we also witness telecommunication operators starting to adopt cloud, orchestration, and virtualization technologies that are enablers for SDMN. SDN standardization organizations such as the Open Network Foundation (ONF) have dedicated groups for wireless and mobile network that are actively working on SDMN use cases and standards.

A traditional security model may not work with SDMN and may require consideration of an integrated security architecture. Security should not be restricted to selected components but should be equally applied to all layers of an SDMN architecture such as infrastructure, SDN, management, orchestration, automation, and applications.

A security architectural approach for SDMN will help gain better visibility and control. Security for individual layers is discussed in the following sections.

17.6.1   Securing SDN Controller

The SDN controller is an important component of the SDMN architecture and requires necessary security hardening to protect it from any threat that can affect its availability. Hackers can leverage its position of the central control and visibility for their nefarious objectives. Gaining unauthorized access to control can enable hacker to manipulate network functions, capture packets, divert traffic, and misuse the network functions.

A controller is usually installed on the top of an operating system (OS) platform such as Linux. Like any other OS, we can harden the underlying OS for the controller by installing necessary patches and fixes, enabling role-based access control (RBAC), enabling accounting and logging, and disabling unnecessary services, ports, and protocols.

A controller software usually ships with basic security management protocols such as SSH, HTTPS, and a proper RBAC to manage controller resources.

17.6.2   Securing Infrastructure/Data Center

SDMN separates the control and data plane. The control plane is centralized in a controller software, while the data plane resides on hardware devices such as Serving Gateway (S-GW) and PDN Gateway (P-GW) or network components such as routers and switches. Depending on what SDN model (basic SDN, hybrid SDN, or full SDN) is used in the environment, necessary security controls can be enabled on the infrastructure devices that are controlled by the SDN. It is common to keep the management plane and data plane function on the device, while the control plane function can be placed on box or off box per the SDN model.

Security controls for infrastructure supporting SDMN include authentication, authorization, and accounting (AAA), secure management protocols, logging, and monitoring controls. Data plane-specific security controls such as port security, access control lists, and private VLANs can be enabled to protect the data plane.

17.6.3   Application Security

SDMN offers an open interface for software applications to call or manage different control plane functions; such interfaces are called Application Programming Interfaces (APIs).

It is required to ensure that applications accessing the SDMN environment are authenticated using a digitally signed code and certification process. Such applications should be developed following secure application development lifecycle and principles of least privilege and fail safe and tested against possible threats such as buffer overflow and resource leakage. A code analysis is performed to ensure applications are secure.

17.6.4   Securing Management and Orchestration

Management and orchestration are the key components of the SDMN architecture, and it is required to apply necessary security controls to protect these components. It is preferred to put management and orchestration in a secure zone that is protected by a firewall and a proper role RBAC system is in place to ensure least privilege and access to authorize users and to monitor and account activities.

17.6.5   Securing API and Communication

As discussed previously, SDMN offers an open interface for applications to call or manage control plane functions; such interface is also called as an API. Access to an API needs to be properly authenticated and authorized. API access also needs to be monitored and revoked as needed. Encryption is required when an open communication channel is used to send and receive API access requests.

17.6.6   Security Technologies

The SDMN environment can be further protected through physical or virtual security technologies such as firewalls, security gateways, deep packet inspections, and intrusion prevention systems.

SDMN can also be used to provision these security technologies as a service for customer tenants or create security service chain for mobile applications.

17.7 SDMN Security Applications

SDMN helps solve security issues in mobile networks. It not only simplifies the network and services provisioning in a mobile network, but it can also be used to solve security problems in a mobile network such as encrypting selected traffic, creating on-demand network segmentation, applying necessary access control, protecting infrastructure in real time, mitigating security threats, and enhancing visibility and telemetry for the network (Fig. 17.5).

ONF [4] has shared two use cases of OpenFlow-based SDN in LTE network. One of the use cases discusses how SDN can be used to centrally manage the radio resources and resolve the interference issues that are traditionally resolved using techniques applied in a distributed fashion.

A similar approach can further be leveraged to apply end-to-end security policies from the eNB all the way up to the EPC using a centralized SDN controller.

17.7.1 Encryption: eNB to Network

In the new all-IP-based mobile network architecture, it is difficult to protect traffic between the eNB and EPC, especially when the eNB can be installed in an untrusted environment or when H(e)NB is located at a customer premise. It is not that difficult to intercept control plane or user plane communication between the eNB and EPC. User and control traffic is at risk without an encryption in place. IPSec is used today to protect eNB-to-EPC communication, but it has its limitations of introducing scalability and complexity to the network.

SDMN can be leveraged to encrypt traffic for all or selected eNBs residing in a trusted or an untrusted environment. A selected traffic identification, policy, and encryption can be applied from an SDN centralized controller.

An SDN controller with centralized policy and control to a mobile network can be used to verify and enforce necessary control and encryption for user plane data. It can also be used to further encrypt traffic from the eNB to EPC, which is unencrypted today.

17.7.2 Segmentation

During a security event such as malware and DoS, DDoS network gets paralyzed. Access to all parts of the network is affected including the critical assets. During such a network state, the security administrator wants to have their critical assets reachable during a network attack; meanwhile, it needs to make sure that no one can access the critical asset until the incident is handled and the attack is mitigated.

An SDN-based application can make sure that critical assets (i.e., Home Subscriber Server (HSS) DB) are accessible during a breach or network attack. System can also make sure that only authorized people can access the assets until the incident is fully resolved.

Such SDN applications can build a tier, a zone, or a network segment in real time by doing a requirement analysis (application criticality, classification, network trust level, risk focused) and create an appropriate type segment.

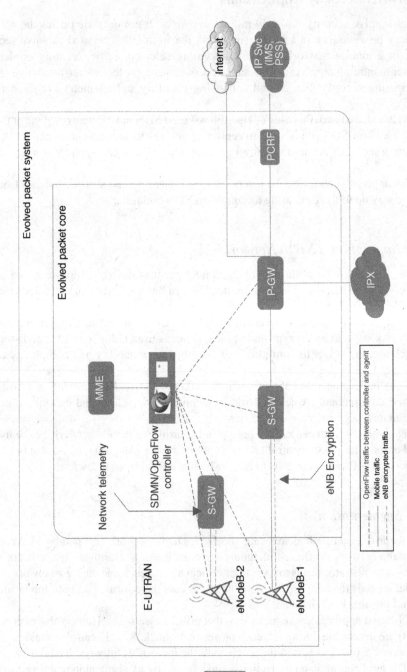

Figure 17.5 SDMN/OpenFlow security applications.

17.7.3 Network Telemetry

Network intelligence can be developed by enabling network visibility and knowing your network. An effective threat protection and mitigation are not possible without having capability to see what is going inside your network.

A network telemetry system can provide information about the origin, destination, nature, and other attributes of the traffic from various network components and help identify and mitigate an ongoing threat.

SDN technology can leverage its capabilities to develop a telemetry system of mobile components in a mobile network where eNBs, S-GW, and P-GW can collaborate and develop an intelligent network telemetry system. This will help introduce network intelligence and visibility capabilities to take informed security decisions.

References

[1] Wikipedia Infocommunication. Available at http://en.wikipedia.org/wiki/Infocommunications (accessed February 18, 2015).

[2] 3GPP. 3GPP System Architecture Evolution (SAE); Security architecture. TS 33.401. Available at http://www.3gpp.org/DynaReport/33401.htm (accessed February 18, 2015).

[3] 3GPP. 2G Security; Security architecture. TS 33.102. Available at http://www.3gpp.org/DynaReport/33102.htm (accessed February 18, 2015).

[4] Open Network Foundation. OpenFlow™-Enabled Mobile and Wireless Networks document. Available at https://www.opennetworking.org/images/stories/downloads/sdn-resources/solution-briefs/sb-wireless-mobile.pdf (accessed February 18, 2015).

18

Security Aspects of SDMN

Edgardo Montes de Oca and Wissam Mallouli

Montimage EURL, Paris, France

18.1 Overview

This chapter presents the security issues introduced by software defined networking (SDN), network function virtualization (NFV), and future mobile networks that integrate these technologies to become software defined mobile networks (SDMNs). Even though existing fault management and network security solutions used in traditional networks are sometimes also applicable in SDMN, the concepts introduced by these technologies bring new opportunities, challenges, and vulnerabilities that need to be investigated or addressed.

The introduction of centralized controllers, network virtualization, programmability, and NFV; the separation of the control plane and the data plane; the introduction of new network functions; and even the introduction of new stakeholders such as mobile virtual network operators (MVNO) will all have impact on how security needs to be assured and managed.

To better understand these issues, in Section 18.2, we present an overview of the state of the art; in Section 18.3, we give a more detailed analysis of security monitoring techniques; and in Section 18.4, other important issues are presented: reaction and mitigation techniques, economic viability, and secure services.

18.2 State of the Art and Security Challenges in SDMN Architectures

Existing security techniques applied or applicable in SDMN will be presented in this section, including techniques for end-to-end security and privacy, monitoring techniques (IDS, IPS, behavior, QoS statistics, etc.), security of virtual and physical network elements (NEs) and interfaces, and reaction and mitigation techniques.

SDMNs impose new challenges on network security involving LTE-EPC mobile network security, cloud security, Internet security, and SDN security.

Software Defined Mobile Networks (SDMN): Beyond LTE Network Architecture, First Edition.
Edited by Madhusanka Liyanage, Andrei Gurtov, and Mika Ylianttila.

18.2.1   Basics

Security in networks involves making sure that the network provides the services expected from it and that the subscribers can rely on them without prejudice. Several issues need to be considered that include the following main categories:

Identification: Users need to be identified in a unique manner. In LTE-EPC, the International Mobile Subscriber Identity (IMSI) is provided via the USIM card and is stored in the Home Subscriber Server (HSS) database.

Mutual authentication: Users (e.g., subscribers, administrators) and NEs need to be able to interact with the assurance that all parties involved are who they claim to be. LTE-EPC provides similar security features as its predecessors (UMTS and GMS).

Access control: Prevents unauthorized use of the network and services by maintaining a user equipment (UE) profile in the HSS database.

Integrity: Interactions include the communication of control plane and user plane data that should not be modified in an unauthorized or undetected manner. In LTE-EPC, this is assured for control plane data only. For the Nonaccess Stratum (NAS) network, both encryption and integrity are provided.

Confidentiality: Privacy, or the ability to control or restrict access so that only authorized individuals or elements can view or understand sensitive information, also needs to be assured. LTE-EPC defines mechanisms to ensure data security during its transmission over the air interface and through the LTE-EPC system by encryption of both user plane and control plane data (e.g., in the Radio Resource Control (RRC) layer). LTE and SDN security will be presented in this section.

Privacy: Keeping identity and location confidential. In LTE-EPC, the MME provides a Globally Unique Temporary Identity (GUTI) to the EU to temporarily replace the IMSI.

Availability: The users need assurance that the network and services are available when required. There is no LTE-EPC integrated feature that deals with this. LTE networks must be strongly safeguarded and proactively monitored from end to end in order to avert casual as well as advanced persistent threats. Monitoring and cyberthreat mitigation will be presented in the next section.

18.2.2   LTE-EPC Security State of the Art

Global System for Mobile Communications (GSM) mobility networks were designed to address mainly privacy and authentication. Encryption and authentication were improved in UMTS and LTE-EPC, and most important, mutual authentication was introduced.

The security model adopted in mobile LTE-EPC networks integrates different security mechanisms at different levels. First of all, it reuses the authentication mechanisms from UMTS, in other words USIM cards in the mobiles, mutual authentication with the network, and key generation (e.g., Ck, Ik). LTE introduces new mechanisms, such as key derivation during mobility to and from LTE (KASME), high-level protection of signaling (including NAS integrity control and ciphering, end-to-end security from mobiles to MME), protection of radio interfaces (Packet Data Convergence Protocol (PDCP) frames, user session ciphering, RRC radio signaling integrity control, and ciphering); and use of HMAC-SHA-256 for successive key derivations. These mechanisms will continue to be used in future 4G and 5G networks, but how they are impacted in NFV contexts is yet to be studied.

EPS adapts GSM and 3G security mechanisms for obtaining an optimized architecture by embedding confidentiality and integrity mechanisms in the EPS protocol stack (as shown in Fig. 18.1). It also needs to interwork with legacy systems. The UE is identified by the Mobility Management Entity (MME) in the serving network that uses authentication data from the home network and triggers the Authentication and Key Agreement (AKA) protocol in the UE. This allows to share a Key Access Security Management Entity (KASME). Further keys can be derived for confidentiality and integrity protection at the NAS level. More keys are derived for confidentiality and integrity protection of the signaling data between the eNB and the UE at the Access Stratum (AS) level. AS signaling integrity and encryption protects the RRC protocol. Confidentiality protection between the UE and the eNB is embedded in the PDCP that performs IP header compression and decompression. No layers below PDCP are confidentiality protected. Integrity protection is not applied between the UE and the eNB, but IPsec can optionally be used to encrypt user data. Likewise, signaling and user data between the eNBs and the core network can be protected using IPsec on the X2, S1-MM2, and S1-U interfaces.

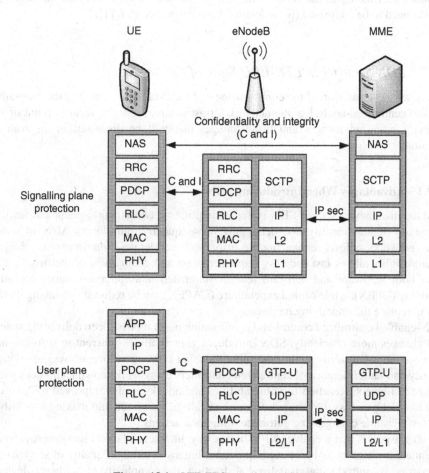

Figure 18.1 LTE-EPC security architecture.

To resume, the role of the different protocol layers is as follows:

- NAS (i.e., all functions and protocols used between the UE and the core network): Performs NAS key handling and integrity and confidentiality protection of NAS. NAS is the layer in charge of managing the establishment of communication sessions and for maintaining continuous communications with the UE as it moves.
- AS (i.e., all functions and protocols used between the UE and the access network): The RRC messages rely on integrity and confidentiality protection from key handling and security activation in PDCP. AS is the layer responsible for carrying information over the wireless portion of the network. PDCP also performs confidentiality protection in the user plane.

The main vulnerabilities in the LTE-EPC security framework concern the system architecture, the access procedures, the handover procedure, and the security mechanism of IP Multimedia System (IMS), Home eNodeB (HeNB), and Machine-Type Communications (MTC). Many vulnerabilities existing in the security framework and the security mechanisms of 4G LTE networks need to be addressed (for a detailed description, see Ref. [1]).

18.2.3   SDN Security in LTE-EPC State of the Art

SDN allows the separation of the control plane and the data plane, enabling the programmability and centralized control of the network infrastructure. From the security point of view, this brings many advantages and disadvantages that will be discussed in the following subsections.

18.2.3.1   Advantages When Introducing SDN

One of the main advantages of SDN is that it simplifies network management and facilitates the upgrade of functionality and debugging. Consequently, introducing SDN in wireless mobile networks allows enhancing security and accelerates innovation in the area. Programmability allows fast and easy implementation and deployment of the new functionality at both hardware and software levels. Automated management reduces operational expenditure (OPEX), while capital expenditure (CAPEX) can be reduced by making it unnecessary to replace the underlying hardware.

SDN-enabled centralized control and coordination make it possible to deliver the state and policy changes more efficiently. SDN introduces vulnerabilities inherent to software-based systems, as we will describe in the next subsection, but at the same time allows improving the resiliency and fault tolerance of centralized controllers using well-known techniques such as automated failovers. Reaction to vulnerabilities and attacks is also improved by giving the ability to quickly assess the network from a centralized viewpoint and making it possible to apply fast dynamic changes and automate mitigation actions.

Another aspect is that it enables NFV. In this way, Internet and cloud service providers can differentiate themselves and propose improved solutions in terms of quality of service (QoS) and security. By introducing virtualized abstraction, the complexity of hardware devices is hidden from the control plane and SDN applications. Furthermore, managed network can be divided into virtual networks (VN) that share the same infrastructure but are governed by different policy and security requirements. SDN and NFV make possible the sharing, aggregation, and management of available resources; enable dynamical reconfiguration and changes of policy; and provide granular control of network and services through the abstraction of the underlying hardware.

The introduction of open SDN standards, such as OpenFlow, not only promotes research and collaborations between different operators and providers but improves the possibility of interoperability in multiservice and multivendor environments and with the legacy systems.

18.2.3.2   Disadvantages When Introducing SDN

One of the main security issues introduced by SDN is that the controllers act as centralized decision points and, as such, become potential single points of attack or failure. Also, the southbound interface (e.g., OpenFlow) between the controller and data-forwarding devices is vulnerable to threats that could degrade the availability, performance, and integrity of the network.

Controllers become a security concern and where they are located and who has access to them needs to be managed correctly. Communications between the controllers and NEs need to be assured by encryption techniques (e.g., SSL), and the keys need to be managed securely. But these techniques are not sufficient to assure high availability because denial-of-service (DoS) attacks remain difficult to detect and counter. Controllers are vulnerable to these types of attacks, and guaranteeing that they are available at all times is a complex task that requires guaranteeing resilience using redundancy and fault tolerance mechanisms. Furthermore, every change and access needs to be monitored and audited for troubleshooting and forensics; and this is more complicated in virtual environments where visibility is often reduced. Thus, the following challenges need to be addressed:

- Secure the controller: Contrary to traditional network architectures where the security functions and mechanisms are orchestrated in a distributed manner, the controller in SDMN architecture is the centralized decision point. Access to such controller needs to be tightly secured and monitored to avoid that an attacker takes control of the NEs.
- Protect the controller: If the controller goes down (e.g., because of a DDoS attack), so goes the network, which means the availability of the controller needs to be maintained.
- Establish trust: Protecting the communications throughout the network is critical. This means ensuring the controller, the applications loaded on it, and the devices it manages are all trusted entities that are operating as they should.
- Create a robust policy framework: What's needed is a system of checks and balances to make sure the controllers are doing what you actually want them to do.
- Conduct forensics and remediation: When an incident happens, you must be able to determine what it was, recover, potentially report on it, and then protect against it in the future.

In Ref. [2], the authors identify the main threat vectors in an SDN-type architecture (depicted in Fig. 18.2).

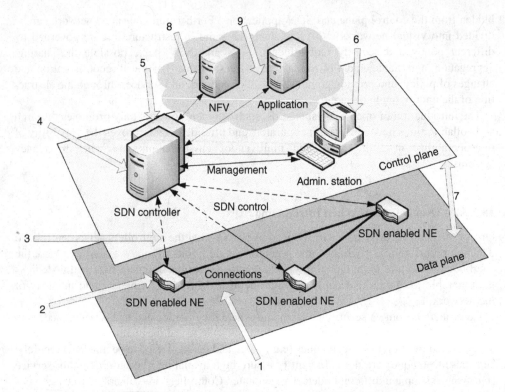

Figure 18.2   Main SDN architecture threat vectors [2].

The seven threat vectors that may enable the exploit of SDN vulnerabilities are:

1. Forged or faked traffic flows: Encryption is not completely reliable and not always possible.
2. Attacks on switches: Introduction of programmability makes them more vulnerable.
3. Attacks on control plane communications: Same issues regarding encryption.
4. Attacks on controllers: Introduction of a new NE or a set of controllers organized hierarchically that need to act in a secure concerted manner.
5. Lack of mechanisms to ensure trust between the controller and management applications: Public key management can be vulnerable.
6. Attacks on administrative stations: Same issues regarding encryption.
7. Lack of trusted resources for forensics and remediation.

To these "classical" vulnerabilities, we need to add vulnerabilities specific to NFV and network programmability:

8. Attack on virtualized network functions.
9. Programmability of network via the controller by untrusted applications.

NFV Specific

Implementing network functions in the cloud introduce vulnerabilities typical in cloud computing. The main security challenges introduced are:

- Introduction of new elements that need to be trust assured, such as virtual machines (VM), virtual switches (VS), hypervisors, controllers, and management modules
- Reduced isolation of network functions
- Resilience dependencies due to resource pooling and multitenancy
- Control of cryptographic keys of hosted network functions

In a cloud environment, multitenancy drives the need for logical separation of virtual resources among tenants. Through orchestration, certain virtualized NFs can be deployed on separate compute nodes, and they can be further segregated by using separate networks. In addition, the use of security zones allows virtualized NFs to be deployed on, or migrated to, hosts that satisfy security-pertinent criteria such as location and level of hardening (e.g., some hosts may employ trusted computing technology).

Automated incident response should include rapid and flexible reconfiguration of virtual resources. If a virtualized NF is suspected of having been compromised (e.g., through unauthorized access via a back door), an uncompromised version can be instantiated to replace it, and the compromised version can be deactivated and be saved for forensic analysis.

In this context, encryption allows protecting the integrity and confidentiality of the signaling and transmitted data, but it is not enough for the following reasons: software is vulnerable and encryption algorithms themselves can be vulnerable (e.g., the OpenSSL Heartbleed bug and backdoors that bypass built-in computer security); the cost of encrypting everything is too high, making it necessary to limit security according to the cost and the risk involved; encryption doesn't mitigate all types of attacks (e.g., DoS attacks); if public keys are used, then how they are managed and stored becomes critical; and a compromised SDN controller potentially allows eavesdropping, exfiltration of data, and unwanted network behavior. Traditional network management tools didn't allow the flexibility to dynamically change the behavior of a network on a node-by-node basis as is possible with SDN.

The Open Networking Foundation (ONF) has identified the southbound communications between controllers and data-forwarding devices as vulnerable. Southbound interface protocols such as OpenFlow have authentication technology that prevents spoofing flow commands from a controller to a switch, but this can be vulnerable if the authentication certificates between controllers and SDN switches are not implemented correctly. Furthermore, authentication cannot prevent DoS attacks from saturating the interface between the control and data planes. To assure secure interactions, they need to be ciphered and monitored, but also, software and hardware need to be kept up to date and also monitored, and unusual behavior that potentially implies a certain level of risk needs to be detected, analyzed, and dealt with.

Programmability Specific

The ONF Northbound Interface Working Group has also been investigating vulnerabilities in northbound communications between the applications and the controllers. Programmability allows installing security applications on the controller's northbound interface to easily introduce new ways to apply security policies on a network. These applications instruct the controller to use the switches and routers that it controls as policy enforcement points. However, the programmable northbound interface is also a potential vulnerability. Here, applications can reprogram the network through the controller, and these can be compromised or contain exploitable vulnerabilities.

Furthermore, as in the case of traditional incompatibilities in routing tables, OpenFlow-type applications have the ability to insert rules that, when combined, may have unexpected results. SDN controllers generally lack the sophistication to understand that security applications should have priority over other applications that communicate with it. Even a harmless application can break security policies if the controller doesn't understand how to handle application requests that contradict security policies. For instance, a security application might quarantine an infected machine, but a load balancing application might still divert traffic to it.

18.2.3.3 Conclusion

To resume, the security issues in SDN are concentrated around the following main areas: (i) application plane, (ii) control plane, (iii) data plane, and (iv) communication security including controller-data path (southbound) and the controller-application (northbound) communication security.

Application Plane Security
SDN enables applications to interact with and manipulate the behavior of network devices through the control layer. SDN has two properties that can be seen as attractive to malicious users and problematic for operators. These properties are, first, the ability to control the network by software and, second, the centralization of network intelligence in network controllers. Since there are no standards or open specifications to facilitate open APIs for applications to control the network services and functions through the control plane, applications can pose serious security threats to the network resources, services, and functions. Although OpenFlow enables deploying flow-based security detection algorithms in the form of security applications, there are yet no compelling OpenFlow security applications [3, 4].

Control Plane Security
In SDNs, the controllers are a particularly attractive target of attack for unauthorized access and exploitation. Without robust and secure controller authentication platform, it is possible to masquerade the controller to carry out malicious activities. Mechanisms to deal with DoS and distributed denial-of-service (DDoS) attacks in large networks are not yet proved viable. Similarly, the controller can become a single point of failure or bottleneck, since the controller southbound and northbound interface securities are also not confirmed. In OpenFlow, most of the complexity is pushed toward controller where forwarding decisions are taken in a logically centralized manner [5]. A challenge for the currently available controller implementations is specifying the number of forwarding devices to be managed by a single controller to cope with the delay or latency constraints. In multiple OpenFlow infrastructures, inconsistency in the controller configurations will result in potential interfederated conflicts [6].

Data Plane Security
In SDNs, switches are most often considered as the basic forwarding hardware accessible via an open interface, while the control logic is moved to the control plane as opposed to the legacy networks where decisions are based on the local configuration of the devices. There are

many security challenges for such architectures. For example, if the control plane is compromised, the data plane is handicapped. The data plane is also prone to saturation attacks since it has limited resources to buffer flow initiation (e.g., using TCP/UDP mechanisms) until the controller issues flow rules. Thus, the failure of the control plane has direct implications on the data plane [4]. Recognizing and differentiating genuine flow rules from false rules is another challenge for the data path elements.

Communication Security

The OpenFlow specification defines Transport Layer Security (TLS) and Datagram Transport Layer Security (DTLS) for the controller–switch communication. The switch and controller mutually authenticate by exchanging certificates signed by a site-specific private key. The switches must be user configurable with one certificate for authenticating the controller and another for authenticating to the controller. Similarly, in the case of the User Datagram Protocol (UDP), the security features are optional and the TLS version is not specified. OpenFlow implementations that use TLS 1.0 may be subjected to man-in-the-middle attacks, as well as other existing attacks against TLS 1.0. OpenFlow implements nonsecure control channel connectivity to ensure interoperability among different systems. However, the standard does not describe how to fall back in case of an authentication failure. Similarly, no mechanisms are demonstrated for application plane and control plane communication.

SDMN Security

SDMNs, carrying the security issues of SDN, have its own set of security concerns. The end user devices in this case often do not have enough processing capabilities, memory, and battery power. Since the communication is IP based, these user devices are prone to the same security threats as their fixed counterparts. The air interface is open to the feats of hacks and thefts; hence, securing the air interface to counter malicious programming of open and programmable network devices is a real challenge. Since the mobile users are mostly on the move and topological changes are frequent, updating the security procedures according to mobility and topological changes is very important. The security between the controller and switches specified in the OpenFlow switch specification is using TLS to secure the channel between the controller and the switch. Similarly, in the case of the UDP, DTLS is described but no mechanism for its usage is currently available. Since there is yet no mobility option available for the use of OpenFlow in mobile networks, SDMNs must develop mobility architectures and the required security mechanisms. SDMNs however must not be limited to an OpenFlow-based architecture, since it is not the only alternative available even in fixed SDN architectures.

18.2.4   Related Work

The concept of a centralized control plane, together with the control channel that is used to exchange information with network devices, introduces new security issues that need to be characterized. From an attacker's perspective, the network controller is attractive due to its important role and so requires specific protection mechanisms. This is an example of a case where network security solutions are more application specific and less dependent on specialized hardware solutions. The scope of such concepts and concrete application scenarios requires a more complete understanding by the research community and stakeholders. As examples, we briefly describe some of the recent research work that is being done.

In Ref. [7], the authors focus on how to utilize SDN to enhance network security. They categorize the target environments into four groups: enterprise networks, cloud and data center, home and edge access, and general design. They analyze different existing security solutions (e.g., OF-RHM, NetFuse, CloudWatcher, AVANT-GUARD, FRESCO, OpenWatch, NIDS Arch., FleXam) and identify the main challenges that include the following: "mobility and roaming" adding dynamicity and therefore complexity to the diagnosis and detection of anomalous activities and security credential exchanges; the "monitoring overhead of OpenFlow-based systems" limiting the effectiveness in the case of high bandwidth and incomplete sample information; "multiaccess and multioperator environment" leading to complex negotiation process, privacy concerns, and potential conflicting policy and QoS requirements that pose a challenge to the security enforcement; and challenges related to the deployment, backward compatibility, interoperability (e.g., between 3G and 4G), and intercommunication with other providers.

Particularly interesting is FleXam [8] that takes into account the optimizations and the dynamics required by a mobile environment. It proposes a flexible sampling extension for OpenFlow to promote the development of security applications such as monitoring. Inspired by this and other works, Ding et al. [7] propose an architecture with local agents that are deployed close to the wireless-edge access to meet the requirements of responsiveness, adaptation, and simplicity. These agents include flow sampling, tracking client records, and mobility profile, and instead of inserting actuation triggers in the data plane, they allow to adaptively query information from the underlying devices and report to the controller, hence alleviating the monitoring load on the central controller.

This architecture has similar objectives as the one proposed in the SIGMONA [9] (a more detailed description can be found in Section 18.3.3).

Besides studying the security vulnerabilities introduced by SDN itself, the authors of Ref. [2] propose a security-by-design technique to achieve secure and dependable SDN platforms based on replicated controllers. Several studies propose solutions based on redundancy. To address the issues of scalability and reliability of centralized controllers, Dixit et al. [10] propose ElastiCon, an elastic distributed controller architecture, in which the controller pool is dynamically grown or shrunk according to traffic conditions and the load is dynamically shifted across controllers.

On the other hand, Araújo et al. [11] study how SDN can be used to guarantee network transport resilience by maintaining multiple virtual forwarding planes that the network assigns to flows. This could be used to mitigate certain types of attacks that provoke path failures. Similarly, Reitblatt et al. [12] present FatTire, a language for writing fault-tolerant network programs based on regular expressions that allows developers to specify the set of paths that packets may take through the network as well as the degree of fault tolerance required. This is implemented using fast-failover mechanisms provided by OpenFlow.

In Ref. [13], the authors propose a hierarchical model of SDN that reduces the number of points of serious failure. Hierarchical deployment of both public key and shared key protocol mechanisms has so far been abstract and largely limited to scalability of cryptographic technology. For the authors, SDN provides an environment with a real need for hierarchical security and raises the question of whether we can use delegation with public key mechanisms, or hierarchical Kerberos mechanisms, to support tiered security in networks. The authors also explain the need for a monitoring service that in turn feeds relevant data back into the management service, completing the loop by connecting to the root-level controllers.

In case TLS is used, a public key infrastructure (PKI) manager is needed, but an alternative could be a Kerberos-type system. Note that the PKI, management, and monitoring services represent concepts and are not necessarily physically separate from the main hierarchy.

From another perspective, introducing SDN also makes it possible to define high-level configuration and policy statements, which can then be translated to the network infrastructure via OpenFlow-type switches. This eliminates the need for individually configuring network devices each time an endpoint, service, application, or policy changes. Thus, SDN controllers can provide improved visibility and control over the network to ensure access control and security policies are enforced end to end. On the other hand, SDN controllers become single points of failure that need to be integrated into the threat and security model.

A number of challenges need to be addressed related to the introduction of SDN controllers that act as important centralized decision points. They need to be secured, their availability needs to be assured, they need to be integrated in the policy framework, and it needs to be assured that they are acting as expected (e.g., via monitoring as in Fig. 18.4) but also enabling forensics, troubleshooting, and remediation.

Furthermore, security should be deployed, managed, and controlled in an SDN environment. For this, we need to add the possibility of both virtualizing the security functions (i.e., NFV of security) and allowing the security functions to act on virtualized networks and functions. This can be called software defined security (SDS), which is to provide network security enforcement by separating the security control plane from the security processing and forwarding planes. This will result in a dynamic distributed system that virtualizes the network security enforcement functions, scales like VM, and can be managed as a single logical system. In Figure 18.3, a possible architecture that implements SDS is represented. Here, all the functions above the southbound interface can be in the cloud where we have VM that are created by an Orchestrator where the virtualized network functions or virtualized network elements (VNE) are run. These VM are connected via VS forming VN. The software defined network and software defined monitoring controllers (SDN/SDM CTRL) translate the requests from the applications (including network management applications) and configure the physical NEs (e.g., switches, appliances, load balancers).

In this architecture, security analysis and monitoring can be done in the cloud where the virtualization is visible and encryption can be managed (e.g., by supplying the security application with the necessary keys). Nevertheless, security analysis and monitoring can also be done, at least in part, by hardware-specific security appliances, even though doing this might no longer be necessary.

With the logically centralized control plane, SDN enhances network security through global visibility of the network state where a conflict can be easily resolved from a remotely monitoring device. The logically centralized SDN architecture supports highly reactive security monitoring, analysis, and response systems to facilitate network forensics, security policy alteration, and security service insertion [3]. For network forensics, SDN facilitates quick and adaptive threat identification through a cycle of harvesting intelligence from the network to analyze network security, update the security policies, and reprogram the network accordingly. Following this logic, Yu et al. [14] propose a software defined security service (SENSS) solution to facilitate attack diagnosis and mitigation. SENSS has three key features: it is victim oriented, that is, the users have access to network information that concerns their address space and can request security services from multiple remote ISPs (such as statistics gathering, traffic filtering, rerouting, or QoS guarantees); a simple detection/mitigation interface that

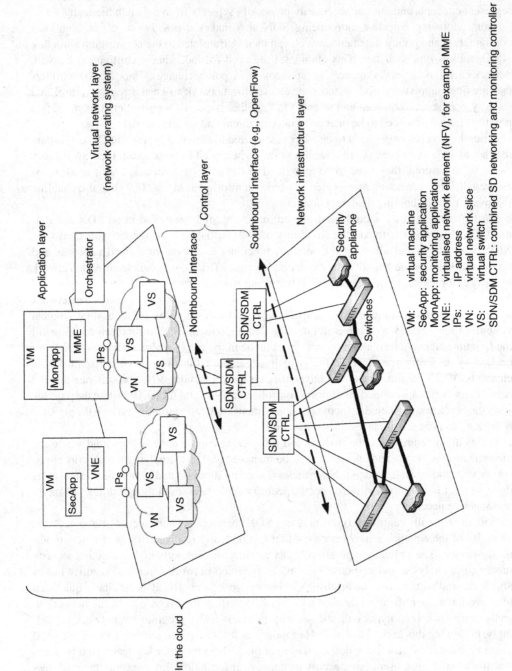

Figure 18.3 Software defined security.

Application layer

Virtual network layer
(network operating system)

Control layer

Northbound interface

Southbound interface (e.g., OpenFlow)

Network infrastructure layer

VM

MonApp

MME

IPs

VS

VS

VS

VN

Orchestrator

VM

SecApp

VNE

IPs

VS

VS

VS

VN

SDN/SDM
CTRL

SDN/SDM
CTRL

SDN/SDM
CTRL

Security
appliance

Switches

In the cloud

VM: virtual machine
SecApp: security application
MonApp: monitoring application
VNE: virtualised network element (NFV), for example MME
IPs: IP address
VN: virtual network slice
VS: virtual switch
SDN/SDM CTRL: combined SD networking and monitoring controller

needs to be implemented by the ISPs makes these requests possible; and a programmable attack detection and mitigation across ISPs allows users to program their own attack detection and mitigation solutions across autonomous systems.

An example of an SDS solution is Catbird (www.catbird.com). Here, all security "devices" are managed and controlled by a common security policy language in which the underlying rules are translated by software. The policy is tied to an asset, with potential for many different policies within the same organization depending on the particular requirements of the people and resources within that organization. Security policies are automatically executed, allowing for quick response time while significantly reducing human error. In an SDS environment, it is easy to imagine assets of different "scopes" safely coresiding in the same virtualized host but subject to very different security policies centrally controlled.

There are several key attributes of software defined network security:

- Abstraction: Security is abstracted away from physical constructs such as stateful port firewalls (FW) and wire sniffers and replaced by a set of flexible controls in the form of policy envelopes blanketing the virtualized (or physical) assets. Abstraction is the foundation for establishing common security models that can be deployed repeatedly without concern for underlying physical hardware capabilities.
- Automation: As each asset is redeployed, its security policy trails it. Concerns about inadvertent operator error are eliminated, as SDS can ensure that no asset can be created without being automatically put into a security trust zone. Role-based controls assure that only properly-privileged administrators can make modifications. SDS automation also means wire-speed reaction to anomalous security events, instantly alerting and quarantining as policy would indicate. By contrast, traditional security is still heavily dependent on manual detection, action, and administration.
- Scalability and flexibility: Eliminating dependencies on physical hardware and expense means security can be deployed on a scale appropriate to each host hypervisor, growing in scope commensurate with business needs. Because this is software only, security policy is elastic and can extend across a cluster or a data center. It also means that security is available "on demand."
- Control orchestration: SDS is designed to integrate a range of network security controls (intrusion detection and prevention, vulnerability management, network segmentation, monitoring tools, etc.) into a single coordinated engine for intelligent analysis and action. Unlimited sources of security input can be funneled into a policy-driven orchestration system, greatly improving the accuracy of the data and attendant action. Orchestration is critical for successful compliance enforcement, as all major compliance standards dictate a variety of controls as parts of the specifications.
- Portability: In a data center governed by SDS, assets carry their security settings with them as they move or scale.
- Visibility: By virtue of being software and thus living within the virtualized infrastructure itself, SDS dramatically improves visibility of network activity. Network administrators and security personnel can detect anomalous behavior that would be blind to them with physical devices and can therefore thwart and protect with a greater degree of accuracy. Network informatics are augmented by this additional data, and NetFlow mapping becomes more extensive and precise.

• Economically viable: The virtualization of the security functions allows dynamically deploying them on existing network infrastructure with minimum CAPEX costs; and the adoption of the SDMN makes their management more flexible (dynamic configuration, countermeasures, etc.) and as a consequence reduces the OPEX costs. These characteristics are unique to SDS and are difficult to attain with traditional security appliances.

18.3 Monitoring Techniques

Network monitoring is required for the verification and validation of SLAs, managing performance (QoS) and user experience (QoE), troubleshooting, assessment of optimizations, and use of resources. In the context of SDMN, network virtualization, and NFV, monitoring needs to be rethought to be able to deal with requirements introduced by virtualization and profit from the flexibility obtained from NFV.

Performance, resource, and security monitoring can be viewed as complementary. Monitoring can provide the knowledge necessary to assure the network's QoS and security. To be able to detect certain types of security issues, performance analysis is necessary. On the other hand, both security breaches and security enforcement mechanisms will have impact on the performance.

LTE-EPC connectivity management of extremely large amounts of devices with various capabilities and intelligence (e.g., mobile phones, ePads, M2M, IoT, etc.) requires automated security services to assure confidentiality and integrity. This leads to high signaling and processing costs and the need for new strategies for cost-effective adaptive security. For this, it is necessary to have a clear view of what is happening in the network and the devices used and how they are used. Monitoring is instrumental for understanding the network traffic and how the services and applications are being used, enabling improved and automated security assurance.

Existing security solutions (e.g., SIEM, IDS, IPS, FW) need to be adapted and correctly controlled since they were meant mostly for physical and not virtual systems and boundaries and do not allow fine-grained analysis adapted to the needs of LTE-EPC and SDN network management. The lack of visibility and controls on internal VN created and the heterogeneity of devices used make many security applications ineffective.

On one hand, the impact of virtualization on these technologies needs to be assessed. For instance, security applications need to be able to monitor virtual connections. Virtualization can help isolate systems but can also be used to make malicious systems that are difficult to detect; for instance, virtualization creates boundaries that could be breached by exploiting vulnerabilities and bugs in the virtualization code (e.g., hypervisors), and the whole systems actually become files that can more easily be stolen.

On the other hand, the security technologies need to cope with ever-changing contexts and trade-offs between the monitoring costs and risks involved. Here, virtualization and SDN facilitate changes, making it necessary for security applications to keep up with this dynamicity.

Security Information and Event Management (SIEM)-type solutions are necessary in order to gain security and status awareness. If an incident happens, the system should be able to determine the source, recover, and protect against it in the future. It should be verified that everything that comes out of the system is logged. Managers have centralized control over the network, and it is necessary to log every change and treat it accordingly in a management solution. Log analysis and event correlation in SDN will fast become a "big data" issue. Tools also are needed that can address all the forensics and compliance requirements.

With SDN, it is possible to create network monitoring applications that collect information and make decisions based on a network-wide holistic view. This enables centralized event correlation on the network controller and allows new ways of mitigating network faults.

Many types of network monitoring techniques exist today that offer different capabilities. First, we have router-based monitoring protocols that allow gathering information supplied by the NEs:

- Simple Network Monitoring Protocol (SNMP): Management of NEs and high-level information on resource use (e.g., monitor bandwidth usage of routers and switches port by port, device information like memory use, CPU load, etc.)
- Remote Monitoring (RMON): Exchange of network monitoring data
- NetFlow or sFlow: Collect information on IP network flows and bandwidth usage

These protocols are dedicated more for performance analysis and network management; but they have also been used for detecting some security problems, as, for instance, NetFlow [15].

We also have packet sniffing, deep packet inspection (DPI), deep flow inspection (DFI), virus scanners, malware detectors, and other techniques for analyzing network packet headers, complete packets, or packet payloads. They are used by Network Intrusion Detection Systems (NIDS), Intrusion Detection and Prevention Systems (IDPS), FW, antivirus scanning appliances, content filtering appliances, etc. and combined with different methods (e.g., statistics, machine learning, behavior analysis, pattern matching, etc.) to detect security breaches (i.e., passive security appliances) or prevent/block detected security problems (i.e., active security appliances).

The adoption of all-IP-type networks introduces vulnerabilities and attacks inherent to the Internet that can be passive or active (i.e., attacks affecting the behavior of the network and services or just dedicated to recuperate information as in the case of scanning and eavesdropping), localized or global (i.e., attacks targeting the network or specific entities or services), and, though less common, insider attacks (i.e., compromised NEs). Also, SDMN introduces new vulnerabilities since it combines mobile phones, the Internet, SDN, network virtualization, and NFV. Some examples are:

- Internet based:
  - DDoS, Smurf, and cyberattacks
  - Spoofing, man in the middle, and ARP poisoning
  - Buffer and heap overflow
  - Format string attack and SQL injection
  - Malware distribution and phishing
  - Data exfiltration
  - Wiretapping and port scanning
- LTE-EPC and mobile network based:
  - Radio and femto-based jamming and saturation of the wireless interface
  - NE vulnerabilities (e.g., in eNodeB, MME), LTE-EPC signaling, and saturation-based attacks

- M2M-based attacks
- Infected mobile phones (e.g., same types of attacks as for the Internet)
- Theft of Service (ToS) (e.g., access to unauthorized services, billing avoidance)
- Protocol misbehavior
- Interoperability between network providers and with legacy networks
- SDN based (see Fig. 18.2)
- NFV based (same as cloud computing vulnerabilities)

Furthermore, amplification effects inherent to the operation of mobility networks are made possible due to centralized authentication nodes and HSS [16]. Some research studies have identified the potential risks that amplification attacks brings to LTE-EPC. For instance, a single event triggered on the phone (a state transition in the RRC state machine) implies a substantial number of messages exchanged among several LTE-EPC nodes. This could be exploited to become a DDoS attack by infecting many phones as explained in Ref. [17].

Many other studies have been made to identify and find solutions to the different types of attacks. In the following, we briefly describe some of them addressing LTE-EPC and SDN vulnerabilities:

- Bassil et al. [18] investigate the effects of signaling attacks that consist of malicious users who take advantage of the signaling overhead required to set up and release dedicated bearers in order to overload the signaling plane by repeatedly triggering dedicated bearers requests.
- Jover [16] analyzes attacks that can affect the LTE-EPC network availability. Common DoS and DDoS attacks could have a severe effect on network performance as already demonstrated, for instance, by a fortuitous error in an android application that created havoc in one of the mobile networks [19]. In Ref. [16], the authors identify that advanced persistent threats, which are well organized and financed, can have very negative effects and provoke both general and very targeted attacks. They propose enhancing mobility network security particularly by improving attack detection techniques and constructing a mobile security architecture based on the following main areas:
  1. Introduction of multiple antennas at the eNodeB to enable the possibility of advanced antijamming techniques [19].
  2. Analysis of traffic and signaling load to modify the network configuration to mitigate DDoS attacks. Common NAS operations, such as idle-to-connected and connected-to-idle RRC state transition, can provoke signaling overloads and are potentially a way of attacking mobile networks and M2M.
  3. Introduction of software defined cellular networks allowing deploying network functions in the cloud. This makes it possible to obtain flexible and adaptable security to counter attacks.
  4. Enhancing SDMN standards and architecture to include other security techniques besides encryption and authentication. Interconnectivity and heterogeneity need to be properly addressed. In particular, SDMN needs to take into account M2M and IoT that require being addressable from the Internet in order to deploy services. This opens new attack vectors, especially for multihomed devices.

Monitoring is an important function that is required for addressing points 2 and 4. To be effective, this function needs to be improved in the following main areas:

- Information extraction: Understanding how to deal with virtualization to obtain information on traffic flows, profiles, and properties by means of extracted protocol metadata, measurements, data mining, and machine learning techniques.
- Scalability and performance issues: The design of the monitoring architecture and the location of the observation points need to be done in such a way as to assure scalability and different monitoring use cases that need to be studied to obtain the best balance between performance, cost, and completeness of the results. Furthermore, hardware acceleration and packet preprocessing technologies need to be integrated and controlled by applications and functions to obtain highly optimized solutions.
- Analysis of different control and user plane traffic flows over the network domains and new interfaces between SDMN and existing networks and identification of related flows in different network domains.
- Dynamicity: Changes in virtualized networks and applications become more easy and frequent. Monitoring solutions need to be able to adapt to these changes.

18.3.1 DPI

DPI is a form of network traffic analysis that involves the act of examining the header and payload content of a packet. Initially, DPI was used to help tackle harmful traffic and security threats and to throttle or block undesired or "bandwidth hog" applications. This role has evolved very fast, including in the mobile sector, where DPI can be deployed for a wide range of use cases aimed at helping to assure and improve the performance of individual customer services and to improve the customer quality of experience.

The key function of the DPI is traffic flow identification and classification. To achieve this, the DPI engine can internally utilize various classification methods from explicit layer information to pattern matching, behavior analysis, and session-level correlation. The classification methods make it possible to support a wide range of protocols and applications without the extensive use of resources in the inspection phase.

Moreover, DPI has the capability to extract traffic information (i.e., metadata) from the inspected packet and the related data stream or application sessions. Typically, this includes:

- Extraction of application and quality metrics such as packet loss, jitter, burstiness, and MOS
- Extraction of protocol details such as IP addresses, HTTP URIs, and RTP audio codecs used

KPI values can be calculated based on DPI extraction results and then integrated with predefined threshold values or optionally with trend analysis that helps identify abnormal changes in the traffic profile.

Thus, the functionality of DPI engines and IDS remains the same, essentially the classification of traffic, metadata extraction, data correlation, and identification of malicious or unwanted traffic. The question is how DPI and IDS need to be adapted to deal with SDN, mobile networks, VN, and VNF.

One critical aspect in all of this is that the applications and associated control elements need a holistic view of infrastructure conditions. This is a central and something that DPI, in principle, can provide, by gathering information throughout the network and feeding it back to the control layer (i.e., the controller) and to the applications so as to ensure that the right resources and capabilities are made available and that the security requirements are met.

It is important to note that at all times legal aspects need to be considered. This includes the storing of information as required by law and protecting the privacy of citizens and organizations. As a legally sanctioned official access to private communications, lawful interception is a security process in which a service provider or a network operator collects and provides law enforcement officials with intercepted communications of private individuals or organizations.

18.3.2   NIDS

The nature of mobile networks and virtualization creates new vulnerabilities that do not exist in fixed wired networks. Currently, many of the proven security techniques (e.g., FW, encryption, IDS) are ineffective in these types of networks and applications since they rely on protecting localized physical assets and interfaces. This is not the case when mobility and virtualization are used since they introduce the need for wireless connections open to eavesdropping and active interfering, virtual boundaries not visible and more vulnerable, virtual applications with remote storage and execution, and mobile devices that connect to unprotected networks. Thus, new architectures, techniques, and tools need to be developed to protect the virtual wireless networks and mobile applications; mobile nodes and the infrastructure must be prepared to operate in a mode that trusts no peer [21].

In this article, the authors argue that intrusion detection can complement intrusion prevention techniques (such as encryption, authentication, secure MAC, secure routing, etc.) to secure the mobile computing environment. However, new techniques must be developed to make intrusion detection work better for wireless networks. This is also true for virtualized networks and functions. In SDN, the SDN-enabled switches can make a first evaluation to detect suspicious traffic. The controller can then mirror this traffic so that it can be analyzed by the IDS appliance (i.e., off-path detection). In this way, one avoids interfering with the traffic flow, but there will be a delay in the blocking of unwanted traffic. In the case of online detection, the IDS would intercept real traffic and act as an FW. This will have impact on the latency. The controller could interact with the SDN switch, the FW, or the IDS to filter unwanted traffic.

The use of IDS appliances in the core network has several drawbacks that reduce its effectiveness. First, encryption will make it harder or impossible to analyze the traffic. This means that IDS appliances are effective only at the edge or at the users' premises where the traffic is lower and the encryption is easier to manage. Another problem is the amount of rules that need to be managed by the appliances. Today, the order of magnitude is in the hundreds of thousands where the requirements are more in the order of millions.

The authors of Ref. [21] show that an architecture for better intrusion detection in mobile computing environment should be distributed and cooperative. Anomaly detection is a critical component of the overall intrusion detection and response mechanism. They stipulate that trace analysis and anomaly detection should be done locally in each node and possibly through cooperation with all nodes in the network. Furthermore, intrusion detection should take place in all networking layers in an integrated cross-layer manner. This type of solutions has been studied for ad hoc networks and is also relevant for VN. This collaboration could be done in the SDN context where the SDN controllers exchange data supplied by the IDSs in order to correlate it and make the appropriate decisions.

18.3.3   Software Defined Monitoring

Different architectural possibilities are studied and proposed in the SIGMONA [9] project where an extension of OpenFlow-type interfaces, SDN CTRL INTERFACE (referred to as SDM CTRL INTERFACE in Fig. 18.4), allows obtaining the packet and flow data and metadata needed by the security applications (e.g., the modules referred to as Management/ Monitoring/Security, Applications, and Network Services) from either the switches or the

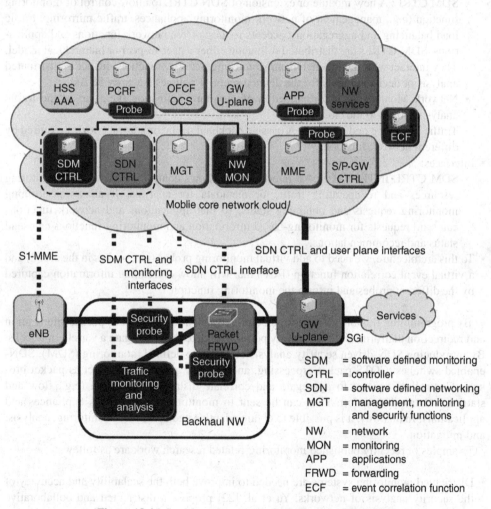

Figure 18.4   Security enhancement framework for SDMN [9].

probes (i.e., agents). The probes can be passive (e.g., the module Traffic Monitoring and Analysis that analyzes mirrored traffic) or active (e.g., the module Active Probe acting as an FW that filters traffic). The SDM CTRL acts as a controller for the software and hardware security devices and could be integrated to the SDN CTRL or separate. If separate, then it will interact with the SDN CTRL via an OpenFlow-type interface. The architecture of the devices and controllers can be hierarchically organized or distributed (e.g., with peer-to-peer communications between the controllers).

The added modules and interfaces are:

- Modules:
  - Security Sensor: An active monitoring probe for the detection of security- and behavior-related information (e.g., security properties and attacks) and mitigation (e.g., filtering). It can be installed on the NEs or in network taps (passive network observation points).
  - SDM CTRL: A new module or extension of SDN CTRL to allow control of monitoring function (e.g., management of network monitoring appliances, traffic mirroring, traffic load balancing and aggregation); accepts requests from network functions and applications. SDM CTRLs are distributed following either a peer-to-peer or hierarchical model. They interact with the Management/Monitoring/Security function and act as distributed analysis or decision points for the defined security policies (security SLAs).
  - Network Monitoring: A virtualization of monitoring function (i.e., part of the traffic analysis moved to the cloud).
  - Traffic Mirroring and Analysis: A passive backhaul traffic monitoring device required by different network functions.
- Interfaces:
  - SDM CTRL INTERFACE: An interface that allows controlling the use of monitoring resources and recuperating traffic or metadata for analysis. It allows performing monitoring requests and obtaining status, so that applications and network functions can send requests for monitoring-based information and monitoring functions can send status and recommendations.
- To this architecture, we need to add virtual monitoring probes (i.e., probes in the cloud) and a virtual event correlation function (ECF) that will allow correlating information captured by the different probes and inform the monitoring function.

By programming flexible switches and other network devices to act as packet interception and redirection platforms, it is potentially possible to detect and mitigate a variety of attacks. By introducing SDN-driven security analysis or software defined monitoring (SDM), SDN-enabled switches, COTS packet processing, and security appliances can act as packet brokers. Controllers can act to aggregate and correlate distributed metadata (e.g., flow and statistical data). This information can be sent to monitoring and analysis appliances and applications. In this way, it is possible to obtain adaptive and optimized monitoring, analysis, and mitigation.

Examples of recently published monitoring-related research work are as follows:

- Distributed monitoring systems are needed to improve both the scalability and accuracy of the security analysis of networks. Yu et al. [22] propose a distributed and collaborative monitoring (DCM) system that allows switches to collaboratively achieve flow monitoring tasks, balance measurement load, and perform per-flow monitoring. It relies on a two-stage Bloom filter to represent monitoring rules using small memory space and centralized SDN control to manage it; but only two rather basic functionalities have been evaluated: flow size counting and packet sampling.

- Choi et al. [23] study scalability problems introduced by centralized SDN control that leads to excessive control traffic overhead to obtain the needed global network visibility. To solve this, they propose a software defined unified monitoring agent (SUMA) that acts as a management middlebox to provide intelligent control, management abstraction, and a filtering layer. Choi et al. [24] also propose a layered control and monitoring management abstraction and filtering solution: software defined unified virtual monitoring function (SuVMF) for SDN-based networks.
- Niels et al. [25] propose a monitoring solution for capturing per-flow metrics (e.g., delay and packet loss) in OpenFlow networks. But the adaptive polling rate technique used that increases when flow rates differ between samples and decreases when flows stabilize to minimize the number of queries could be applied to detecting end-to-end performance problems due to DoS and DDoS attacks. Similarly, Bianchi et al. [26] introduce SDN techniques to improve programming and deployment of online (stream-based) traffic analysis functions that can also be used for detecting security breaches.
- Adapting monitoring techniques to effectively deal with virtualized context is also a major research area. In Ref. [27], SDN, NV, and traditional methods are adapted for gathering evidence and auditing activities on a per-tenant basis, allowing monitoring tenants' VN. Zaalouk et al. [28] study how to adapt the SDN architecture for security use cases. OrchSec, an Orchestrator-based architecture that utilizes network monitoring and SDN control functions to develop security applications, is proposed. Even though limited to sFlow-type data analysis, it shows the benefits that can be derived from SDN-enabled flexibility.
- Wenge et al. [29] outline the current research areas in Security as a Service (SaaS), especially SIEM.

18.4 Other Important Aspects

18.4.1 Reaction and Mitigation Techniques

Cloud computing attack mitigation has been studied by several research teams (e.g., Refs. [30, 31]), but little or no research has been published on mitigation of attacks on LTE and SDN networks.

The work done is mainly how SDN can be used to detect and mitigate attacks on the network. For instance, in the NOVI project Networking Innovations over Virtualized Infrastructures [32], the authors study extending SDN functionalities for performing anomaly detection and mitigation. It is based on flow statistics that may be used to reveal massive DDoS attacks. They demonstrate that OpenFlow statistics collection and processing overloads the centralized control plane and propose a modular architecture for the separation of the data collection process from the SDN control plane with the employment of sFlow monitoring data. The results show that the sFlow-based mechanism is more effective than the native OpenFlow approach and that the OpenFlow protocol can effectively mitigate attacks via flow table modifications.

Similarly in Ref. [33], SDN is used to mitigate attacks detected on virtual appliances. Vizváry and Vykopal [34] present a very succinct analysis of current and future possibilities for the detection and mitigation of DDoS attacks in SDN environments but says little on the vulnerabilities introduced by them.

All the studies are in laboratory settings and are not necessarily adapted to operating network environments. Nevertheless, it has been shown that with SDN flexibility, it is possible to more rapidly react to detected DDoS attacks. But many issues need to be studied, as, for instance, how to block these attacks without blocking legitimate traffic, how to balance risk and cost to obtain efficient solutions, how to scale to large networks, and how to limit the number of false positives detected. Also, work has mainly been oriented toward mitigating DDoS attacks, but many more mitigation types need to be addressed, including attacks on signaling and control plane, more localized DoS attacks difficult to detect using global performance statistics, advanced persistent threats, and compromised network functions.

18.4.2   Economically Viable Security Techniques for Mobile Networks

Economic viability of the solutions and their estimated cost levels is an important aspect that needs to be studied in the context of SDMN.

For instance, securing the controller forcibly has a cost. First of all, it is necessary to know and audit who has access to the controller and where it resides on the network, and security between the controller and end nodes (routers or switches) needs to be assured; likewise, high availability, changes that need to be logged and controlled, existing security devices, and applications need to be configured and integrated correctly.

Many studies have been made to evaluate cost and optimize security mechanisms for the cloud and data centers (e.g., the recent studies: Refs. [35, 36]) and others for mobile applications (e.g., the recent studies: Refs. [37, 38]). In Ref. [39], the authors study new strategies that need to be devised for cost-effective security provision and propose a context-aware security controller for LTE-EPC networks to minimize the overall security cost that activates security mechanisms according to the contextual information such as the application type and the device capabilities. More specific to SPAM mitigation but applicable in general, Bou-Harb et al. [40] identify the high cost of centralized security solutions in LTE-EPC and proposes a distributed architecture made cost-effective by utilizing commercial-of-the-shelf (COTS) low-cost hardware in the distributed nodes to mitigate SPAM flooding attacks.

In Ref. [41], the authors study how to obtain virtualized cost-effective DPI monitoring solutions based on genetic algorithms. They argue that any network function (e.g., DPI, FW, caching, ciphers, load balancers) can be virtualized, but deployment and operation costs due to licensing and power consumption need to be optimized. The genetic algorithms allow this by minimizing the number of deployed DPI engines and determining their location and at the same time minimizing network load introduced by DPI.

Overall, there are two types of costs that need to be considered: CAPEX (i.e., investments costs) and OPEX (i.e., operation costs). In the following, we give some observations that need to be considered when trying to obtain the best cost-effective security.

18.4.2.1   CAPEX

First of all, NFV provides lower total cost of ownership, reducing CAPEX by migrating functions from proprietary to commodity hardware and from dedicated NE to VM, including security appliances and functions.

To illustrate this, for example, for their data centers, Microsoft has developed an OpenFlow-based network tap aggregation platform (Distributed Ethernet Monitoring) for analyzing huge volume of traffic within the cloud network. Traditional network packet brokers that do tap and SPAN port aggregation or mirror ports did not scale. By introducing OpenFlow, it was possible to reduce CAPEX costs since it allowed using single merchant silicon switches (switches using of-the-shelf chip components), replacing more expensive specialized tap/mirror aggregation appliances. OpenFlow controllers allowed easy tailoring of the monitoring and aggregation to adapt to the requirements in an optimized way.

Nevertheless, licensing costs and deployment of network functions need to be optimized for the best results.

Security based on policies requires the ability to deploy large number of rules, for example, when using FW to filter network traffic. Existing FW, even when based on very sophisticated hardware like Content-Addressable Memory (CAM) [42], are limited in the number of rules they can handle without introducing exorbitant costs or affecting traffic latency.

18.4.2.2   OPEX

NFV can ease the operational impact of deploying security updates. An upgraded instance of the virtualized NF can be launched and tested while the previous instance remains active. Services and customers can then be migrated to the upgraded instance. The older instances with security flaws can be deactivated and analyzed once this is complete.

Deployment of new or modified security policies is also facilitated. Security has been difficult to implement even in the traditional networks because of difficulty in enforcing the required security policies in a continually changing environment. SDMN provides new ways of dealing with this problem by enabling introduction of sophisticated network architecture that allows network administrators to dynamically enforce and control fine-grained security mechanisms by relying on NFV. Logically centralized SDN control can help simplify security policy deployment and prevent conflicts and inconsistencies in the security procedures by providing a global view of the configurations of different network devices. Formal methods and proof techniques can also be applied more easily to detect misconfigurations.

The introduction of standards (not all yet completely adopted in the case of SDMN) is also a determining factor to reduce CAPEX and OPEX costs by allowing improved competition between stakeholders, eliminating the need for gateways, allowing the use of commodity hardware, and reducing the learning curves.

18.4.3   Secure Mobile Network Services and Security Management

Mobile services are vulnerable and have become a main target for attacks that include unreliable authentication mechanisms, nonencrypted or poorly secured communications, inauspiciously installed malware, lack of security applications, out-of-date systems and applications, and unauthorized modifications (e.g., jailbreaking, rooting). Virtual and software defined network techniques will facilitate modifying and configuring network functions using centralized controllers, making it easier to adapt security functions to the needs of the mobile services and their users.

Future mobile networks will be a composite of multiple architectures and infrastructures to cover different geographic locations and support different set of services with high data rate. In the current mobile networks, DoS attacks, authorization vulnerability, service degradation attacks, location tracking, and bandwidth stealing are common threats, and more will appear as the usage of the Internet in mobile networks increases. With the migration of IP service, the security challenges will also migrate to mobile devices and networks with the IP services. Mobile devices, though, have lesser resources to counter attacks than fixed stable devices. Therefore, it is widely accepted that security measures incorporated to the network itself must be strengthened first to protect the network infrastructure and architecture and then the mobile devices and its users. The network operators thus will face challenging security issues since security will soon be a key differentiator in their commercialization efforts. Stable and robust security policy deployment would require global analysis of policy configuration of all the security devices in networks to avoid conflicts and inconsistencies in the security procedures. These policies diminish the chances of serious security breaches and network vulnerabilities. Therefore, SDMNs will be strong candidates for future secure mobile networks, since in SDMN the logically centralized control can provide a global view of the configurations of different network devices and hence mitigate the risks of security breaches.

Service chaining is not a new concept, but the trend has taken on a new importance with the rise of SDN and NFV. A service chain is a carrier-grade process for continuous delivery of services based on network function associations, such as FW or application delivery controllers (ADCs) interconnected through the network to support an application. SDN and NFV make the service chain and application provisioning process faster and easier [43, 44]. In the past, building a service chain to support a new application required installing specialized hardware and tailored configurations. Furthermore, service chains did not adapt easily to changes in application needs, so they needed to be overprovisioned and made as generic as possible to support multiple applications. By separating management functions from the infrastructure, SDN and NFV allow standardized automated reconfiguration. Network security functions to support mobile services execute as VM under control of a hypervisor and can be easily adapted to the context and needs of the applications.

For instance, a service chain can consist of an edge router at the customer premises, followed by a DPI service that determines the type of traffic, which in turn informs the controller to create a service chain specific for the customer and the traffic. Another example is an email or web service chain that includes virus, spam, and phishing detection, routed through connections offering the required performance. Thus, service chains allow automated tailoring of network security functions adapted to the needs of the services and customers.

18.5 Conclusion

On the one hand, the introduction of NFV, network virtualization, and SDN facilitates the security assessment and mitigation in future mobile networks. On the other hand, these techniques introduce new vulnerabilities inherent to software- and Internet-based systems and the addition of new elements, for example, SDN controllers. In this chapter, we have presented both the advantages and disadvantages of introducing these technologies in future mobile networks (4G/5G). We also presented ongoing work and possible solutions. We also briefly addressed other issues: mitigation, economic viability, and mobile service security. Other topics that were not covered but are important are standardization, open source solutions, and legal and network neutrality considerations.

References

[1] J. Cao, M. Ma, H. Li, Y. Zhang, Z. Luo; A survey on security aspects for LTE and LTE-A networks; IEEE Communications Surveys and Tutorials 16(1):283–302 (2014).

[2] D. Kreutz, F. Ramos, P. Verissimo; Towards secure and dependable software-defined networks; in Proceedings of the Second ACM SIGCOMM Workshop on Hot Topics in Software Defined Networking. ACM, New York, 2013, pp. 55–60; ACM SIGCOMM 2013, Hong Kong, August 12 and August 16, 2013.

[3] S. Sezer, S. Scott-Hayward, P. K. Chouhan, B. Fraser, D. Lake, J. Finnegan, N. Viljoen, M. Miller, N. Rao; Are we ready for SDN? Implementation challenges for software-defined networks; Communications Magazine, IEEE 51(7):36–43 (2013).

[4] S. Shin, V. Yegneswaran, P. Porras, G. Gu; Avant-guard: scalable and vigilant switch flow management in software-defined networks; in Proceedings of the 2013 ACM SIGSAC Conference on Computer & Communications Security. ACM, New York, 2013, pp. 413–424; CCS 2013, November 4–8, 2013 Berlin, Germany.

[5] J. Naous, D. Erickson, G. A. Covington, G. Appenzeller, N. McKeown; Implementing an openflow switch on the NetFPGA platform; in Proceedings of the Fourth ACM/IEEE Symposium on Architectures for Networking and Communications Systems. ACM, New York, 2008, pp. 1–9; ANCS 2008, November 6–7, 2008, San Jose, California, USA.

[6] E. Al-Shaer, S. Al-Haj; Flowchecker: configuration analysis and verification of federated openflow infrastructures; in Proceedings of the Third ACM Workshop on Assurable and Usable Security Configuration. ACM, New York, 2010, pp. 37–44; CCS 2010, October 4–8, 2010, Chicago, IL, USA.

[7] A. Yi Ding, J. Crowcroft, S. Tarkoma, H. Flinck; Software defined networking for security enhancement in wireless mobile networks; Computer Networks 66:94–101 (2014).

[8] S. Shirali-Shahreza, Y. Ganjali; Efficient implementation of security applications in OpenFlow controller with FleXam; in Proceedings of IEEE Symposium on High-Performance Interconnects. ACM, New York, 2013, pp. 167–168; ACM SIGCOMM 2013, Hong Kong, August 12 and August 16, 2013.

[9] http://celticplus.eu/project-sigmona/ (project where the authors participate in defining the SDMN architecture) (accessed January 24, 2015).

[10] A. Abhay Dixit, F. Hao, S. Mukherjee, T. V. Lakshman, R. Rao Kompella; Towards an elastic distributed SDN controller; Computer Communication Review 43(4):7–12 (2013).

[11] J. Taveira Araújo, R. Landa, R. G. Clegg, G. Pavlou; Software-defined network support for transport resilience; in Proceedings of Network Operations and Management Symposium (NOMS), 2014 IEEE, pp. 1–8; 5–9 May 2014, Krakow.

[12] M. Reitblatt, M. Canini, A. Guha, N. Foster; FatTire: declarative fault tolerance for software-defined networks; in Proceedings of the second ACM SIGCOMM workshop on Hot topics in software defined networking, pp. 109–114, ACM New York; HotSDN '13, Hong Kong, August 12 and August 16, 2013.

[13] Y. W. Chen, J. T. Wang, K. H. Chi, C. C. Tseng; Group-based authentication and key agreement; Wireless Personal Communications, Springer US; February 2012, Volume 62, Issue 4, pp 965–979.

[14] M. Yu, Y. Zhang, J. Mirkovic, A. Alwabel; SENSS: software-defined security service; SIGCOMM Workshop ONS 2014, in Proceedings of the ACM conference on SIGCOMM; Pages 349–350; ACM New York, NY, USA 2014; ONS 2014, March 2014, Santa Clara, CA.

[15] M. Scheck; Cisco's whitepaper: "netflow for incident detection"; http://www.first.org/global/practices/Netflow.pdf (accessed January 24, 2015).

[16] R. Piqueras Jover; Security attacks against the availability of LTE mobility networks: overview and research directions; in Proceedings of 16th International Symposium on Wireless Personal Multimedia Communications (WPMC), 2013, IEEE, pp 1–9; 24–27 June 2013; Atlantic City, NJ.

[17] R. Bassil, A. Chehab, I. Elhajj, A. Kayssi; Signaling oriented denial of service on LTE networks; in Proceedings of the 10th ACM International Symposium on Mobility Management and Wireless Access. ACM, New York, 2012, pp. 153–158; MobiWac '12, October 21–25 2012, Paphos, Cyprus Island.

[18] R. Bassil, I. H. Elhajj, A. Chehab, A. I. Kayssi; Effects of signaling attacks on LTE networks; in Proceedings of the 27th International Conference on Advanced Information Networking and Applications Workshops (WAINA), 2013, IEEE, pp 499–504; Barcelona, Spain, 25–28 March 2013.

[19] M. Dano; The Android IM app that brought T-Mobile's network to its knees; Fierce Wireless, October 2010, http://goo.gl/O3qsG (accessed January 24, 2015).

[20] R. Jover, J. Lackey, A. Raghavan; Enhancing the security of LTE networks against jamming attacks; EURASIP Journal on Information Security 2014:7 (2014).

[21] Y. Zhang, W. Lee, Y.-A. Huang; Intrusion detection techniques for mobile wireless networks; Mobile Networks and Applications; 2003, Volume 9 Issue 5, September 2003 Pages 545–556.

[22] Y. Yu, Q. Chen, X. Li; Distributed collaborative monitoring in software defined networks; in Proceedings of the third workshop on Hot topics in software defined networking SIGCOMM'14 ACM Conference, ACM New York, pp. 85–90; August 17–22, 2014, Chicago, IL, USA.

[23] T. Choi, S. Song, H. Park, S. Yoon, S. Yang; SUMA: software-defined unified monitoring agent for SDN; in Proceedings of Network Operations and Management Symposium (NOMS), 2014 IEEE, pp. 1–5; 5–9 May 2014, Krakow.

[24] T. Choi, S. Kang, S. Yoon, S. Yang, S. Song, H. Park; SuVMF: software-defined unified virtual monitoring function for SDN-based large-scale networks; in Proceedings of CFI '14 Ninth International Conference on Future Internet Technologies, Article No. 4, ACM New York; CFI'14, June 18–20 2014, Tokyo, Japan.

[25] N. L. M. van Adrichem, C. Doerr, F. A. Kuipers; OpenNetMon: network monitoring in OpenFlow software-defined networks; in Proceedings of Network Operations and Management Symposium (NOMS), 2014 IEEE, pp. 1–8; 5–9 May 2014, Krakow.

[26] G. Bianchi, M. Bonola, G. Picierro, S. Pontarelli, M. Monaci; StreaMon: a data-plane programming abstraction for software-defined stream monitoring; in Proceedings of 26th International Teletraffic Congress (ITC), 2014, IEEE, pp. 1–6; 9–11 Sept. 2014, Karlskrona.

[27] A. TaheriMonfared, C. Rong; Multi-tenant network monitoring based on software defined networking; in Proceedings of On the Move to Meaningful Internet Systems (OTM 2013) Conferences, Lecture Notes in Computer Science Volume 8185, 2013, Springer Berlin Heidelberg, pp. 327–341; OTM 2013, September 9–13, 2013, Graz, Austria.

[28] A. Zaalouk, R. Khondoker, R. Marx, K. M. Bayarou; OrchSec: an orchestrator-based architecture for enhancing network-security using network monitoring and SDN control functions; in Proceedings of Network Operations and Management Symposium (NOMS), 2014 IEEE, pp. 1–9; 5–9 May 2014, Krakow.

[29] O. Wenge, U. Lampe, C. Rensing, R. Steinmetz; Security information and event monitoring as a service: a survey on current concerns and solutions; Praxis der Informationsverarbeitung und Kommunikation 37(2):163–170 (2014).

[30] J. Szefer, P. A. Jamkhedkar, D. Perez-Botero, R. B. Lee; Cyber defenses for physical attacks and insider threats in cloud computing; in Proceedings of the 9th ACM symposium on Information, computer and communications security, pp. 519–524, ACM New York; ASIA CCS '14, Kyoto, Japan, June 4–6, 2014.

[31] S. S. Alarifi, S. D. Wolthusen; Mitigation of cloud-internal denial of service attacks; in Proceedings of the 8th International Symposium on Service Oriented System Engineering (SOSE), 2014, IEEE, pp. 478–483; SOSE'14, 7–11 April 2014, Oxford, UK

[32] K. Giotis, C. Argyropoulos, G. Androulidakis, D. Kalogeras, V. Maglaris; Combining OpenFlow and sFlow for an effective and scalable anomaly detection and mitigation mechanism on SDN environments; Computer Networks 62:122–136 (2014).

[33] G. Carrozza, V. Manetti, A. Marotta, R. Canonico, S. Avallone; Exploiting SDN approach to tackle cloud computing security issues in the ATC scenario, in: M. Viera, J.C. Cunha (eds), Dependable Computing Springer-Verlag, Berlin/Heidelberg 2013, pp. 54–60.

[34] M. Vizváry, J. Vykopal; Future of DDoS attacks mitigation in software defined networks; in Proceedings of the 8th IFIP WG 6.6 International Conference on Autonomous Infrastructure, Management, and Security, Springer Berlin Heidelberg, pp. 123–127; AIMS 2014, Brno, Czech Republic, June 30 – July 3, 2014.

[35] F. Malecki; The cost of network-based attacks; Network Security 2014(3):17–18 (2014).

[36] Y. Chen, R. Sion; Costs and security in clouds; Secure Cloud Computing:31–56 2014, ISBN 978-1-4614-9277-1.

[37] G. Moody, D. Wu; Security, but at what cost?—An examination of security notifications within a mobile application; in Proceedings of the 15th International Conference Human Interface and the Management of Information. Information and Interaction for Health, Safety, Mobility and Complex Environments, Springer Berlin Heidelberg, 2013, pp. 391–399; HCI 2013, Las Vegas, Nevada, USA, July 21–26, 2013.

[38] N. Vrakas, D. Geneiatakis, C. Lambrinoudakis; Evaluating the security and privacy protection level of IP multimedia subsystem environments; IEEE Communications Surveys and Tutorials 15(2):803–819 (2013).

[39] S. B. H. Said, K. Guillouard, J-M. Bonnin; On the benefit of context-awareness for security mechanisms in LTE-EPC networks; 24th IEEE Annual International Symposium on Personal, Indoor, and Mobile Radio Communications, PIMRC 2013, London, United Kingdom, September 8–11, 2013. IEEE 2013; pp 2414–2118.

[40] E. Bou-Harb, M. Pourzandi, M. Debbabi, C. Assi; A secure, efficient, and cost-effective distributed architecture for spam mitigation on LTE 4G mobile networks; Security and Communication Networks 6(12):1478–1489 (2013).

[41] M. Bouet, J. Leguay, V. Conan; Cost-based placement of virtualized deep packet inspection functions in SDN; Military Communications Conference, MILCOM 2013—2013 IEEE, San Diego, CA, 2013, pp. 992–997.

[42] A. X. Liu, C. R. Meiners, E. Torng; Packet classification using binary content addressable memory; in Proceedings of INFOCOM, 2014, IEEE, pp. 628–636; INFOCOM'14, April 27 2014–May 2 2014, Toronto, ON.

[43] W. John, K. Pentikousis, G. Agapiou, E. Jacob, M. Kind, A. Manzalini, F. Risso, D. Staessens, R. Steinert, C. Meirosu; Research directions in network service chaining; in Proceedings of SDN for Future Networks and Services, 2013, IEEE, pp. 1–7; SDN4FNS'13, 11–13 Nov. 2013, Trento.

[44] Y. Zhang, N. Beheshti, L. Beliveau, G. Lefebvre, R. Manghirmalani, R. Mishra, R. Patney, M. Shirazipour, R. Subrahmaniam, C. Truchan, M. Tatipamula; StEERING: a software-defined networking for inline service chaining; in Proceedings of the 21st IEEE International Conference on Network Protocols (ICNP 2013), Gottingen, Germany, October 2013, pp. 1–10.

19

SDMN
Industry Architecture Evolution Paths

Nan Zhang, Tapio Levä, and Heikki Hämmäinen
Aalto University, Espoo, Finland

19.1 Introduction

Global mobile data traffic is expected to increase at a compound annual growth rate of 61% between 2013 and 2018 [1], which can make the current centralized gateway system a bottleneck. Software defined networking (SDN) [2] based on, for example, the OpenFlow protocol [3] is one suggested solution to dissolve this bottleneck by separating the network into centralized control functions and distributed forwarding switches.

Separating the control plane functions from the user plane elements creates more signaling traffic [4]. However, the network could also see cost savings [5, 6] from capacity sharing and economies of scale benefits from shared cloud platforms. In addition, acquiring and maintaining standardized general-purpose switches are assumed to be cheaper than the costs of proprietary specific-purpose components currently used in mobile networks [5]. Before the net benefit can be quantified, the industry architectures mapping the technical and business relationships between the network elements and the market actors need to be identified.

Industry architectures and business models for virtualized mobile networks are not an entirely new research topic. However, previous studies have typically focused only on one business model or use case. For example, Fischer et al. [7] describe a simple infrastructure-as-a-service business model, where only infrastructure providers, service providers, virtual mobile operators, and virtual mobile providers are considered. Dramitinos et al. [8] discuss video-on-demand use cases over virtualized Long-Term Evolution (LTE) networks and present two industry architectures covering many business roles.

In contrast to the existing studies, this study provides an overview of multiple potential industry architectures for software defined mobile networks (SDMN). The analysis describes how SDN could improve the performance of mobile Internet service provisioning and how the

---

Software Defined Mobile Networks (SDMN): Beyond LTE Network Architecture, First Edition.
Edited by Madhusanka Liyanage, Andrei Gurtov, and Mika Ylianttila.
© 2015 John Wiley & Sons, Ltd. Published 2015 by John Wiley & Sons, Ltd.

industry structure could change due to the introduction of SDMN. The industry architectures are illustrated with Casey et al.'s [9] method that maps the technical components onto business roles and roles further to actors. The industry architectures are based on the current Finnish LTE network structure, and they are identified by interviewing ten technical and business experts representing academia, mobile network operators (MNOs), and network equipment vendors.

New technologies can be sustaining or disruptive [10]; thus, two deployment approaches for SDMN are discussed: (1) evolutionary SDMN and (2) revolutionary SDMN. *In the evolutionary SDMN*, the network elements (i.e., base stations, routers and switches, and gateways) are separated into their control plane functions and user plane elements. The user plane elements are assumed to stay in the same physical locations, though their control plane functions are moved into the cloud. For example, Basta et al. [11] discuss a functional split of Serving/Packet Data Network Gateways (S/P-GWs), where parts of gateway are moved into a cloud platform. *In the revolutionary SDMN*, the control plane functions can be optimized to operate more efficiently or simply by dividing them into subfunctions and forming new functional groups. From the economic perspective, evolutionary SDMN would bring incremental improvements to the operational efficiency of the currently existing industry architectures, whereas revolutionary SDMN could disrupt the current market situation through new industry architectures, where the value is redistributed among the existing and potential new actors in the market.

The rest of this chapter is structured as follows. Section 19.2 gives an overview of the current LTE network and defines the evolutionary and revolutionary SDMN from the technical perspective. The business roles of SDMN are defined in Section 19.3. These roles are used in Sections 19.4 and 19.5 to illustrate the identified industry architectures for evolutionary and revolutionary SDMN, respectively. Finally, Section 19.6 summarizes the findings and discusses the factors that influence which SDMN industry architectures are most likely to succeed.

19.2   From Current Mobile Networks to SDMN

For forming the industry architectures, a thorough understanding of the underlying technical architecture is needed. This section provides the technical background for the work by briefly explaining the technical evolution from the current mobile network to evolutionary and revolutionary SDMN architectures together with the simplifications and assumptions employed in this study. The sources used include the 3GPP specifications 23.002 on mobile networks [12], Pentikousis et al. [13], Penttinen [14], and interviews with technical and business experts in the Finnish market.

Ten semistructured interviews were conducted during spring 2014 with each lasting on average one hour. The interviewees included technology directors from Finnish MNOs, senior research scientists, and business unit representatives from network equipment vendors and senior researchers from academia. Topics discussed covered the current mobile network topology and structure in Finland and how SDN would change the structure.

19.2.1   Current Mobile Network Architecture

A simplified version of the current LTE network is shown in Figure 19.1. The end user's traffic is sent to an evolved NodeB (eNB), that is, an LTE base station, through the radio interface. The eNB contacts the Mobility Management Entity (MME) and Home Subscriber Server (HSS) for subscriber and authentication information. The traffic is then forwarded through the network routers and switches to the S/P-GW, which decides where to route the traffic. P-GW is the point where the traffic leaves the core network and enters the public IP network through the external interfaces (i.e., Internet exchange points, roaming, etc.). Additionally, P-GW acts as a firewall and S-GW serves as the mobility anchor, when the user is changing eNBs. The Policy and Charging Rules Function (PCRF) keeps track of the network usage and handles the billing for each user. In addition, traditional mobility management typically uses the GPRS Tunneling Protocol (GTP) for transporting the IP packets through the mobile network from eNB to P-GW. However, the tunnel is not drawn into the architecture figure, because it is only a logical connection.

As an example of the quantities of each network element, Figure 19.1 shows the numbers for the aggregate Finnish LTE network combining all the MNOs in Finland. Finland has three MNOs with approximately equal number of subscribers. The main challenge is a scarcely populated country with only 18 inhabitants per square kilometer [15], which increases the relative need for eNBs compared to more densely populated areas. For example, all three MNOs in Finland together currently have roughly 10,000 eNBs that serve the 5.4 million population [16], whereas the German mobile operators have 25,000 eNBs serving a customer base of 55 million [5].

In the Finnish LTE network, MME, HSS, and PCRF have already been centralized into data centers. However, virtualization has not happened yet and the elements still run on dedicated MME, HSS, and PCRF servers. In addition, the border between the user plane and the control plane runs through the eNB, S/P-GW, and routers and switches. Routers and S/P-GWs are also running on dedicated hardware in the traditional Evolved Packet Core. This means that the hardware providers have higher control over the market dynamics and they can charge high prices for the maintenance and updating of the network elements.

19.2.2   Evolutionary SDMN Architecture

Figure 19.2 shows the same LTE network with the eNBs, routers and switches, and S/P-GWs split into their control plane functions and user plane elements. The user plane elements are marked with a U, and control plane functions with a C in their names. In addition to the split,

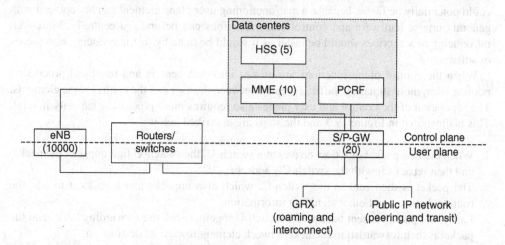

Figure 19.1   Current mobile network architecture.

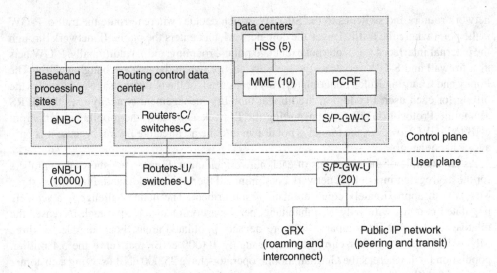

Figure 19.2 Evolutionary SDMN architecture.

the control plane functions of eNBs, routers and switches, and S/P-GWs are also centralized and put into a data center. The degree of centralization depends on the delay sensitivity of the function. For example, eNB-C and routers-C/switches-C cannot be too far away from their user plane equivalents, because it would increase the latency. Thus, eNB-C and routers-C/switches-C are placed in baseband processing pooling sites, which are typically within few kilometers from the user plane equipment (i.e., eNB-U), and the sites have to be connected with a fiber connection [17]. In this evolutionary SDMN architecture, GTP is assumed to function as in the traditional mobile network.

After the control plane functions are stripped from the user plane elements, the user plane can run on general-purpose hardware instead of dedicated servers or routers. This will decrease the costs on building and maintaining the mobile network. In addition, the repair process could potentially be faster, because a malfunctioning user plane element can be replaced with general-purpose hardware and control plane functions can be updated centrally. Moreover, introducing new services should be faster as it would be done by just introducing new pieces of software.

When the control plane functions are moved into data centers and baseband processing pooling sites, more signaling traffic is created between the user and the control plane elements. The separation of the control and user planes also requires more processing capacity in total. This is illustrated in Figure 19.3 and the steps are described below:

1. When the first packet of a flow arrives at a switch-U, the switch-U first unpacks the packet and then repacks it with the switch-C's address.
2. The packet is then sent to the switch-C, which also unpacks and repacks it to add the routing decision and other signaling information.
3. Lastly, the packet is sent back to the switch-U together with the forwarding table, and the packet is then forwarded to the next network element toward its destination.

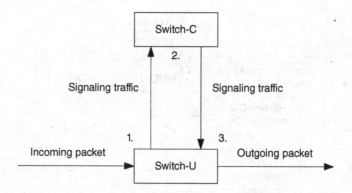

Figure 19.3 First packet processing of a flow and signaling traffic in SDN.

With the traditional switch, the packet would be unpacked and packed only once at the switch and no additional signaling traffic is needed. How much the signaling traffic increases depends on the length of the flows, because only the first packet of a flow is sent to the switch-C. With the following packets of a flow, the switch-U already knows where to forward them. Thus, the scalability of a network with shorter flows is an uncertainty. However, 80–90% of all consumer traffic in the Internet is estimated to be video traffic by 2017 [1], which is a significant optimization opportunity for SDMN. In addition, the amount of the signaling traffic and processing capacity can be decreased with proactive SDN. This means that the switches have predefined routing decisions and rules on how to process certain types of flows. In reactive SDN, each new flow (or its first packet) arriving to a switch-U causes the switch-U to trigger a signaling message, which would cause scalability issues. The number of controllers in the network also affects the performance; the performance decreases with each added controller [4].

On the other hand, the flexibility of the split architecture may reduce the risk of scalability bottlenecks in the mobile networks. In addition, it may also bring cost savings through more efficient resource sharing including both the user plane (spectrum, forwarding) and the control plane (cloud platforms). On the other hand, more advanced resource sharing in the radio access network (RAN) is already under discussion by the 3GPP, for example, the RAN sharing enhancement standards [18], which reduces the resource sharing benefits of SDMN compared to traditional non-SDN mobile networks.

19.2.3 Revolutionary SDMN Architecture

A more radical longer-term step—called revolutionary SDMN—is to reoptimize the control plane elements to function more efficiently. This means that current control plane functions are split further into subfunctions, that is, functional split is performed. These subfunctions can be regrouped into new functional groups, and these new functional groups are marked with a + sign in Figure 19.4, which shows the revolutionary SDMN architecture.

The longer-term revolutionary approach of SDMN technically assumes a standard scalable fiber-based packet network, where control plane processing resides both close to base stations (e.g., cloud RAN) and in large centralized data centers. Mobility anchors can be distributed

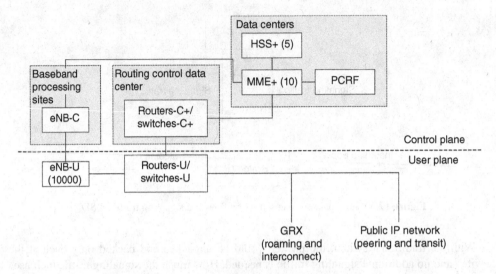

Figure 19.4   Revolutionary SDMN architecture.

and commoditized in all switches, that is, the traditional centralized S/P-GWs can be removed from the architecture. An example of a cloud-optimized control plane function is mobility management based on a standardized tunneling capability in all switching elements, instead of few specialized S/P-GWs. This idea matches with the SDN Enhanced Distributed P/S-GW use case of the Open Networking Foundation's Wireless & Mobile Working Group [19], where distributed S/P-GWs are communicating with a centralized MME to reduce redundant traffic between eNBs and the centralized S/P-GWs.

Removal of centralized S/P-GWs also means changes in GTP. The earlier versions of OpenFlow do not support GTP, but the user plane elements can be realized with different technologies, such as VLAN tagging, or by implementing GTP directly into the OpenFlow switch [20]. The control plane element of GTP could also be moved into the cloud. The role of GTP can be important for a cost-optimized transition from the traditional GTP to a more lightweight mobility management. For example, another use case from the same working group (SDN-based mobility management in LTE [19]) discusses the potential of removing GTP tunneling and, thus, eliminating the signaling traffic and GTP overhead during handovers. The other functionalities of the S/P-GWs are divided between MME+ and routers-C+/switches-C+.

19.3   Business Roles of SDMN

Introduction of new technology also affects the value networks and business models, that is, the industry architectures, of mobile Internet service provisioning. In order to explore the impact of SDMN systematically, the required functionalities for providing mobile Internet over SDMN are divided into business roles, which are then allocated to different actors. The business roles are identified, and the industry architectures are constructed by using the value

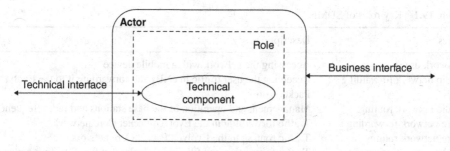

Figure 19.5 Industry architecture notation.

network configuration method of Casey et al. [9]. The components of the notation are illustrated in Figure 19.5 and they are defined as follows:

- **Technical component:** a collection and realization of technical functionalities, including the technical interfaces to other technical components
- **Role:** a set of activities and technical components, the responsibility of which is not divided between separate actors
- **Actor:** an actual market player that takes roles and establishes business interfaces (contracts and revenue models) between other actors

Technical architecture consists of technical components, whereas industry architecture (also value network configuration in the paper of Casey et al. [9]) describes how the roles are divided among actors and how the actors are connected to each other. The notation allows illustration of technical and business architectures in the same figure, which is beneficial for understanding the dependencies between them. Since the technical architectures were introduced already in Section 19.2, from this on, the focus is on industry architectures.

Role analysis forms the basis for constructing and analyzing the industry architectures. Table 19.1 describes the key roles related to SDMN, which can be mapped to the technical components of the traditional mobile network architecture, as is shown in the generic role configuration presented in Figure 19.6. However, the distinct feature of SDMN compared to the traditional mobile networks is the breaking of the tight integration of multiple roles in the same technical components, allowing different actors to control these earlier integrated roles. Therefore, the user plane and control plane functionalities of base stations, routers, and gateways are separated into different roles. Also, the functionalities of S-GW and P-GW are defined separately, even though they are integrated in the existing mobile networks. In order to focus the analysis on the basic functionalities of mobile networks, some advanced functionalities may be missing or they are assumed to be included in the identified key roles.

In addition to the roles related to the internal structure of SDMN, the two key external roles related to network usage and interconnection to other networks are also defined and included in the industry architecture. Even though these two roles are always assigned to end users and interconnection providers, respectively, their inclusion is highly relevant, since a significant amount of revenues and costs are transferred over the business interfaces to these two actors.

Table 19.1   Key roles of SDMN

Roles	Description
Network usage	Accessing the network with a mobile device
Radio network forwarding	Receiving the user traffic in eNBs and forwarding it to the Evolved Packet Core
Radio network routing	Management and operation of the base stations and radio frequencies
Core network forwarding	Traffic forwarding in the Evolved Packet Core network
Core network routing	Traffic routing in the Evolved Packet Core network
Public network forwarding	Traffic forwarding and filtering (i.e., firewall functionality) between the public network and the core network
Connectivity management	Management of connectivity (i) between the public network and the core network and (ii) in the Evolved Packet Core, including situations of inter-eNB handover. Can be divided into (i) public network connectivity management and (ii) mobile network connectivity management, respectively
Mobility management	Management of control plane signaling between the eNB and other network elements like HSS
Subscriber management	Management of the user- and subscription-related information, including user authentication, access authorization, and home network information
Policy and charging	Brokering quality of service and charging policy on a per-flow basis
Interconnection provisioning	Providing the interconnection to public IP networks and other mobile networks through transit, peering, and roaming agreements

Particularly, the business interface to end users, the customers using the mobile networks, is highly valuable as connecting to the source of revenues provides negotiation power against other actors.

19.4   Industry Architectures of Evolutionary SDMN

This section explores the business opportunities of the evolutionary SDMN by analyzing three industry architectures, where SDMN could be introduced. The presented industry architectures are technically feasible with the current technology and existing deployments of the mobile operators in the Finnish market. Since the identified industry architectures already exist in the market, this section focuses on analyzing how SDMN could increase flexibility and improve operational efficiency of the mobile Internet service provisioning.

19.4.1   Monolithic MNO

A very natural evolution path would be one, where an MNO drives the SDMN deployment due to their ownership of the current network infrastructure and the business relationship with the end users (Fig. 19.7). Thus, in the first scenario, MNO operates its own mobile cloud platform and runs the control plane functions on it. The MNO also negotiates with the interconnection providers for the roaming, transit, and peering agreements.

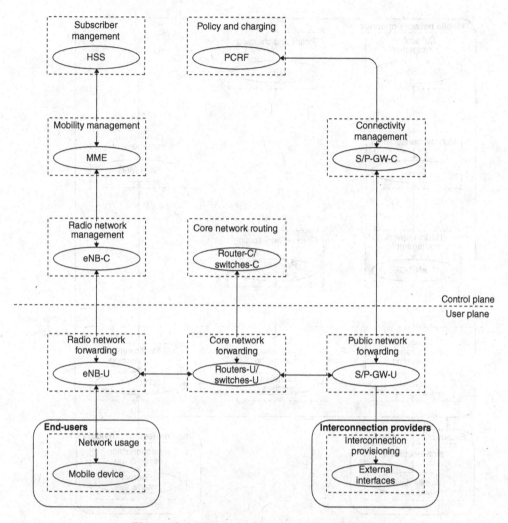

Figure 19.6 Generic role configuration of SDMN.

This scenario's benefit lies in the potential cost reduction from using general-purpose and standardized hardware in both the user plane and in the control plane cloud platform. For example, Naudts et al. [5] show that an SDN-enabled mobile network in Germany presents substantial capital expenditure savings, especially in the preaggregation sites (i.e., the routers). In addition, network configuration can be handled more dynamically, which enables new services to enter the market faster.

However, cost savings from future capital expenditure as the only incentive are not enough for the MNOs to switch to SDMN immediately. The SDMN deployment would most likely happen only when the existing network infrastructure is being renewed. For immediate deployment, SDMN should show more benefits, for example, a decrease in operational expenditure or new revenue potentials, which remain to be determined.

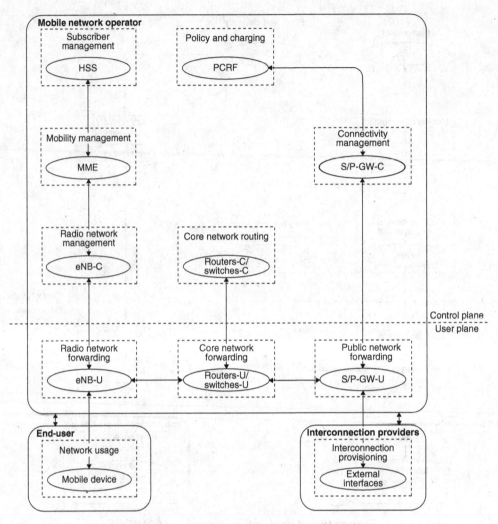

Figure 19.7 Evolutionary SDMN with monolithic MNO.

19.4.2 Outsourced Subscriber Management

Another common business arrangement in the current mobile market is such that a mobile virtual network operator (MVNO) handles the business relationship with the end users and partially or fully uses the MNO's network infrastructure [21]. For example, Virgin Mobile in the United States is a MVNO running over a network provided by the Sprint Network [22], and Lycamobile is a United Kingdom-based MVNO operating in 17 countries and partnering with local MNOs in each country [23].

When a MVNO is entering the market, it can choose different levels of investment into the network equipment [21]. Figure 19.8 shows a scenario, where a service provider MVNO manages a front-end HSS and PCRF. The MNO in this scenario retains control of the critical

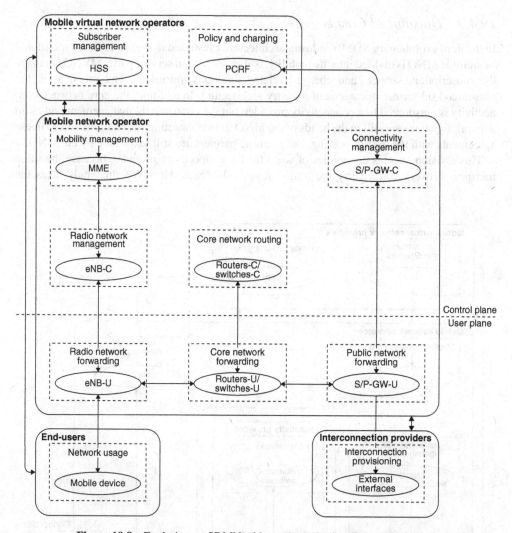

Figure 19.8 Evolutionary SDMN with outsourced subscriber management.

functions of the network, such as MME, S/P-GW, and HSS database. In addition, the service provider MVNO might not wish to own any infrastructure and could be managing the operations of the subscriber management and policy and charging over a leased cloud provided by, for example, a data center or a mobile infrastructure provider.

The SDN-related benefits and challenges for the MNO in this industry architecture are the same as in the monolithic MNO industry architecture. Some additional benefits could be obtained if the MNO would charge the MVNOs the same price as before but still enjoy the cost savings from SDN. In addition, a more agile network could potentially enable the MVNOs to offer better services to the end users and the MNO could use this argument to charge more from the MVNOs.

19.4.3   Outsourced Connectivity

In the third evolutionary SDMN industry architecture illustrated in Figure 19.9, the traditional monolithic MNO is divided into the mobility, connectivity, and service parts. MVNOs manage the subscription, service, and charging-related functions toward the end users, as in the outsourced subscriber management industry architecture. In addition, the core network connectivity is outsourced to a connectivity provider, but the connectivity management and radio network functions are still in the hands of the MNO. Interconnection negotiations and business agreements with the transit, peering, and roaming partners are still taken care by the MNO.

This division enables economies of scale for the connectivity provider, because the same transport network could carry the traffic of several MNOs. However, this study takes the

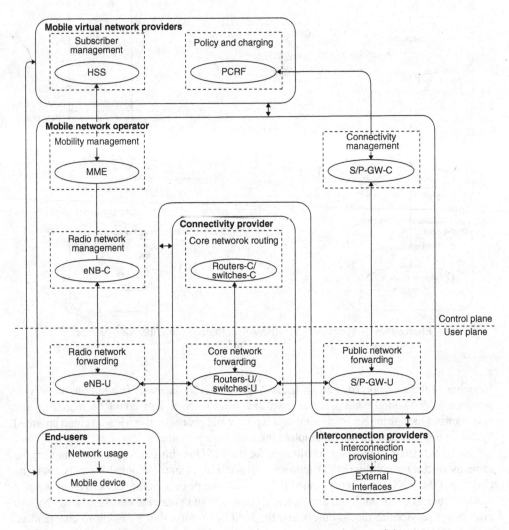

Figure 19.9   Evolutionary SDMN with outsourced connectivity.

perspective of one MNO; thus, only one MNO is illustrated in Figure 19.9. Consequently, the MNO loses some control of the transport network and behaves rather like an overlay network operator. Whether the cost savings from not owning a transport network are enough to compensate the lost control remains to be determined by market forces.

In addition, the performance of a virtualized network is assumed to be weaker due to more signaling and longer distance between the control plane functions and user plane elements. However, to decrease latency and improve service quality, for example, caching can be deployed at the centralized baseband processing sites.

From the MNOs' perspective, the capital expenditure savings from SDMN in this industry architecture are not significant, because according to Naudts et al [5], substantial cost savings come from preaggregation sites, which are now owned by the connectivity provider. Thus, MNOs need more incentives to deploy SDMN, when it outsources its connectivity. On the other hand, the connectivity provider may have an incentive for bringing SDN into its own network.

19.5   Industry Architectures of Revolutionary SDMN

This section explores the potential industry architectures for the drafted revolutionary architecture. The scenarios of evolutionary SDMN have their continuations in the revolutionary phase, but the focus here is on discussing the changes in the market structure due to SDN as well as new actors taking the roles. A big change that revolutionary SDMN enables is the outsourcing of the different network elements, which reduces the control of the MNO. On the other hand, as MNOs control the radio frequencies and the base station infrastructure, their position in the market is still strong. The three revolutionary scenarios discussed are (1) MVNO (2) outsourced interconnectivity, and (3) outsourced mobility management.

19.5.1   MVNO

On the other end of the MVNO spectrum, a full MVNO manages the whole mobile core network and leases just the frequency and connectivity from the MNO, as is shown in Figure 19.10. This scenario is enabled by the division of user plane and control plane. The actor that takes the MVNO role in this scenario has to be big enough to accommodate all the network elements and potentially act internationally to enjoy economies of scale benefits.

The public network forwarding and, as a consequence, the interconnection agreements are also controlled by the MVNO in this scenario. This makes sense for a global MVNO, whose users are highly mobile and who wants to control its own roaming agreements instead of relying on the MNO's roaming partners. In addition, the MVNO could operate on different MNO's networks in different countries. For example, Volvo's connected cars service enabled by Ericsson [24], which offers data connection to Volvo's new cars in any country, could benefit from a more flexible SDMN architecture. Ericsson would then be taking the MVNO role in Volvo's case.

In this scenario, the MNO loses control over the network operations and becomes a mere radio network and connectivity provider. In addition, giving up the management of the baseband processing pooling sites (i.e., eNB-C) means losing control over access management

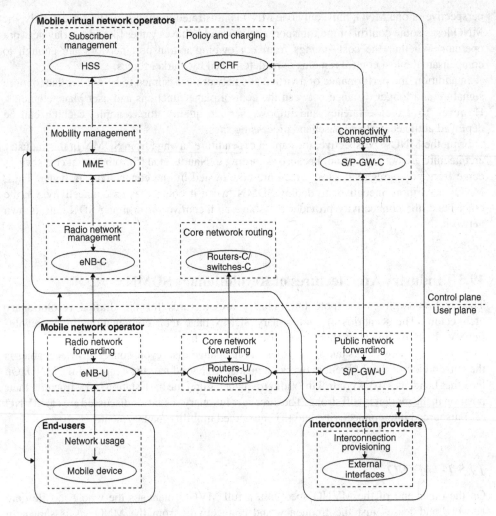

Figure 19.10 Revolutionary SDMN with mobile virtual network operator.

and resource optimization in the radio and core network, which might not be for the best interest of the MNO. However, if the MVNO is a subsidiary of the MNO and targets a different market segment, this industry architecture could be more feasible.

19.5.2 Outsourced Interconnection

Compared to the first revolutionary industry architecture, also, the scenario in Figure 19.11 has outsourced the management of the interconnection agreements but to the connectivity provider. This move of responsibilities is the result of the removal of the S/P-GW technical components from the network (see empty role boxes in Fig. 19.11), which enables a more flexible division of the control plane functions. The roles related to S/P-GWs are divided

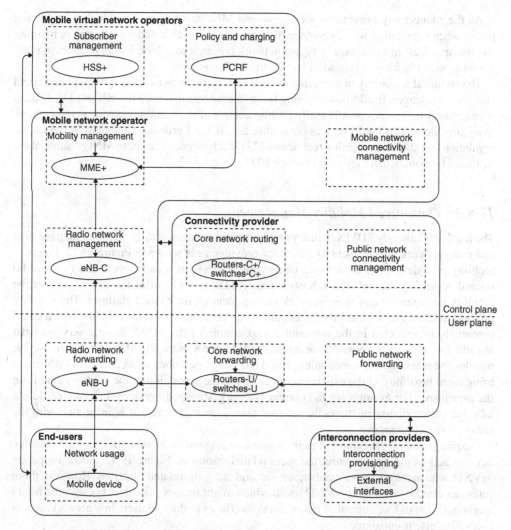

Figure 19.11   Revolutionary SDMN with outsourced interconnection.

between the MNO (mobile network connectivity management role) and the connectivity provider (public network connectivity management role). In addition, the functionalities are now performed by the MME+, HSS+, and routers-C+/switches-C+, which are collections of functionalities rather than traditional technical components.

As a consequence, mobility management could be done in a simpler and lighter way. For example, if the MVNO is big enough (e.g., Google) or has a trustworthy reputation (e.g., banks), it could partner with several MNOs in different countries and local breakout could be handled more efficiently. Mechanisms for local breakout are defined already in the current mobile network; however, due to the lack of trust and inequality in sizes between the MNOs in different countries, its popularity still remains low [25]. In addition, the lowering roaming fees in Europe [26] may become a driving force for simpler mobility management.

As the connectivity provider is serving several MNOs, it might have a higher bargaining power when negotiating with interconnection providers, which could be reflected in roaming and transit prices. In addition, as a bigger network player, it may have higher chances of peering with other big players instead of buying transit from them.

The technical feasibility of removing the S/P-GW from the network is not tested and could still pose challenges. In addition, roaming brings significant revenues for MNOs [27]. Thus, it is uncertain whether they would really give up control of the roaming agreements to connectivity providers. This could be feasible within EU, if the European Commission passes the regulation that abolishes roaming fees across EU [28]. However, European MNOs' subscribers still need to roam, when they travel outside EU.

19.5.3   Outsourced Mobility Management

The third revolutionary SDMN industry architecture outsources also the mobility management and radio network management to a mobility provider, as is shown in Figure 19.12. Potential mobility providers could be network equipment vendors, such as Ericsson and Nokia, who instead of selling the network infrastructure to MNOs could offer the service to run the mobility management and radio network management on their cloud platform. The mobility provider could serve several MNOs and, thus, gain economies of scale benefits, which could potentially be reflected in the operational expenditure of the MNO. Similar services exist already in the current network, for example, Ericsson's Network Managed Services [29] handles the planning, implementation, and day-to-day operations of an MNO. SDMN could bring more flexibility to the existing services due to the separation of the control plane from the user plane. For example, in the current Network Managed Services, Ericsson could not take full responsibility of the radio network management, because it is integrated with the radio network forwarding.

Despite giving up the control of radio network management, MNOs still own the frequency licenses and would, thus, control the radio-related resources. Figure 19.12 shows a separate MVNO, who manages the subscriber, service, and charging-related functions. However, these roles can also be taken by the MNO itself, which might be more feasible, because the MNO might not wish to lose control of the business interfaces to the end users in a network, where everything else is outsourced.

19.6   Discussion

This chapter discussed the evolution path from the current mobile network to SDMN and identified three evolutionary industry architectures and three revolutionary industry architectures. The analysis shows that in the first phase, the SDMN improves the flexibility and operational efficiency of the industry architectures already existing in the market. For example, evolutionary SDMN sees improvements in network management including faster maintenance as well as more efficient and easier software updates, which enable faster entrance of new services into the market. In addition, an SDN-enabled mobile network introduces cost savings due to the use of general-purpose hardware. In the outsourced connectivity industry architecture, the connectivity provider may also experience economies of scale, if it serves several MNOs. However, the deployment of evolutionary SDMN can be slow, because cost

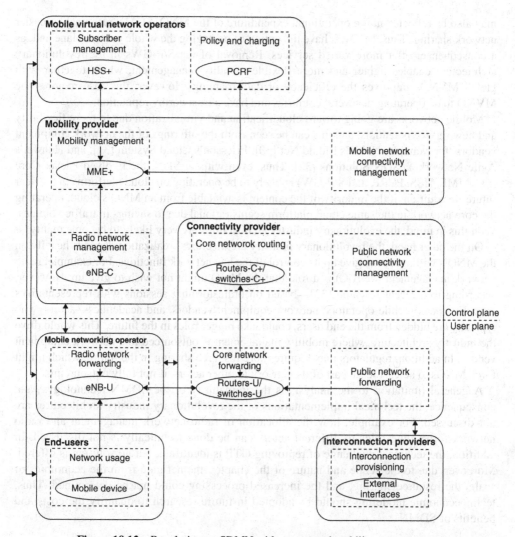

Figure 19.12 Revolutionary SDMN with outsourced mobility management.

savings from future capital expenditure mean that the MNO would implement SDN into the mobile network only when the current infrastructure is at the end of its lifecycle. More incentives, such as lower operational expenditure or additional revenue enabled by SDMN, are needed for faster deployment of SDMN.

However, SDMN also enables new, disruptive industry architectures, where the MNO outsources more functions and new industry actors enter the market. The main benefit from outsourcing is the more efficient operation of the network by more specialized actors. For example, in the outsourced mobility management industry architecture, the mobility provider enjoys economies of scale, which may be reflected in the pricing toward MNOs. At the same time, the connectivity provider, due to its large scale, potentially does more peering than a single MNO and has higher bargaining power over transit and roaming agreements, which may also be reflected in the operational expenditure of the MNO. SDMN also enables easier network sharing. Thus, MVNOs have the option of managing the whole control plane and, as a consequence, offer more varied services. Removal of the S/P-GWs in the revolutionary architecture enables lighter and more flexible mobility management, which together with global MVNOs improves the efficiency of data roaming. However, this case requires the MVNO to be operating in several countries and have a trustworthy reputation.

Mobile networks are going toward cloudification and virtualization due to more flexibility and new service potentials, and this can be seen from the offerings of the network equipment vendors, for example, Nokia's Liquid Net [30], Ericsson's Cloud System [31], and Huawei's Agile Network & SDN Solutions [32]. Thus, even without SDN, the Evolved Packet Core (i.e., MME, HSS, PCRF, and S/P-GW) is likely to be operating on cloud platforms in the near future. In addition, if the majority of the content is available from an MNO's cloud, operating the core network in the same cloud platform seems rational due to savings in traffic volumes. With this in mind, the evolutionary industry architectures are very likely to become reality.

On the other hand, the revolutionary industry architectures' realization depends heavily on the MNO's willingness to give up its control of the key network functions. For example, radio network management (eNB-C) is a role that the MNO might not wish to give up. However, according to a recent research [33]—based on brainstorming sessions with representatives from European mobile operators, network equipment vendors, and academics—actors, who typically are hidden from the end users, could take bigger roles in the future. This would drive the industry architecture, where mobility management is outsourced to a network equipment vendor. In addition, regulators may require more resource sharing in the future, which might force MNOs to at least lease part of its network to other actors, if not fully outsourcing.

A general limitation to the analysis is the immaturity of the SDMN technology. As a consequence, the technical implementations of the revolutionary industry architectures are not discussed. For example, how the allocation of radio network management and radio network forwarding roles to different actors can be done technically is not discussed. In addition, the technical challenge of removing GTP is identified, but no solution is offered. Moreover, due to the scope and nature of the chapter, the net gain from the economies of scale, the resource sharing, and the increased processing could not be determined. Thus, technoeconomic modeling should be adopted in future research to quantify the costs and benefits of SDMN.

References

[1] Cisco (2014). Cisco visual networking index: forecast and methodology, 2013–2018. Updated June 10, 2014, Accessed April 1, 2015, at: http://www.cisco.com/c/en/us/solutions/collateral/service-provider/ip-ngn-ip-next-generation-network/white_paper_c11-481360.pdf.

[2] Raghavan, B., Casado, M., Koponen, T., Ratnasamy, S., Ghodsi, A., and Shenker, S. (2012). Software-defined internet architecture: decoupling architecture from infrastructure. Proceedings of the 11th ACM Workshop on Hot Topics in Networks, October, 29-30, 2012, Redmon, VA, USA, pp. 43–48.

[3] McKeown, N., Anderson, T., Balakrishnan, H., Parulkar, G., Peterson, L., Rexford, J., Shenker, S., and Turner, J. (2008). OpenFlow: enabling innovation in campus networks. ACM SIGCOMM Computer Communication Review, 38(2), pp. 69–74.

[4] Valdivieso Caraguay, A.L., Barona Lopez, L.I., and Garcia Villalba, L.J. (2013). Evolution and challenges of software defined networking. Proceedings of 2013 IEEE SDN for Future Networks and Services (SDN4FNS), November 11–13, 2013, Trento, Italy, pp. 49–55.

[5] Naudts, B., Kind, M., Westphal, F.-J., Verbrugge, S., Colle, D., and Pickavet, M. (2012). Techno-economic analysis of software defined networking as architecture for the virtualization of a mobile network. Proceedings of 2012 European Workshop on Software Defined Networking, October 25–26, 2012, Darmstadt, Germany, pp. 67–72.

[6] GSMA (2012). Mobile infrastructure sharing. Accessed January 24, 2015, at: http://www.gsma.com/publicpolicy/wp-content/uploads/2012/09/Mobile-Infrastructure-sharing.pdf. Accessed February 18, 2015.

[7] Fischer, A., Botero, J.F., Beck, M.T., de Meer, H., and Hesselbach, X. (2013). Virtual network embedding: a survey. IEEE Communications Surveys & Tutorials, 15(4), pp. 1888–1906.

[8] Dramitinos, M., Zhang, N., Kantor, M., Costa-Requena, J., and Papafili, I. (2013). Video delivery over next generation cellular networks. Proceedings of the Workshop on Social-aware Economic Traffic Management (SETM) at the 9th International Conference on Network and Service Management, October 18, 2013, Zurich, Switzerland, pp. 386–393.

[9] Casey, T., Smura, T., and Sorri, A. (2010). Value network configurations in wireless local area access. Proceedings of 9th Conference on Telecommunications Internet and Media Techno Economics (CTTE), June 7–9, 2010, Ghent, Belgium, pp. 1–9.

[10] Christensen, C. (2003). *The Innovator's Dilemma*. HarperBusiness Essentials, New York, p. xviii.

[11] Basta, A., Kellerer, W., Hoffmann, M., Hoffmann, K., and Schmidt, E.-D. (2013). A virtual SDN-enabled LTE EPC architecture: a case study for S-/P-gateways functions. Proceedings of IEEE SDN for Future Networks and Services (SDN4FNS), November 11–13, 2013, Trento, Italy, pp. 8–14.

[12] 3GPP TS 23.002, "3rd Generation Partnership Project; Technical Specification Group Services and System Aspects; Network architecture (Release 12)", Updated June 2013, Accessed April 1, 2015, at: http://www.3gpp.org/ftp/Specs/archive/23_series/23.002/23002-c20.zip.

[13] Pentikousis, K., Wang, Y., and Hu, W. (2013). MobileFlow: toward software-defined mobile networks. IEEE Communications Magazine, 51(7), pp. 44–53.

[14] Penttinen, J. (2012). LTE and SAE architecture. In: J. Penttinen (ed.), *The LTE/SAE Deployment Handbook*. John Wiley & Sons, Ltd, Chichester, pp. 63–77.

[15] The World Bank Group (2012). World DataBank: population density. Accessed January 24, 2015, at: http://databank.worldbank.org/data/views/reports/tableview.aspx.

[16] Statistics Finland (2014). Official Statistics of Finland (OSF): preliminary population statistics. May 22, 2014. Accessed January 24, 2015, at: http://www.stat.fi/til/vamuu/2014/04/vamuu_2014_04_2014-05-22_tie_001_en.html.

[17] NSN Whitepaper (2013). Nokia Solutions and Networks, Liquid Radio, let traffic waves flow most efficiently. Accessed, April 1, 2015, at: http://br.networks.nokia.com/file/26241/liquid-radio-let-traffic-waves-flow-most-efficiently.

[18] Costa-Perez, X., Swetina, J., Guo, T., Mahindra, R., and Rangarajan, S. (2013). Radio access network virtualization for future mobile carrier networks. IEEE Communications Magazine, 51(7), pp. 27–35.

[19] Open Network Foundation (2014). Wireless & Mobile Working Group charter. Accessed January 24, 2015, at: https://www.opennetworking.org/images/stories/downloads/working-groups/charter-wireless-mobile.pdf.

[20] Kempf, J., Johansson, B., Pettersson, S., Lüning, H., and Nilsson, T. (2012). Moving the mobile evolved packet core to the cloud. Proceedings of the Fifth International Workshop on Selected Topics in Mobile and Wireless Computing, October 8–10, 2012, Barcelona, Spain, pp. 784–791.

[21] Smura, T., Kiiski, A., and Hämmäinen, H. (2007). Virtual operators in the mobile industry: a techno-economic analysis. NETNOMICS: Economic Research and Electronic Networking, 8(1–2), pp. 25–48.

[22] Virgin Mobile (2014). Check coverage. Accessed January 24, 2015, at: http://www.virginmobileusa.com/check-cell-phone-coverage.

[23] Lycamobile (2014). Lycamobile across 17 countries. Accessed January 24, 2015, at: http://www.lycamobile.com/lycamobile.php.

[24] Ericsson (2012). Press release: Connected Car services come to market with Volvo Car Group and Ericsson. December 17, 2012. Accessed January 24, 2015, at: http://www.ericsson.com/news/1665573.

[25] van Veen, M. (2013). Local breakout—a new challenge for networks. LTE World Series Blog, August 7, 2013. Accessed January 24, 2015, at: http://lteconference.wordpress.com/2013/08/07/local-breakout-a-new-challenge-for-networks/.

[26] European Commission (2014). Roaming tariffs. Accessed January 24, 2015, at: http://ec.europa.eu/digital-agenda/en/roaming-tariffs.

[27] Bhas, N. (2012). Press release: Mobile roaming revenues to exceed $80bn by 2017, driven by data roaming usage. Juniper Research, October 3, 2012. Accessed January 24, 2105, at: http://www.juniperresearch.com/viewpressrelease.php?pr=341.

[28] European Commission (2014). EU plans to end mobile phone roaming charges. Updated March 6, 2014. Accessed January 24, 2105, at: http://ec.europa.eu/news/science/130916_en.htm.

[29] Ericsson (2014). Network managed services. Accessed January 24, 2015, at: http://www.ericsson.com/us/ourportfolio/telecom-operators/network-managed-services?nav=marketcategory004fgb_101_127.

[30] Nokia Solutions and Network (2014). Liquid Net. Accessed January 24, 2015, at: http://nsn.com/portfolio/liquidnet.

[31] Ericsson (2014). Ericsson Cloud System. Accessed April 1, 2015, at: http://www.ericsson.com/ourportfolio/products/cloud-system.

[32] Huawei (2014). Agile network & SDN solutions. Accessed January 24, 2015, at: http://enterprise.huawei.com/en/solutions/basenet/agile-network/index.htm.

[33] Bai, X. (2013). Scenario analysis on LTE mobile network virtualization. Master's thesis, Department of Communications and Networking, Aalto University School of Electrical Engineering, Espoo, Finland.

Index

AAA, 6–7, 229, 295, 326, 349
access
  App, 89–93
  control, 9, 30
  control lists *see* ACL
  interface, 198
  internet, 89
  list, 265
  non-3GPP, 195
  point, 7, 29, 46, 201
  policing, 67
  provider, 197
  resources, 103
  rights, 292–3
  services, 155
  technology, 68, 74, 226
  web, 4
access network, 7, 48, 71, 84, 286
  infrastructure, 294
  security, 332
access point name, *see* APN
access stratum, *see* AS
ACL(s), 319, 329
active networking, 24–5, 29, 292
  community, 25
active queue management, *see* AQM
active
  APN, 194 *see also* APN
  applications, 25
  control, 24

interfering, 348
management, 14
monitoring probe, 350
monitoring, 14
nodes, 24
probes, 350
security appliances, 345
services, 345
VMs, 111 *see also* VM
VNF applications, 172 *see also* VNF
adaptation, 340
  content, 46, 201
  layer, 47
  of network virtualization, 9
  of OF, 49
  of SDN, 5
  real-time, 72
ADC, 251
adequate security, 318, 320
ADSL, 26, 78
advance persistent threat *see* APT
aggregation, 92–3, 254, 322, 350
  data, 69
  element, 92
  layer, 158
  network, 7, 73, 89
  of data paths, 93
  of flows, 34
  of forwarding rules, 103
  of resources, 335

Software Defined Mobile Networks (SDMN): Beyond LTE Network Architecture, First Edition.
Edited by Madhusanka Liyanage, Andrei Gurtov, and Mika Ylianttila.
© 2015 John Wiley & Sons, Ltd. Published 2015 by John Wiley & Sons, Ltd.

aggregation (*cont'd*)
  of tracking areas, 93
  platform, 353
  point, 89, 178
  pre-, 7, 367
  switch, 253
  traffic, 92
  user, 201
ALTO, 40, 189
  applications, 235
  client, 202–3, 205
  guidance, 203
  in SDMN, 201
  information, 203
  network information provision, 205
  network map, 206
  problem, 201
  protocol, 202
  SDN architecture, 204
  SDN use case, 202
  server, 202–3, 205
  service, 202
API, 6, 18, 109, 112–15, 226
  abstraction, 74
  access, 326
  collection of, 113
  features, 114
  heterogeneous, 292
  HTTP, 115 *see also* HTTP
  network, 25, 74
  northbound, 29, 104
  open, 25, 49, 338
  OF, 27, 234
  programmable, 223
  REST, 114, 124
  RRM, 72 *see also* RRM
  SDN, 72, 223
  securing, 326
  southbound, 29, 204
  standardized, 109
  vendor-specific, 198
  web service, 112, 114
APLS, 239
APN–AMBR, 194–5
APN, 194 *see* also APN-AMBR application, 110, 230,
    243, 291
  delivery controllers, 354
  detection and control *see* ADC
  domain security, 323
  domain, 37
  driven networks, 63
  function, 196, 253
  label switching *see* APLS
  layer network service, 217
  layer, 6, 29, 231

level signaling, 197
  model, 308
  plane security, 338
  plane, 65
  programming interface *see* API
  SDN, 235
  security, 326–7
  service subsystem *see* ASS
  services, 211, 290
  stratum, 322
  supported traffic management, 190
application detection and control *see* ADC
application layer traffic optimization *see* ALTO
APT, 318
AQM, 30
architecture
  of current mobile network, 359
  of evolutionary SDMN, 360
  of revolutionary SDMN, 364
  SDMN 7, 295
  SDN, 28–9
AS, 333–4
ASS, 291
  services, 291
Aster*x, 235, 236
asymmetric digital subscriber line *see* ADSL
authentication, authorization, and accounting *see* AAA
automated
  packet flow shaping and dropping, 268
  router configuration, 251, 265
automation, 210
availability, 213, 321
average
  case latency, 134, 136, 143
  latency, 135, 144, 145
  of minimum propagation latency, 134
  propagation latency, 134
  business tariff per service, 296–9

backhaul
  App, 88, 90
  devices, 8
  network, 6, 84, 343
  provisioning, 89
  scaling, 91
BalanceFlow, 236–7
base station, 86
  App, 88, 89
BBERF, 196–9, 251
bearer, 122
  binding and event reporting function *see* BBERF
  EPS, 193
  services, 292
best placement, 144, 145
BGP, 29, 98, 115

border gateway protocol *see* BGP
Bowman, 25
brute-force
  algorithm, 142
  approach, 142
BSS, 112, 171
BSS, 17, 171, 212
BTS, 21, 83, 220
buffer
  depletion, 255
  estimation, 262, 263
  fill level, 259, 263
  OF, 345
business
  architectures, 365
  framework, 290
  interfaces, 365
  model, 289, 293, 302
  relations, 310
  roles, 364
  service subsystem *see* BSS
  strategies, 396
  support system, 112
business service system *see* BSS
business support system *see* BSS

caching, 39, 56, 205, 216
  content, 100
  dynamic, 84
  location of, 102
  network, 67
  nodes, 101
  packet classifiers, 103
  relocated, 102
  servers, 10, 215
  techniques, 71
CAP, 296
DPI EX, 5, 65, 98, 334, 353
carrier grade
  ethernet, 93, 96, 97
  NAT *see* CG-NAT
  networks, 66
  process, 354
  services, 4
CCN, 234
CDMA, 8, 12, 226, 308 *see also* WCDMA
CDN, 7, 40, 56
CDNI, 40
cell, 63, 201, 220, 226
  edge, 227
  ID, 259
  load condition, 222
  serving, 240
  shrinking, 63
  splitting, 227

cellular, 61
  4G architecture, 130
  access, 53
  architecture, 37
  connection, 15
  coverage, 55
  data networks, 130
  devices, 37
  modems, 15
  networks, 3, 29, 31, 37–8, 71
  operator, 53
  SDN architecture, 132
  traffic, 19
centralization, 109, 231
centralized
  control plane, 233
  controlling, 7, 244
  DPI, 254 *see also* DPI
  MLB, 241 *see also* MLB
  policy, 321
  servers, 241
CES, 89, 97, 98
CGE, 90, 101
CG-NAT, 215, 218
CISCO, 12
classification, 253, 327
  flow, 254
clean-slate approach, 131
cloud
  based gateway, 40
  carrier, 62
  hybrid, 154
  infrastructure, 66
  mobile, 63
  network, 14
  operator, 4
  platform, 360
  private, 154
  provider, 4
  public, 154
  services, 41, 45
cloud computing, 5, 107, 111, 127
  abstraction levels of, 153
  deployment models of, 154
cloud-RAN, 66
CoA, 268, 275, 276, 281
code division multiple access *see* CDMA
cognitive
  access points, 305 *see also* CAP MVNO
  network management and operation, 76
  OAM, 77 *see also* OAM
  process, 77
  radio management *see* CRM
  radio, 50, 78 *see also* software defined
    radio

cognitive (*cont'd*)
  strategies, 296
  switching, 291
collaborative
  approach, 318
  attack, 318
  network, 55
  security system, 320
communication, 137
  application plane, 339
  architectures, 6, 131
  car-to-car, 69
  channels, 67
  control plane, 339
  cross-layer, 238
  cross-controller, 237
  D2D, 219
  data, 87
  HTTP, 204
  interface, 22
  M2M, 158
  mobile, 4
  multihop relay, 71
  multipath, 71
  overhead, 301
  path, 94
  satellite, 77
  seamless, 130
  secure, 293, 323, 327
  security, 339
  services, 289
  short distance, 289
  southbound, 35 *see also* southbound interface
  system, 109, 289
  user plane, 327
  wireless, 62
complexity
  device, 38, 71
  network, 18, 109
  of the network infrastructure, 13
confidentiality, 320–321
connectivity management, 366
content
  adaptation, 201
  caching, 100
  centric networks *see* CCN
  centric operation, 131
  consumption, 100
  delivery networks interconnection
    see CDNI
  delivery networks *see* CDN
  delivery, 190
  filtering, 200
  providers, 104, 190
  resource selection, 189

storage, 153
user-generated *see* UGC
control
  logically centralized, 22, 27, 29, 31–3
  physically centralized, 32–3
control and data plane
  separation, 22
  wireless, 31
control plane, 130, 137, 214
  distributed, 236, 237
  load balancing, 236
  protocols, 146
  scalability, 230, 231
controller failures, 135
controller placement, 132–44
  algorithm, 132–6
  decision, 134
  for SDN, 133
  metric, 134
  problem *see* CPP
  related challenges, 130
  scheme, 132
  strategies, 136
controller, 7, 51–6
  DevoFlow, 30, 35
  DIFANE, 30
  FlowVisor, 32
  HyperFlow, 32
  Kandoo, 32
  Maestro, 32
  McNettle, 33
  NOX, 32
  Onix, 32
  performance, 32–5, 37
  SANE/Ethane, 27, 34
converged intelligence, 317
convergence, 73
COTS, 215
CPP, 129, 140
CRM, 305
customer
  edge switching *see* CES
  facing service, 169
cyber-attackers, 317

D2D, 219, 244
data center, 37, 171
data path, 233
Data Plane Development Kit *see* DPDK
data plane, 129, 130, 132, 232, 238
  forwarding, 223
  load balancing, 238–9
data replication, 46
DDoS, 319, 321, 338 *see also* DoS
dead zone, 59

dedicated bearers, 250–255
deep packet inspection *see* DPI
 *see*demonstrator, 261, 265
depletion events, 258
device-to-device *see* D2D
DevoFlow, 238
DHCP, 89, 185
differentiated services code point *see* DSCP
DiffServ, 255, 263, 268
distributed
 computing, 33
 control plane, 236
 controller architecture, 236
 denial of service *see* DDoS
 forwarding switches, 359
 mobility management *see* DMM
 monitoring systems, 350, 353
 MVNO, 292 *see also* MVNO
 P/S-GW, 364
 path optimization, 9
 routing protocols, 30
 SDN model, 77
 security, 325
DMM, 72, 269–72
DNS, 105, 270–271
domain name service *see* DNS
domain name, 271
DoS, 231, 335, 337, 338
DPDK, 117
DPI, 72, 347–8, 352, 354
DSCP, 120, 255, 263
dynamic
 network control, 182
 network information provision, 205
 network map, 205
 network structure, 136
 operation, 214, 215
 PCC, 196, 206
 QoS enforcement, 198
 topology, 146, 213
Dynamic Host Configuration Protocol *see* DHCP

ECMP, 193–5
economic viability, 352
EDGE, 63, 300
E-function, 260, 261
element manager *see* EM
EM, 174
EMS, 175, 177, 183–9
eNB *see* eNodeB
encryption, 323, 326–7
 IPsec, 323
 SSL-based, 221
enforcement database, 255
enhanced data rates for GSM evolution *see* EDGE

enhanced
 visibility, 317
 business strategies, 304–5
 PCRF *see* ePCRF
eNodeB(s), 85–98, 106–10, 130
EPC, 61–7, 107–28
ePCRF *see* PCRF
EPS, 253
equal cost multipath, 228
 estimate, 261
ethernet, 67, 87, 252
 monitoring, 353
 services, 40
 over Wi-Fi, 57
 frames, 64
 VLAN, 92
 switch, 8, 27, 64
ETSI, 170–177, 184–90, 227
E-UTRAN, 86, 192, 328
evolution
 generic technology, 150
 of mobile networking, 11, 12, 149
 of programmable networks, 25
 of the mobile industry, 12
 of the NFV system, 108, 180
 of the wireless communication, 62
evolutionary approach, 131
evolved packet core *see* EPC
evolved packet system *see* EPS
evolved UMTS terrestrial radio access network *see*
 E-UTRAN
evolving threat landscape, 318
experimentation, network, 27
Extensible Messaging and Presence Protocol *see*
 XMPP

Facebook, 253, 267
fault tolerance, 132, 133, 136
femtocells, 46, 48
fiber
 -base, 361
 connection, 17
 optical, 74
5th generation networks *see* 5G
Firewalls, 90, 319, 323
5G, 8, 61, 134, 332
 cellular network, 77
 mobile backhaul, 92
 mobile networks, 61, 104
 network architecture, 78, 89
 network infrastructure, 75
 network, 61, 88, 134
 systems, 68–9
 technology, 61, 72
 wireless networks, 62

five tuple, 191, 255
fixed-mobile convergence *see* FMC
flash video, 257
flexibility, 8, 267, 342
  of mobility management, 285
  SDN-enabled, 351
flow, 232
  detection, 257
  matching, 233
  packet, 24, 30
  rules, 233
  table, 132, 232
FlowVisor, 47, 51
FLV, 257
FMC, 254
ForCES, 26–9
forum
  broadband, 40
  metro ethernet, 40
  optical internetworking, 40
forwarding
  in core network, 366
  in public network, 366
  in radio network, 366
  plane, 133, 137
  rule, 28, 30
  table, 30, 37
4D project, 27
4G, 38, 191, 303 *see also* LTE
4th generation networks *see* 4G

gateway GPRS support node *see* GGSN
gateway tunneling protocol *see* GTP
gateway, 3, 5, 78
  cloud-based, 40
  home, 64, 74
  media, 64
  packet (P-GW), 84, 115–17 *see also* PDN GW
  packet data network *see* PDNWAP
  residential, 69
  serving (S-GW), 84, 115–17
generic routing encapsulation *see* GRE
generic technology evolution, 150–152
GGSN, 12, 191, 211, 212
Gi interface, 192, 216
Gi-LAN, 182–5
global system for mobile communications *see* GSM
GPRS 12, 291
  roaming exchange *see* GRX
  tunneling protocol *see* GTP
GRE, 197
Greedy
  algorithm, 138, 141
  approach, 301
  controller placement algorithm, 141

  placement, 138
  routing tree algorithm, 139
green networking, 46
GRX, 361–3
GSM, 8, 12, 62, 332
GTP, 72, 86–90, 106–8
Gx, 124, 196
Gxx, 196

HA, 267–9, 271–3, 284–5
handover, 46, 53
  horizontal, 46
  low-latency, 84
  procedure, 46
  process, 49
  soft, 54
  vertical, 46, 53
HAS, 254
heterogeneous network, 8
heuristic
  higher-level, 142
  lower-level, 142
  meta-, 142
  method, 132
Hierarchical Mobile IPv6 *see* HMIPv6
HIP, 270–271, 277
history
  mobile network, 11
  of programmable networks, 23
  SDN, 23
HMIPv6, 268, 271–2
HoA, 268–9, 275–6, 281
home subscriber subsystem *see* HSS
host identity protocol *see* HIP
HSS, 85, 115–16, 118, 126
HTTP, 253, 254, 258, 269
  adaptive streaming services *see* HAS
hypervisor, 178

I2RS, 40, 210, 216
identification
  bearer, 125
  flow, 255
  functions, 215
  state, 219
identifier/locator network protocol *see* ILNP
identifier/locator split, 268
IDS, 21
IETF, 40, 210, 270
ILNP, 270–271, 283–5
immersive services, 78
impairment mitigation, 263
IMS, 73, 110, 213, 253
industry architecture
  definition, 365

evolutionary, 366
revolutionary, 371
industry specification group *see* ISG
inflexibility, 5
infocommunication, 317, 320, 329
information centric networks, 46, 67
information technology *see* IT
infrastructure as a service, 153
initial buffering, 260
integrity, 321
interactive
   applications, 71
   games, 194
   services, 78
   voice, 91
interconnection provisioning, 366
intercontroller latency, 136
interface
   mitigation, 48
   northbound, 65, 109
   southbound, 65, 109
interface to the routing system *see* I2RS
Internet Engineering Task Force *see* IETF
internet multimedia subsystem *see* IMS
internet of things *see* IoT
interoperability, 8
intrusion detection systems *see* IDS
IoT, 63
IP multimedia subsystem, 253 *see also* IMS
IP security *see* IPsec
IP, 4, 211–12
IPsec, 192, 264
IPv4, 98, 108, 119 *see also* IP
IPv6, 98, 119 *see also* IP
ISAAR, 249–54
   framework, 250–252
ISG, 212, 213
isolation, 8
IT, 5, 11, 149, 209
ITU-T, 40

key
   encryption, 48
   performance indicator *see* KPI
KPI, 250, 256

LAN, 217, 218
languages, SDN programming, 23, 36–7, 40
latency, 134
layer-2 switch, 215
LISP, 270–271
load balancers, 227
load balancing, 225–46
  as SDN applications, 235
  challenges of, 244

flow-based, 239
future directions, 244
in case of controller failure, 238
in heterogeneous networks, 227
in OF, 233
in wireless networks, 226
inter-technology, 226
mobility, 227
SDN-enabled, 233
server, 230, 234
shortcomings in, 227, 228
load imbalance, 136
location aware service flow observation, 268
Locator Identifier Separation Protocol
   see LISP
Long Term Evolution *see* LTE
LR-WPANs, 69
LTE, 4, 62, 83–9
  advanced, 63, 252
  architecture, 131
  EPC security architecture, 333
l-w-greedy
  algorithm, 141
  controller placement, 141

M2M, 17–18, 62, 210
MaaS, 92, 97, 98
MAC, 58, 100–101
machine-to-machine *see* M2M
MAG, 269, 284
management and orchestration
   see MANO
management plane, 130
MANO, 171, 175–80
MAP, 268
matching rules, 265
MCP, 140–142
mean opinion score, 250, 259
measurement point, 258, 261
memory effect, 260, 261, 266
metaheuristic, 142
metro, 12
microflow, 233
mininet, 281, 283
MIP, 267–73
  care-of address *see* CoA
  home address *see* HoA
  home agent *see* HA
  Mobile IPv6 *see* MIPv6
  triangle routing, 269–71, 277–8, 285
MIPv6, 268, 272
mitigation techniques, 351–2
MLB, 227, 239, 241
MME, 6, 84–93, 230
  in cells, 239–40

MNO, 104, 113
mobile
  cloud, 12
  IP *see* MIP
  network operator *see* MNO
  network security, 318
  networks, 12, 130–132
mobility, 227
  as a service, 92
  data center, 228
  handover, 227
  load balancing in cells, 239
  load balancing, 227, 239
  management App, 90–92
  management entity *see* MME
  virtual machine, 228
mobility anchor points *see* MAP
model
  capsule, 25
  proactive control, 34
  programmable router/switch, 25
  reactive control, 34
monitoring, 249–55
  App, 91, 92
  techniques, 334
MOS, 250, 259–61
MP4, 257
MPLS, 108, 119–21, 127, 193
  labels, 119–21
  LSPs, 264
MSC, 220
multihoming, 46
multihop networks, 62
multiple controller placement *see* MCP
multiprotocol label switching
      see MPLS
multi-tenancy, 210
MVNO developments
  mobility support, 303–4
  management schemes, 304
  enhancing business strategies, 304–5
MVNO, 50, 64
  aggregator MVNO, 296
  business strategies, 307
  centralized and distributed, 292
  challenges, 293
  features, 291–2
  functional aspects, 292–3
  hierarchical MVNOs, 294–6
  module type and services, 302
  multiple MVNO, 294
  pricing schemes and tariffs, 309
  service model and features, 308
  single MVNO, 294
  types, 294

NAS, 100, 332–4
NAT, 39, 220, 323
Negative impact, 266
Netflix, 221
NetFlow, 345
network
  as a service, 74
  control, 229
  devices, 5
  disruption, 135
  management protocol, 109
  management, 5
  operating system, 229 *see also* NOS
  service chaining, 66
  telemetry, 329
  usage, 366
network configuration
  automatic, 8
  manual, 5
network function virtual infrastructure, 175
network interface card *see* NIC
network intrusion detection systems
      see NIDS
network service chaining, 66
network service subsystem *see* NSS
next generation network *see* NGN
NFV, 6, 41, 62
NGN, 252
NIC, 117
NIDS, 348–9
NNSF, 241, 243
NodeOS, 24–5, 29
non-access stratum *see* NAS
northbound API, 91, 113
NOS, 5, 6, 229, 230
NSS, 291

OF, 6, 24, 27–8
  controller, 118–19, 121–3, 232
  counters, 108
  GTP gateway, 121
  protocol, 6, 108, 199, 232
  QoS provisioning, 200
  switch, 108, 117
  wildcard rules in, 238
OFDM, 211
offloading, 46, 49, 50, 57, 66
  in mobile networks, 200
ONF, 40, 198–9, 210, 337
Open Networking Foundation *see* ONF
open signaling *see* OpenSig
Open Virtual Switch *see* OVS
OpenFlow *see* OF
OpenRoads, 51–4, 60, 66
OpenSig, 24

OpenStack, 111–17, 124–5, 154
  community, 156
  components of, 155–6
  design and architecture, 155
operating system
  network, 29, 31, 41
  node, 29
operation and management, 76
operational support subsystem see OSS
operators, 46, 49, 50
OPEX, 105, 109, 234, 352, 353
optical fibers, 74
optimal
  algorithm, 138
  placement, 138
  reliability, 145
  result, 142
  solution, 136, 140–142
optimization, 210, 220
  path, 9
optimizer, 210, 217
orchestration, 39, 209, 213, 326
  history, 14
OSS, 112, 175–7
OTT, 16, 74
  providers, 16
over-the-top see OTT
OVS, 108, 117

packet
  capture, 261
  core network, 223
Packet Data Convergence Protocol see PDCP
packet gateway see P-GW
PCAP, 261
PCC, 194–8, 253, 254, 267, 269
  architecture, 196
  rule and QoS rule, 197
  service, 195
PCEF, 250, 255, 264, 267
PCRF, 6, 7, 84, 115–16
  enhanced, 124–5
PDCP, 332, 334
PDN, 192, 193
per APN aggregate maximum bit rate
  see APN–AMBR
performance, 210, 213
perimeter, 319
  security, 319
PFB, 250, 262, 264
P-GW, 72, 85–8
phone hacking, 318
phreaking, 318
physically decentralized, 32–3
PIN, 318, 323

PLAN, 25
platform as a service, 153
play out buffer, 258, 259, 268
play out timestamp, 259
playback stalls, 258
PMIPv6, 266, 268, 271, 275, 283–5
  mobile access gateway see MAG
policy and charging control see PCC
policy and charging enforcement function
  see PCRF
policy and charging rules function see PCRF
policy
  sased, 262
  control, 191–8
POS, 15
POTS, 220
power control, 46
Pox, 281
predefined packet handling, 250, 265
priority markings, 268
proactive
  capability, 77
  control model, 34
  control, 34
  load balancing, 236
  monitor, 332
  SDN, 363
programmability, 8, 21, 209
  control plane, 23
  data plane, 23
  in ATM networks, 24
  network, 8, 21
programmable network, 25–6
progressive download, 258
  streaming services, 258
propagation
  delay, 133, 135, 140
  latency, 134
provision, 9
Proxy Mobile IPv6 see PMIPv6

QCI, 253, 263
QEN, 250, 255
QMON, 250, 255
QoE, 62, 249–56
  enforcement, 255
  estimation, 256, 258, 268
  monitor, 255, 257, 265
  rules, 255
QoS characteristics, 191–5
  for EPS bearer, 191–5
  for non-3GPP access, 195
QoS control
  network-initiated, 197
  user-initiated, 198

QoS enforcement, 195
  in EPS, 195
  in SDN, 199
QoS, 5, 15, 46, 51
  class IDs, 253
  enforcement, 268
  provisioning, 75
QRULE, 249–50
Quality
  enforcement, 250
  estimation, 257, 259–61
  monitoring, 249
  score, 262
quality of experience *see* QoE
quality of service *see* QoS
quality rules *see* QRule

radio access technologies *see* RAT
radio network controller *see* RNC
radio
  access network *see* RAN
  access technologies *see* RAT
  network, 7, 211
  signaling, 212
radio resource control *see* RRC
radio resource management *see* RRM
RAN, 63, 116, 211
random placement, 141–3
RAT, 84
reaction, 249, 351–2
reactive
  control model, 34
  function, 89
  load balancing, 236
  SDN, 363
  security monitoring, 341
real-time, 114
  adaptation, 71
  games, 194
  monitoring, 38
  services, 252
  threat mitigation, 317
  vision, 58
reliability, 133–7
remote monitoring *see* RMON
remote procedure call *see* RPC
representational state transfer *see* REST
resilience, 135–8
resource negotiation and pricing, 309
REST, 109, 114, 122–4
RMON, 345
RNC, 210
routing
  in core network, 366
  in radio network, 366

RPC(s), 122–3, 232
  GTP_routing_update, 123
  P-GW, 123
RRC, 100, 332, 334
RRM, 68, 72

S/P-GW, 84–5, 99
S10 interface, 100
SA, 142, 143
sandvine, 251, 254
satellites, 63, 77
scalability, 15, 30–32
  limitation, 4
scaling, 66
SCP, 137
SCTP, 85, 193
SDM, 349–50
  control interface, 349–50
  controller, 349–50
SDMN, 3–7, 83–7
SDN, 3–5, 12, 23, 108
  applications, 84, 91, 97
  controller, 7, 109
  enforcement, 266
  protocols, 109
  security *see* SDS
  service-provider, 112–14, 123
  switch, 7, 29–31
SDNRG, 40
SDR, 47, 50, 78
SDS, 334–5, 342–4
SDWN, 68
secure
  mobile network services, 353–4
  service delivery App, 91, 95
securing
  API, 326
  infrastructure, 325
  perimeter, 319
security, 210, 213
  applications, 327
  architecture, 322–5
  controls, 318–23
  management, 353–4
  mechanisms, 332
  technologies, 326
security information and event management
  see SIEM
segmentation, 8, 327
self-organizing networks, 227
separation of networks, 212
server(s), 225, 230
  centralized, 230
  load balancing, 230
  logically centralized, 230

service
    chaining, 354
    flow class index, 255
    integration, 210
    level agreement see SLA
    oriented architecture see SOA
    provider(s), 11
serving gateway see S-GW
serving/packet gateway, 84
session
    border controller, 65
    initiation protocol, 254
SGi interface, 217
SGSN, 211, 216
S-GW, 192
sharing resources 13
short message service see SMS
SID, 179, 180, 198
SIEM, 344, 351
simple network management protocol see SNMP
simulated annealing see SA
single controller placement see SCP
SIP, 254
SLA, 46, 172
small cells, 18
smartphones, 15
SMS, 15
SNMP, 109
SNMP, 345
SOA, 114–15, 123–5
software
    architecture, 213
    as a service, 153
software defined
    mobile network see SDMN
    monitoring see SDM
    network see SDN
    networking research group see SDNRG
    radio see SDR
    security see SDS
southbound interface, 132
SPR, 253
stalling events, 258–61, 266, 267
standardization, 40–41, 253
STUN, 223
subscriber database, 262
subscriber management, 366
subscriber policy repository see SPR
subscription profile repository, 253
SW, 213

TAI, 86
TCP, 257, 258
    flow, 259
    segments, 258

TDF, 253
TDM, 13
TEID, 86, 119–21, 123
telecommunication standardization sector
        see ITU-T
telecommunication, 169
Tempest, 24, 32
tenant, 170, 195, 196
tesseract, 27
TFT, 86, 92
3rd generation (3G) networks see 3G
3rd generation partnership project (3GPP) see 3GPP
3G, 8, 15, 303 see also UMTS
3GPP, 4, 15, 73, 217
    access, 195
    classifier, 184
    IMS, 70
    legacy, 7
    network architecture, 210
    policies, 184
    QoS, 253, 268
    release 7, 192
    standardized, 227
time division multiplexed see TDM
TMF, 179, 180
TN, 213
topology discovery, 59, 62
traces, 261
tracking Area ID see TAI
tracking area, 262
traditional security, 317
traffic
    class, 255
    detection function, 253
    engineering, 255, 265, 266
    flow template see TFT
    offloading see offloading
    steering, 227
traffic management, 190–191, 198
    application-supported, 190
    in software-defined mobile networks, 198
    macroscopic-level, 190
    microscopic-level, 190
    traffic steering usage models, 191
transformation, 317
transport quality, 250, 254, 256
tunnel endpoint identifier see TEID
tunnel, 85, 121, 264
2G, 15, 191, 304 see also GSM and GPRS
2.5G, 63, 320 see also EDGE

UDP, 67, 99, 221
UDR, 251
UE, 84–6
    multihoming, 117–18

UGC, 221
UMTS, 11–12, 38, 62, 291
user data repository *see* UDR
user equipment *see* UE
user plane traffic, 90, 116
user supportive subsystem *see* USS
USS, 290–291

VDI, 15
VDU, 172, 174
video
  buffer, 258
  playback, 256, 262
  QoE, 258, 264
  quality, 256
  streaming quality, 252
  streaming services, 249, 252
VIM, 175, 176
  migration, 235
virtual
  deployment unit, 174
  LAN *see* VLAN
  link, 172, 178
  machine *see* VM
virtual desktop infrastructure *see* VDI
virtual network function initiation descriptor, 183
virtual network functions, 172, 173
virtualization, 169–70
  computer, 154
  history, 13
  network functions, 38–9
  network, 29, 38–9, 41
virtualized infrastructure manager, 175
visibility, 322
Visual Networking Index *see* VNI
VLAN, 108, 114, 120–121, 195, 196
  identifier, 100
  tag, 93, 98, 100, 120–121
VM, 110–111
VNF(s), 172–3
  manager, 174
VNFD, 173, 175
VNFID, 193
VNFM, 174–6
VNI, 15

VoLTE, 191
VPN, 112, 114
  enterprise, 112, 115
vulnerabilities
  application plane, 338
  communication, 339
  control plane, 338
  data plane, 338
  programmability, 337
  SDMN specific, 339
  SDN specific, 337
  threat vectors, 336
VXLAN, 125

WCDMA, 211, 221
web API *see* API
WebM, 257
wideband code division multiple access *see* WCDMA
Wi-Fi, 12, 49–55
  over UMTS 57
  access, 19
WiMAX, 12, 252, 254, 269, 292 *see also* 4G
wireless LAN *see* WLAN
wireless personal area network *see* WPAN
wireless, 45
  channel bandwidth, 45
  channel status, 46
  LANs, 46, 48
  link capacity, 46
WLAN, 37, 226
worldwide interoperability for microwave access *see*
    WiMAX
worst-case latency, 135, 136, 142
WPAN, 69, 308
  LR-, 69

X2, 86, 93
XMPP, 109, 126
XMPP *see* Extensible Messaging and Presence
    Protocol

YoMo, 257, 269
YouTube, 249, 250

zero level buffer estimates, 262

Printed in the United States
By Bookmasters